Benjamin Outram

Benjamin Outram
1764–1805

An Engineering Biography

R. B. Schofield

MERTON PRIORY PRESS

First published 2000

Published by Merton Priory Press Ltd
67 Merthyr Road, Whitchurch, Cardiff CF14 1DD

ISBN 1 898937 42 7

© Copyright R.B. Schofield 2000

For Rob, Neil and Gille

Printed by
Dinefwr Press Ltd
Rawlings Road, Llandybïe
Carmarthenshire SA18 3YD

Contents

	List of Illustrations	vi
	Acknowledgements	viii
	Preface	x
	Abbreviations used in the Footnotes	xii
1	The Outrams of Alfreton	1
2	Benjamin Outram's Apprenticeship: the Promotion of the Cromford Canal	6
3	The Construction of the Cromford Canal	30
4	The Nottingham and Nutbrook Canals	46
5	The Derby Canal	57
6	The Huddersfield Narrow Canal	76
7	The Ashton and Peak Forest Canals: Design, Promotion and Related Issues	118
8	The Construction of the Peak Forest Canal	129
9	The Completion of the Ashton Canal	149
10	An Incident at Barnoldswick	157
11	Benjamin Outram & Co.: The Ironworks at Butterley	177
12	On Railways and Tramroads	198
13	The Cromford Canal Railways	210
14	The Ashby de la Zouch Canal and Railways	221
15	The Monmouthshire Canal, Brecknock & Abergavenny Canal and their Railways	249
16	Benjamin Outram's other Canals and Railways	274
17	Appraisal	318
	Index	333

List of Illustrations

Plates

1	Interior of Butterley tunnel	33
2	The Amber aqueduct	38
3	The Derwent aqueduct	42
4	The Holmes iron aqueduct	67
5	Balsa model to demonstrate mode of failure of Holmes aqueduct	68
6	Jack o'Darley's bridge	71
7	Marsden entrance to Standedge tunnel	79
8	Interior of Standedge tunnel	85
9	Red Brook engine-house	92
10	Typical bridge at a lock-tail on the Huddersfield Narrow Canal	95
11	Skew bridge at Sparth	96
12	Stakes iron aqueduct at Stalybridge	97
13	Stodhart tunnel, *c.* 1929	125
14	Marple aqueduct	136
15	The inclined plane at Chapel en le Frith, *c.* 1905	138
16	The inclined plane in 1975	139
17	Store Street aqueduct, Manchester	151
18	Cutting and overbridge near South Wood on the Ashby railways	226
19	Central section of tunnel under Calke Abbey carriageway	227
20	Low embankment with original stone blocks, Ticknall line	239
21	Railway bridge over Ticknall main street	242
22	Ashford tunnel on the Brecknock & Abergavenny Canal	254
23	The Long Bridge at Risca on the Sirhowy Railway in 1900	268
24	The Long Bridge during demolition in 1900	269

Figures

1	The Cromford Canal	8
2	Canal feeders at Cromford Bridge	18
3	The Derwent aqueduct: reconstruction stage 1	40
4	The Derwent aqueduct: reconstruction stage 2	41
5	The Cromford Canal construction programme	44
6	The Erewash Canal and connecting waterways	51
7	The River Trent Navigation and connecting canals	58
8	The Derby Canal	62
9	Canal and river crossing in Derby	65
10	The iron aqueduct on the Derby Canal	66
11	Jack o'Darley's bridge	70
12	Narrow canals from Manchester	77
13	Canal tunnel and engine house at Red Brook	86
14	Section of the Huddersfield Canal tunnel	87
15	Construction pits and adits on Pule	88
16	Water engine system at Heathy Lee	89
17	Stakes iron aqueduct	99
18	Standedge tunnel and the summit reservoirs	104
19	Plan of Standedge tunnel showing the deviation of actual from intended alignment	113
20	Stodhart tunnel	124
21	Design of an iron aqueduct	132
22	Marple aqueduct	135
23	The skew arch bridge	153
24	The Coats Hall estate in 1808	161
25	Railway components and rolling stock	205
26	Ashby de la Zouch Canal and railways	225
27	Ashby Canal railways: contract payments	236
28	Canals and railways in South Wales	251
29	The Blisworth Hill railway	284
30	The Somersetshire Coal Canal and railways	292
31	Robert Weldon's caisson lock	293
32	Forest of Dean Railway	311

ACKNOWLEDGEMENTS

I have to thank several friends, as well as acquaintances met on the way, for help in my completing this book but firstly, I am grateful to the Leverhulme Trust for a Research Grant which got me started on this engineering study of Outram's career.

Sir Alan J. Outram Bt, a descendant of Benjamin Outram, is thanked for his interest in this work, and the late Charles Hadfield is remembered for his encouragement and help, as is the late Ronald Bennett of Manchester for assistance on site investigations and his perceptive observations on engineering matters. I acknowledge, too, the help of the late Harold Potter, Hydrologist to the former Trent River Authority, in unravelling the mysteries of the Derby Canal crossing of the River Derwent, and Royston Torrington of Derby for describing the operation of the canal locks adjacent to the river. Simon Stoker is thanked for providing information on the Cromford Canal, as is Douglas Stakes, formerly mining engineer to British Waterways, for details of the Standedge tunnel; also John Blunt and Ernest Upton of Melbourne for their guided tour of the Ashby Railway system, and Geoff Brown of Huddersfield for comments on the Standedge reservoirs. My sons, Robert and Neil, held the tape, sometimes reluctantly, on many site visits and their valuable help is likewise acknowledged.

Members of the Railway and Canal Historical Society who have provided information include Paul Reynolds, Hugh Compton, the late William Skillern and the late Gordon Rattenbury. Particular thanks must go to Peter Stevenson, also a member, who provided a copy of a Butterley Company letterbook, now in private hands, which he had laboriously copied under the most trying conditions—a story in itself. In this country there are many remarkable archives in public libraries and local record offices. The British Transport Historical Records of British Waterways, managed by the Waterways Trust and retained in the Public Record Office at Kew, have been of prime importance and the willing assistance of the staff is acknowledged, as is that of the librarians and archivists of the Institution of Civil Engineers, the House of Lords Record Office and the county record offices of Derbyshire, Gloucestershire, Somerset and Lancashire, and the National Archives of Scotland, to mention but a few. I am especially indebted to Dr Frank Taylor and Glenise Matheson of the John Rylands University of Manchester Library, and also the staff of the public libraries in Manchester, Swinton, Leeds,

Huddersfield, Derby, Sheffield, Salford and elswhere, who have helped over the years.

The photographs are from the my own collection except the following: the iron aqueduct in the Holmes, Derby, by permission of Monks, Contractors, of Warrington; the interiors of the Standedge tunnel on the Huddersfield Narrow Canal and the Butterley tunnel on the Cromford Canal, by permission of Dr Robin Witter; the inclined plane and the Stodhart tunnel on the Peak Forest Canal railway are from the Baxter Collection of the Railway and Canal Historical Society; and the Long Bridge of the Sirhowy railway, by permission of Michael Parker of Risca.

The drawings and maps were prepared from measured site surveys and early Ordnance Survey maps, except that of the Forest of Dean Railway, which was traced, by permission of the Gloucestershire Record Office, from the original survey (QRUm/5). Similarly, the drawing of Standedge tunnel (PRO, RAIL 844/53/7) was copied, by permission of British Waterways and the Waterways Trust, from the original in the Public Record Office. The plan of the feeder system to the Cromford Canal is based on details provided by Mr Simon Stoker.

I would also thank Tony Feenan of Belfast for his care in the development of photographs, some from indifferent negatives. I also acknowledge the help of others who have contributed in some way, especially the Revd Godfrey Higgins, Wilfred Hodgkinson, Dr Jean Lindsay, the late Dr A.R. Griffin, Mary Rea, Raymond Court, and Professor Ronald Pickvance. Finally, I thank my wife Brenda, for her constant interest; also for her diligence in searching archives with me and for assisting on site surveys. The responsibility for the interpretation of documentation and site investigations, as well as the accuracy of the written work and illustrations, however, remains entirely with me.

Wareham, Dorset   R.B. Schofield

Preface

Civil engineers often encounter works of a bygone era whilst surveying and planning for new construction projects. In Great Britain these remains will, most commonly, belong to the time of the Industrial Revolution when the practice of civil engineering was recognised, firstly by the establishment of the Society of Civil Engineers in 1771 and then by the more influential Institution of Civil Engineers, in 1818.

My interest in engineering history began during my early days as a civil engineer in the post-war years whilst surveying for water treatment schemes along the Irwell Valley, north of Manchester. This led to my exploration of the remains of the nearby canal, tramroads, and an ingenious water-power scheme by James Brindley and his engineers. Then again, in the late 1950s, my work on Midland motorways familiarised me with the Grand Union Canal and the interconnected systems. It was of interest to me then—and my colleagues as well for that matter—to consider why and how the routes had been selected for these waterways. We also pondered about the techniques adopted for the control of levels and alignment, all difficult enough problems using modern surveying equipment and aided by Ordnance Survey plans. So many of these schemes were substantial, even by today's standards, and we wondered who had promoted them and how they were financed, designed, and then managed during construction. Moreover, who were the engineers who directed such major projects and what were their origins?

My wider explorations of the canal system then followed but finally, the restoration of the Cheshire ring of canals and in particular the Peak Forest Canal and the Grand Aqueduct at Marple, aroused my interest in Benjamin Outram. Enquiries showed that very little was known of him but over the years since, fieldwork and foraging in primary and secondary source material added to my knowledge and eventually resulted in the present study. Regrettably (as in the case of his friend and colleague William Jessop), very little remains of Outram's personal correspondence and family papers. It was said that much was destroyed in a house fire during the nineteenth century but most of his engineering drawings and data must have been discarded when his wife and children left Butterley Hall, shortly after his death in 1805. Unfortunately too, any engineering drawings that might have survived in the records of his clients have eluded me and but one land survey—that of the Derby

Canal—remains of the many that he must have prepared during his professional lifetime.

Accordingly, of much importance and interest to me were the Chandos-Pole-Gell Manuscripts in the Derbyshire Record Office and the Lords Committee Papers in the House of Lords Record Office, which relate to the early beginnings of the Cromford Canal and to Outram's first experience of engineering construction. These documents provide a rare insight into the parliamentary proceedings of that day and the promotion of a major waterway just before the onset of the Canal Mania of the 1790s; an entire chapter is devoted to these matters. Here we catch rather more than a glimpse of some well-known personalities and their squabbles, intrigues and gossip, but the papers also yield an understanding of the workings-out of a sound engineering scheme by Jessop, enthusiastically supported by the youthful Outram. Exposed too, are the inadequacies of some of the engineering practices of the day, quite apart from the paucity of the scientific concepts then available to these engineers.

Canal company minute books are all too often brief, and omit much of interest on site conditions and any disputes with riparian owners. Some such information is revealed in the minute books and papers of the Huddersfield Canal Company but a more informative source is found in the Bagshawe Muniments (John Rylands University of Manchester Library) from which the entire story of a lengthy local dispute, in which an apparently reluctant Benjamin Outram was involved with his father Joseph, has been derived and described.

Outram is often remembered for his railways and it is fortunate, in a sense, that his biggest contract for such works resulted in a major contractual dispute with the proprietors of the Ashby de la Zouch Canal Company, because much documentation relating to construction, costs, engineering reports and related correspondence survives, from which yet another detailed commentary could be written.

Outram should be remembered too, as the founder and managing partner of the ironworks at Ripley, in Derbyshire, which eventually became known as the Butterley Company. It gradually became obvious to me that he had always regarded this venture as the most important part of his life's work and that civil engineering became secondary to his future plans, particularly after 1801. Chapters have been included to summarise these business interests, although a fuller study of the company can be found in Philip Riden's book on *The Butterley Company 1790-1830* (Derbyshire Record Society, 1990).

And finally, I have frequently criticised both Outram and his colleagues—including the eminent Jessop—for the casual manner in which they always dealt with contracts; for their inadequate planning, inaccurate estimates, and the fact that so much design was prepared only during the progress of the works. Such matters, which seemed to be of secondary importance to them, would never be emulated today. There are sound financial reasons which justify the preparation of complete and accurate contract documentation before work begins on site but, of course, such modern methods are only the result of the mistakes, and their revisions, of two hundred years of professional engineering practice.

ABBREVIATIONS USED IN THE FOOTNOTES

BCL	Birmingham Central Library
DCCR	Derby Canal Company Records
DLSL	Derby Local Studies Library
DRO	Derbyshire Record Office
GRO	Gloucestershire Record Office
HLRO	House of Lords Record Office
ICE	Library of the Institution of Civil Engineers
JRUML	John Rylands University of Manchester Library
NLW	National Library of Wales
PRO	Public Record Office
SRO	Somerset Record Office
WYASK	West Yorkshire Archive Service, Kirklees
WYASL	West Yorkshire Archive Service, Leeds
WYASW	West Yorkshire Archive Service, Wakefield

CHAPTER 1

THE OUTRAMS OF ALFRETON

The family name of Outram is not uncommon in Derbyshire and south Yorkshire Although ostensibly of yeoman stock, the Alfreton branch claimed some relationship with Thomas Outram, rector of Owston Ferry, near Gainsborough, in 1435 and also with William Outram, Doctor of Divinity, who was buried in Poets' Corner in Westminster Abbey in 1695,[1] but direct links with these dignitaries of the past have been assumed, rather than proved.

The name first appears in the Alfreton parish register on the death of Joseph Outram, a gardener, in 1732. His offspring included Benjamin, whose grandson was Joseph, baptised on 5 March 1732. Although the family had been established for many years in Alfreton, Joseph, when writing to a client in August 1798,[2] mentioned that 'I rode to our old house at Ferrybridge to breakfast', thus implying that his branch of the family came from the Pontefract area and that relatives still lived there.

Joseph Outram (1732–1810) was three times married, first to Jane Armfield in 1757 from which union there were no children. After her death he married again, in 1763, to Elizabeth Hodgkinson, the daughter of Edmund Hodgkinson of Ashover,[3] who was to be mother of Benjamin (b. 1764), Edmund (b. 1765), Sally (b. 1769), and Joseph (b. 1771), as well as two other children who died in infancy. Joseph was married, yet again, to Anne Micklethwaite of Alfreton in 1776, and she was mother to Anne (b. 1776) and John (b. 1777).[4] The children were given family Christian names and the eldest son was no exception, although tradition asserts that he was named after the American, Benjamin Franklin, and that the latter stood as the godfather at the infant's baptism. Franklin certainly visited the country in August 1759 because he wrote to his wife of ' ... having spent some time in Derby-

---

[1] M.F. Outram, *Margaret Outram (1778–1863). Mother of the Bayard of India* (1932), p. 91.

[2] JRUML, Bagshawe Muniments, 8/4/2873, J. Outram to J. Bagshawe, 1 Aug. 1798.

[3] F.C. Corfield, *The Family of Outram of Alfreton* (1900) (copy in DLSL). According to W.J. Hodgkinson of St Neots, Corfield is incorrect in stating that Elizabeth was of the Hodgkinsons of Overton Hall.

[4] Private correspondence, W.J. Hodgkinson with author.

shire among the gentry there to whom we were recommended'.[1] He is believed to have stayed in Swanwick, close by Alfreton, and he could have met Joseph Outram there but, even if he had influenced the selection of Benjamin's name, he could not have attended the baptism because he only returned to Derbyshire next in 1770.[2]

Joseph Outram was an unpretentious fellow and although on occasion he was known to describe himself as 'gentleman', even late in life he was content to be known as an 'advanced agriculturist'[3] and to be listed in directories under 'traders etc.', as a maltster.[4] These were far from adequate descriptions of his activities: he was much too able to devote his life to such narrow interests. Joseph Outram was acknowledged to be a leading land agent whose services were frequently in demand. He was particularly well-known as a commissioner for parliamentary enclosures and he held many such appointments from as early as 1773 until his death, 37 years later. Indeed, in his obituary in the *Gentleman's Magazine* (1810), Joseph Outram was said to be 'a gentleman long-known for his judgement and experience as a Commissioner for altering wastelands and an arbitrator in the division of landed property'. Regrettably, such details of his career were lost to later generations of his family, who seemed to regard him as but an eccentric character of little account. For example, a distinguished descendant was to write that 'he was naught to describe beyond being an old gentleman of decided character, and owner of very little property about Alfreton'.[5]

Joseph Outram could be a crusty and difficult man to deal with, as his business correspondence sometimes suggests.[6] He was, nevertheless, able and widely experienced and, probably because of this, he could be self-opinionated and impatient with colleagues, and even clients, although frequent attacks of the gout did not help moderate his temper. For all that, Joseph had a kind and generous streak which commended him to many. 'King' Outram, as he was sometimes known, was a scrupulously honest and reliable man who set the highest ethical standards for himself, as well as for the three sons who assisted him at

---

[1] Yale University Library, Diary of Jonathan Williams jun. (unpublished)

[2] *The Papers of Benjamin Franklin* (ed. L.W. Labarce), viii (1962), p. 431.

[3] DLSL, Tilley Biog. Notes, W. Stevenson to F.D. Outram, 21 Jan. 1909.

[4] Barfoot & Wilkes, *Universal British Directory of Trade, Commerce and Manufactures* (1793 ed.), ii, p. 30.

[5] DLSL, Tilley Biog. Notes, Sir Francis Outram Bt to W. Stevenson (nd, probably June 1909).

[6] JRUML, Bagshawe Muniments, Outram Corres. 8/4/1–4197. A good insight into the work of the Outram family practice can be derived from this source.

various times in his busy practice.

For the most part the practice was concerned with estate management, valuations, building surveying and design, and even mine surveying. But the Outrams were landscape architects too; they could confidently advise on afforestation schemes and were not averse to laying out a bowling green. As capable agriculturalists, they were familiar with farming methods and could deal in livestock, animal feedstuffs and seeds if the opportunities arose. All were competent land surveyors, although Joseph seemed to prefer to put out much of this class of work to local specialists such as the Fairbank family of Sheffield, or to John Nuttall of Matlock, a distant relative, and his son George. The Outrams had earned the respect of clients among the aristocracy, notably the Duke of Leeds and Lord Harrington, besides several county families, including the Gells of Hopton and the Bagshawes of the Oaks, to mention but two.[1] Even so, Joseph Outram was never a wealthy man in spite of his busy practice (his professional charges were only one guinea per day plus expenses) as well as his ownership of sizeable freehold and mineral rights local to Alfreton (there is no truth in the often-repeated story that he had an ironworks in Ripley), and investments in canals and turnpike trusts. The family, nevertheless, seemed to have lived in some style and comfort in one of the biggest houses in the town, in King Street,[2] and Joseph Outram was regarded as second only in standing to Squire George Morewood of Alfreton Hall, in a township which had a population of 2,301 in 1801.[3]

Benjamin (1764–1805) was the first of Joseph's sons to join the practice. Very little is known of his education and formative years. He was listed in directories as a land surveyor and he also described himself as such to a parliamentary committee in 1789,[4] but during his early twenties he was also dealing with briefs which required knowledge and experience of general surveying practice;[5] all the evidence suggests that he was then a competent land agent. Benjamin was an intelligent, self-confident and ambitious young man, who eventually left the practice to become a civil engineer on his own account. He was also soon to take up the managing partnership of the ironworks known as Benjamin Outram & Co. of Butterley, near Ripley. He then lived in Butterley Hall

---

[1] See DRO, D258, and JRUML, Bagshawe Muniments, passim.
[2] DLSL, Tilley Biog. Notes.
[3] R. Johnson, *A History of Alfreton* (1970), p. 171.
[4] HLRO, Evidence on Cromford Canal Bill, 6 July 1789.
[5] DRO, D258/50/15k, J. Wilkinson to P. Gell, 11 March 1789.

and came to describe himself as 'gentleman'.

It is unlikely that the family practice would have satisfied Benjamin's entrepreneural spirit indefinitely but it seems too, that he did not always get on well with his father. Hence it is questionable if any close business relationship between father and son could have survived for long. Joseph, long used to a position of independent authority, was never able to accept with grace the opinions of others, and perhaps more so those of his sons, over matters in which he had long excelled. Perhaps that explains why, in later life, Benjamin, who was not unlike his father in temperament, was rather cautious in any dealings with him but they were, nonetheless, always courteous when conferring on professional matters and clearly respected one another's abilities. Benjamin's wife Margaret, writing many years afterwards, claimed that father and son had not spoken for seven years before her marriage in 1800 and that she had been instrumental in bringing them together;[1] her statement is an exaggeration, judging by the available evidence.

Joseph's second son Edmund (1765–1821), was the very antithesis of Benjamin in character. Mild of disposition but a scholar of real ability, he was the favourite son of Joseph who never lost the opportunity to tell others of the prowess of 'my young man', as he frequently referred to his son Edmund, and also Joseph junior. Edmund was educated at the Manchester Grammar School and was an exhibitioner of St John's College, Cambridge. He graduated as second wrangler, was elected a fellow and proceeded to his doctorate before appointment as Public Orator to the University of Cambridge in 1808. Edmund eventually entered the church and became rector of St Philip's, Birmingham, as well as a magistrate and Master of St John's Hospital, Lichfield, besides other appointments.[2]

Joseph junior (1771–1830), joined his father's practice after Benjamin had departed for greater challenges. He soon proved to be a conscientious and able assistant but he later joined his brother in the commercial management of the ironworks at Butterley. He left to take up an appointment at the Clyde Ironworks in Glasgow about 1803 and eventually, with his father's help, bought into a partnership there. This did not prove to be advantageous and he left in 1808 following disagreements. He later invested successfully in the coastal shipping trade and

---

[1] Outram, *Margaret Outram*, p. 99.

[2] Manchester Grammar School, Admission Register.

held lesser interests in land and property.[1] Joseph was father of George Outram (1805–56), poet and journalist, and of another son who emigrated to Nova Scotia.[2]

Of the remaining children, Sally took care of her father and the family home in Alfreton until she married David Hinckley, a doctor. Her step-sister Anne married Samuel Hodgkinson of Worksop, and her step-brother John was assisting his father in the practice by the turn of the nineteenth century. John's work lacked the style and competence of his older step-brothers but perhaps the career of surveyor held no appeal for him because, in the 1820s, he became a publican at Greasley in Nottinghamshire.[3]

The surveying practice was constantly busy, generally with clients in the Midlands, although Joseph Outram was not averse to accepting commissions in Lancashire and Yorkshire and even occasionally in the south of England. The menfolk were frequently under pressure, often for seven days a week, and their lives included a succession of long and arduous journeys in all weathers by coach, hired chaise or on horseback. Hence Joseph could write one day that ' ... we are close to setting the boundaries between Hallam and Hathersage this day' and ' ... we are directing the planting of 10,000 oak plants 2 years old on the Castleton estate', and, on another, ' ... I am going to his Lordship's estate at Elvaston tomorrow for 3 days'.[4]

During the last decade of the eighteenth century, their round of tasks also began to include business (and also some disputes) with the proprietors of the new canals and turnpike trusts. Hence Joseph was to become embroiled in long-standing battles on behalf of his clients on matters of land-take and evaluations; for example with the Sheffield & Chesterfield Turnpike Trust and also the Leeds & Liverpool Canal Company. But in 1787 there also came an opportunity for him to help promote and invest, on his own behalf, in one of the new waterways: the Cromford Canal.

---

[1] National Archives of Scotland, SC 36/48/22, 26 May 1831, inventory of Joseph Outram's estate.

[2] DLSL, Tilley Biog. Notes.

[3] Information from H.R. Potter, Nottingham, from Kimberley (Notts.) parish register.

[4] JRUML, Bagshawe Muniments, 8/4/2801, J. Outram to J. Bagshawe, 27 May 1793; 8/4/2791, Outram to Bagshawe, 9 Feb. 1793.

CHAPTER 2

BENJAMIN OUTRAM'S APPRENTICESHIP: THE PROMOTION OF THE CROMFORD CANAL

The late eighteenth century in Britain was a time of unprecedented, vigorous industrial growth which was stimulated and nurtured by technical inventions and an increasing population. The turmoil generated brought prosperity for some, but left a wake of much uncontrolled urban development and hardships for many. In the countryside a similar revolution was afoot because industrial growth and the rising population created an expanding market for foodstuffs; hence the modernisation of agriculture had become urgently necessary. A prime result was a reorganisation of land use by means of the enclosure of common land and its redistribution by Acts of Parliament.

Enclosure was not a new phenomen, indeed it had continued unabated since Tudor times, although the process had increased apace during the eighteenth century. Legal though the enclosures were, they were a painful remedy for an archaic problem. Parliamentary commissioners' rulings carried the full force of the law, many rural communities were depleted or destroyed, and landless labourers and their families were often despatched to a life of insecurity and poverty in the new towns. The regions particularly affected at this period extended from the East Riding of Yorkshire through Lincolnshire and Norfolk, as far south as Berkshire and Wiltshire. Irrespective of the harsher realities of the enclosures, however, they created opportunities for the advancement of the agricultural sciences. Thus innovative ideas were applied to systematic livestock breeding, horticulture and good husbandry, and to the design and layout of farm buildings. But at this time other, more subtle, changes were also occurring in the landscape.

Without improved communications, the industrial and agricultural revolutions could not have thrived and, accordingly, exceptional measures were in progress for their provision. During the years 1759 to 1805 165 Acts of Parliament were passed to allow for the construction of a network of inland navigations,[1] the peak period for legislation being during the Canal Mania years of 1791 to 1796 when 51 new waterways

---

[1] J. Phillips, *A General History of Inland Navigations, Foreign and Domestic* (1805), p. vi.

were authorised.¹ The national road system, the maintenance of which had been dependent for generations on the whims of parish surveyors, was also improved by the new turnpike trusts, and some 1,600 Acts were passed between 1751 and 1790.²

At that time, farming and country pursuits were a preoccupation of the wealthy, in part a consequence of the fact that political power was associated with ownership of land and property. Not surprisingly, a professional class of land agents and surveyors had grown around this prosperous agrarian society in order to undertake the skilled tasks of estate management. In the North and Midlands, none was better known than the practice established by Joseph Outram in Alfreton during the 1760s. Such country practices were well-placed to organise and administer the enclosures and also to engage in the design and construction of the new canals and turnpikes, quite apart from the industrial developments which these communications activated. The Outrams were quick to take advantage of these opportunities.

The years following the completion of the Trent & Mersey Canal in 1772 saw the steady expansion of the inland navigation system in the Midlands. The Erewash Canal, for example, a lateral cut which followed the course of the river of that name for nearly twelve miles from Langley Mill to the Trent Navigation at Sawley, was a typical development of the era. Its prime purpose was to export coal from the collieries of the Erewash valley into Leicestershire, via the Loughborough Navigation.³ Although the proprietors had prospered since their waterway opened in 1779, some ten years later the rich coal seams on which it depended were nearing exhaustion. Hence, in May 1787, the Revd D'Ewes Coke of Pinxton proposed a northwards extension of the canal to encourage mining beyond Heanor but this was rejected by the company.⁴ So was a similar recommendation to the committee the following July by a group of businessmen, who included Thomas Hodgkinson, a coal-owner of Pinxton, and his friend, Joseph Outram of Alfreton.

The refusal of the proprietors to adopt these sensible proposals and so extend the life of their waterway probably owed much to complacency resulting from years of monopoly. Whatever the reason, the wishes of local enterprise were not to be frustrated for long and an alternative and more comprehensive scheme was soon to be promoted. This was the

---

[1] C. Hadfield, *British Canals: an illustrated history* (1950), p. 106.

[2] G.M. Trevelyan, *English Social History* (1944), p. 382.

[3] 17 Geo. III c. xxx (1777).

[4] *Derby Mercury*, 10 May 1787.

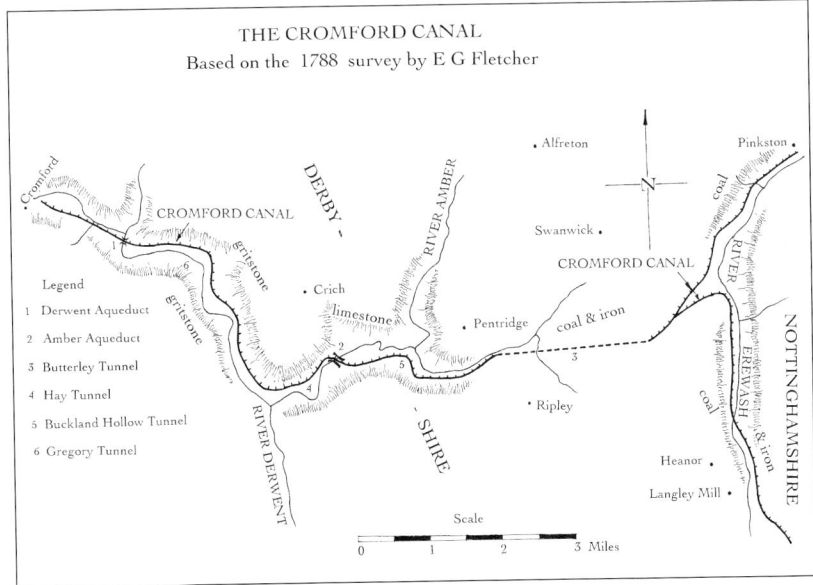

Figure 1 The Cromford Canal.

Cromford Canal, projected and designed by William Jessop (1745-1813), civil engineer of Newark, comprising a canal 'from Langley Bridge to Cromford, with a collateral Branch to Pinckston (sic) Mill' at estimated costs of £42,697 (Fig. 1).[1]

Jessop's waterway consisted of a summit level fourteen miles long, by which means industrial Cromford could be provided with coal and corn from the eastern end at Pinxton. In addition, a link to the Erewash Canal gave access into the Trent valley and permitted the export of lime, stone and timber, as well as cotton goods from the Cromford mills, into the national network. A scheme for an inter-regional line north to Stockport via Bakewell and Whaley Bridge was also examined, but rejected, at a late stage of planning. This was a transitory fancy and the canal remained a cul-de-sac until closure in 1944, although the promotion of the Cromford & High Peak Railway in 1825,[2] providing a link with the Peak Forest Canal at Whaley Bridge, enhanced its prospects in later years. The engineer for that project was to be Josias Jessop (1781–1826), the second son of William.

---

[1] DRO, D258/41/32q, Report of Mr Jessop, engineer, on a design for a canal from Langley Bridge to Cromford, 31 Dec. 1788.

[2] 6 Geo. IV c. xxx (1825).

PROMOTION OF THE CROMFORD CANAL

Jessop's design for the Cromford Canal was practicable, in spite of some severe problems in overcoming the natural obstacles of the rugged terrain. In his report he did not trouble his readers with technicalities, preferring

> only to mention in general, that its Length from Langley Bridge where the present Erewash Canal terminates, to Cromford, is fourteen Miles and an Half; that from Langley Bridge to the summit Level in Codnor Park there is a Rise of 82 feet; and that then, including a Branch of two Miles and an Half to Pinkston, there will be a dead Level of above fourteen Miles in Length ... passing under Butterley Park it will be subterraneous for one Mile, six Furlongs and eighty-eight yards in Length; so that, though Trent Barges may come to the Tunnel, it will only admit, through it, and to Cromford, Boats of seven Feet wide, such as are used on the Western Canal.

Thus the Cromford Canal was to be built to narrow-boat standards at the western end, as was the Trent & Mersey Canal, and to broad standards (78 ft by 14 ft) at the eastern end and down to the Erewash Canal, so as to accommodate Trent barges.

The summit level of the Cromford Canal follows the 300-ft contour closely and the major construction works are situated along this section. These include the Derwent aqueduct, a single segmental stone arch of 79 ft span, rising 19¼ ft above springing level; the Amber aqueduct (an earth embankment originally pierced by three underbridges, one of which contained the River Amber); and the 2,978-yard Butterley tunnel (subsequently extended to its present length of about 3,100 yards) driven through the ridge between the Erewash and Derwent valleys. There are three other tunnels, all with towpaths, at Buckland Hollow (33 yards), Hag (93 yards) and Gregory (76 yards). The link with the Erewash Canal at Langley Mill, now defunct, comprised a three-mile section which fell, via fourteen broad locks, from Codnor Park.

Jessop's report included a rudimentary economic review which was speculative and occasionally inaccurate but argued a strong case for canal transport. It also showed something of the concern Jessop felt for others, particularly for the 'lower Reaches of People which form the great Bulk of Society'. He hoped that 'no one will be influenced by imaginary Injury to oppose the Welfare of Thousands by objecting to a Proposition pregnant with Public Benefit'. He also noted that the scheme would need the support of the Erewash Canal Company and mildly threatened that if they opposed the project, 'another Line might be pointed out which would more effectually Benefit the Public', meaning that this waterway would be in direct competition with the Erewash Canal.

Jessop was aware that his canal's water supply could only be obtained from the River Derwent, on which many factories and mills depended for their power, as well as the River Erewash catchment and various coal mines along the route of the canal. For that reason he expected opposition, hence he played down the problem, stating that the bulk of the trade would be along the summit level 'for which no Water would be used for Lockage'. There is no doubt that the scheme was sound; its potential was evident, though his proposals for the principal water supply were barely adequate. Yet throughout the parliamentary proceedings which followed Jessop was to show a casual unconcern when questioned on this vital issue. Perhaps his experience led him to believe that other sources would be found once the canal was fully operational. If this was the case, events eventually proved him correct.

The Erewash Company lost the opportunity to dictate developments in their valley because initiatives were firmly grasped by a group of Derbyshire gentlemen who were to establish the Cromford Canal Company. The latter was led by Philip Gell, who owned estates at Hopton Hall, Wirksworth, besides quarries and mining rights nearby. Several other land-owning families were involved, including the Eyres and the Milneses, both related to the Gells; also the Hurts and the Fillinghams and, not surprisingly, the Revd D'Ewes Coke of Pinxton as well. Two solicitors were appointed, the distinguished Francis Beresford of Ashbourne and Francis Evans of Nottingham. Captain John Gell of Crich, brother of Philip, was to work on behalf of the group by canvassing and reporting on parliamentary proceedings from his London home in Wimpole Street. Thus the principal shareholders were all known to each other. Investment was by invitation only, so that undesirable speculators, the cause of many of the troubles of other canal companies, were excluded. This was a wise measure which was to ensure a sound management and financial basis for the company.

A name associated with the early history of the canal is that of Sir Richard Arkwright (1732–92), who owned land and cotton mills in Cromford, notably the Masson Mill. Contemporary sources suggest that his was but a minor role in the establishment of the canal company.[1] He was invited to be a shareholder because of his national fame, in order to

---

[1] Arkwright's unreliability is frequently mentioned in correspondence between John and Philip Gell (DRO, D258).

lend credibility to the scheme and to encourage local investment. Arkwright was unpopular with most of his fellow proprietors; he was not of the same social stratum as the Gells and their friends (they often referred to him facetiously, at least among themselves, as 'The Knight') but it cannot be said that dislike was due to any snobbishness on their part. Rather it was his selfish and boorish ways which did not appeal to his gentlemanly associates. Arkwright's intentions were forever uncertain and, throughout the parliamentary proceedings, his colleagues were frequently perplexed and worried by his changes of mind and by his plotting with the opposition parties. He appears to have been a hard, unrelenting man whose principal pursuit was self-interest. Arkwright courted friendship with nobody and certainly not with the professional group with which he might have had interests in common. Such people included Joseph Outram and his 24-year-old assistant and son Benjamin. The latter was a lively and talented youth who was to attract the attention and approval of Jessop, and was soon to be engaged by the company as an expert witness; he eventually became their canal superintendent.

It is uncertain who first invited Jessop to investigate the possibilities for the new waterway but he began fieldwork in September 1787 with the help of a local land surveyor, Edward Green Fletcher, who prepared plans for the recommended route. Jessop then deferred publication of his report for a further fifteen months for a very good reason. An artificial canal relies for its operation on a reliable water supply from river or reservoir and, as noted in his report, Jessop was fully aware that the prime water sources of the River Derwent and its tributaries were already used by several mills and factories between Cromford and Derby. He accordingly took until November 1788 before completing and submitting his report to the promoters, because this allowed him time to make regular observations of flow rates in the Derwent. He was thus able to devise an ingenious method for supplying the canal which he believed was fair and equitable for all, but he noted that 'it would be a singular Case if it met not with Opposition'. He was certainly proved correct in that assumption.

All exceptionally low flows were recorded during this monitoring of the river; the lowest rate occurred on Sunday, 16 November 1788, when the discharge was 570 tons of water per minute (equivalent to two locksfull). Jessop then reasoned that an abstraction of one-twentieth of that flow collected over a 24-hour period from Saturday night through Sunday, even during the driest of weather conditions, could provide 41,040 tons of water. This was just sufficient for the operation of the canal for one week, taking into account lockage, leakage and other water losses. He proposed to excavate the summit pound twelve inches deeper than navigation required, calculating that this would give sufficient storage capacity for the diverted water. Jessop claimed that no one's interests would be harmed because the river water ran to waste on the Sabbath when all the mills and factories were closed.

Here was a nice solution, seemingly enhanced by Jessop's claim that even such quantities would be more than sufficient because most trading on the canal would be along the summit level, where there were no locks. He went on to state in his report that future growth in trade could be met by drawing on the effluent of collieries, and perhaps from the Erewash catchment as well. In the event, the Erewash Canal Company would have none of that; they were to insist that any water taken into the summit level from the River Erewash would only be from ground thirty feet below the canal's water surface! This would prevent the newcomers intercepting streams from ground higher than their canal, although they would be permitted to pump from a lower level, including from coal mines. The Cromford party were to face stiff opposition in Parliament from the Erewash Company; in fact agreement on this issue only followed with their signing of a bond for £10,000 to guarantee acceptance of the Erewash Company's conditions. Clauses to this effect were incorporated into an amended bill.

There were good reasons for the persistent opposition of the Erewash Company. They had evolved as an adjunct of the Loughborough Navigation with which management structures were shared; indeed their waterway was originally planned for the purpose of exporting coal into Leicestershire. The Loughborough Company had since built up a substantial trade in the Trent valley and beyond and their future business prospects looked promising, to the extent that they were even then preparing a bill for a Leicester line.[1] The Cromford Canal would have access to untapped mineral resources of high quality and was seen as a threat to these long-established navigations which carried similar products of a lesser quality. It was hardly surprising that the new neighbours from Cromford were unwelcome.

The Cromford party began preparations for their Act of Parliament with a meeting in Alfreton on 15 December 1788. Jessop was present and although he was to describe himself in Parliament as the 'projector',[2] he had only been given a brief 'in pursuance of the directions

---

[1] 31 Geo. III c. lxv (1791).
[2] HLRO, Evidence on Cromford Canal Bill, W. Jessop, 1 July 1789.

given';[1] the Gells and their colleagues were very much in control. A policy statement was prepared relating, among other matters, to tolls and the mode of applying to the Erewash Company for their assent to a reduction of tolls, as well as the pressing matter of the subscriptions for £100 shares, based on Jessop's estimate of £42,697. The proprietors were then restricted to ten shares each until 15 January 1789 in order to give those unable to attend a chance to invest in the enterprise. Interest was fixed at 5 per cent until works were completed and a call of 2 per cent was made to defray preliminary expenses.

A notable absentee that evening was Sir Richard Arkwright, which was unexpected because he had written beforehand that 'I am fear full of the proposed alteration. In taking the water so soon as it leaves the cotton mill will be an inconvenience and a loss to me'.[2] His statement suggests that the location of the all-important water intake at Cromford was still undecided; in fact it remained so until work began some months later. This was but the beginning of the several disagreements Arkwright was to have with his colleagues; he was generally uncooperative, particularly on issues relating, even marginally, to rights he held in Cromford.

The draft petition and bill were ready for a public meeting held in Nottingham during February 1789, where evidence of the interest aroused was shown by the presence of the Duke of Devonshire, the Duke of Newcastle, Lord Rawdon and Lord Middleton, and also representatives of the Erewash, Loughborough and the proposed Leicester canal, companies. A number of clauses were subsequently revised, as it was desirable to gain local support, but even these were not enough to prevent strong opposition in Parliament, notably from the Derby Corporation and mill-owners on the lower Derwent, as well as from the proprietors of the Derwent Navigation. All objected to this newcomer sharing in the water resources of the Derwent and its tributaries, although the evidence suggests that the Derby party, rather like the Erewash Company, were mainly concerned with the effects the canal would have on their trading monopolies.

Shortly after that meeting Beresford reported to his colleagues, with much satisfaction, that the subscription lists were nearly full. The leader of the enterprise was then seen to be Philip Gell who had subscribed £6,000; his brother Captain John Gell had invested £2,000, Beresford

---

[1] DRO, D258/41/32r, Minutes, 15 Dec. 1788.
[2] DRO, D258/41/32m, Arkwright to P. Gell, 13 Nov. 1788.

also £2,000 and Arkwright but £1,000.[1] Beresford impressed on his colleagues the need to seek the support of all the canal's riparian owners in order to keep control of the company within the hands of local investors and away from undesirable speculators. Several land-owners had been slow to respond to his invitations however, one of whom was Peter Nightingale of Lea Hall (father of Florence). He was not an easy person to deal with; he was well-known to the gentry as an erratic character given to wild and drunken escapades of which they did not really approve. Beresford 'wished that somebody would persuade Mr Nightingale to take £1,000. Pray send somebody to him that bears an influence over him, if any such person there is'.[2] His gracious neighbours were all reluctant to do this but fortunately, for them, young Outram had already been helping the committee by visiting landowners, advising them on investments and collecting subscriptions. Hence he was sent off, innocent and unsuspecting, to meet the eccentric Nightingale. As luck would have it, Benjamin got on well with him and was very successful because he 'was with Mr Nightingale a few days ago. He seemed very anxious for the success of the measure and wishes to meet you'.[3]

Jessop's report had specifically referred to the benefits of the project which would accrue to the poor of the district but this meant, to his wealthy clients at least, by the fruits of their labours rather than as subscribers. At least one investor visited by Benjamin Outram freely admitted his good fortune because he wrote facetiously to Philip Gell that 'Mr B. Outram took my subscription opposite £1,000 for securing a share in the Cromford Canal, undertaken most generously and disinterestedly for the public benefit at the hazard of gaining 40 per cent without fear of loss'.[4]

But Outram was not to wait long before his new acquaintance, William Jessop, diverted his attention to the more interesting and exacting engineering aspects of the project which would have to be addressed very carefully before justifying the design of the canal before the parliamentary committees. Thus 'Mr Jessop is with me and we are going together to get ourselves informed about the mills and reservoirs'.[5] After that journey Jessop had sufficient confidence in his young

---

[1] DRO, D258/41/32u, F. Beresford to P. Gell, 3 Jan. 1789.
[2] Ibid.
[3] DRO, D258/41/32v, B. Outram to P. Gell, 20 Feb. 1789.
[4] DRO, D258/50/15k, J. Wilkinson to P. Gell, 11 March 1789.
[5] DRO, D258/50/13a, B. Outram to P. Gell, 5 March 1789.

colleague to delegate to him entirely the tasks of investigating mill machinery on the Derwent and its tributaries, as well as the monitoring of river flow rates and industrial water usage. In the meantime he went off to supervise the construction of works under his control in other parts of the country. Outram's investigations must have been thorough because both he and Jessop proved to be well-briefed witnesses by the time they were called before the parliamentary committee.

During the spring of 1789, Captain Gell took up residence in Wimpole Street and set about canvassing support for the bill which was soon to be presented to Parliament. He was particularly active in the Lords, where he persevered in his task, though frequently rebuffed. It was reported to Philip Gell that 'you have a most active agent I perceive in London, your brother, who scarce takes rest day or night from his labours in soliciting the Navigation Bill'.[1] In London political circles the Cromford became known as 'Gell's Canal'.[2] Other proprietors also helped and behind all was the reassuring presence of Jessop. The Captain wrote that 'I have much dependence on Jessop if examined before the whole House, they all respect him and attend to what he says'.[3] Outram too, was 'full fraught with knowledge of the Derwent and knows to a fraction what quantities every dam contains'.[4] Clearly the Captain thought that all would go well; 'Jessop wishes much to be examined, perhaps to triumph over his brother engineers, and Outram is full of fire in the same way to show himself. Fletcher promises not to touch a drop of beer till it is all over'.[5]

Captain Gell's letters to his brother also implied that they had other, lesser opponents who, quietly and anonymously, influenced events through their Members of Parliament. Quarry- and mine-owners situated at some distance from the canal, as well as the road hauliers serving them, fell into this category and perhaps this explains the evasive answers, or blunt refusals of support he received from some local Members. A few of the Lords, notably the Duke of Newcastle, openly supported the venture but others, including Devonshire, remained publicly uncommitted throughout, unwilling to offend anyone, yet secretly encouraging the canal proprietors because their own interests stood to benefit.

---

[1] DRO, D258/50/15k.
[2] DRO, D258/50/14sb, J. Gell to P. Gell, 30 April 1789.
[3] DRO, D258/50/13q, J. Gell to P. Gell, – May 1789.
[4] DRO, D258/15/14pa, J. Gell to P. Gell, 9 June 1789.
[5] DRO, D258/50/14sb, J. Gell to P. Gell, 9 May 1789.

Lord George Cavendish (a son of the Duke of Devonshire) agreed to carry up the petition and to promote the entire business, claiming he was always ready to 'obey the commands of the gentlemen of Derbyshire' and also to ensure that 'nothing was carried out against them in a clandestine manner'.[1] At the same time he did not wish to offend the Derby party, hence Gell could only report home that 'he was with us but did not promise absolutely'. The Duke of Devonshire was also visited on one occasion but the Captain found that 'Devon's Duke was not up at two' and on returning later in the day 'at half past three at Devon House, he was but just up. If God does not please to alter the mode of the world there is a certain number of people living in this place who do not see the day or the blessings of the sun'.[2] Devonshire proved to be evasive although their parting was amicable. The duke 'promised to let him know' but he never did so.

Among many others, Gell interviewed Britain's leading sailor and got an encouraging and interesting response, but probably made a rash promise in return:

> This morning [I have] been with Lord Howe. He is not against canals that do not join the seas and cross the island. In his mind he is with us. I mentioned the great number of people employed on the Trent and this could increase and draw many to work on water than had any Idea of it and men may be raised for the Navy, and answered that I would engage to raise 300 men'.[3]

The petition had passed through many hands in the county for signature and its whereabouts were sometimes in doubt. John Gell stressed to his brother Philip the need for signatures by men of substance, preferably freeholders and would 'not have any child support it'. As the parliamentary proceedings drew closer he was anxious to see the document completed and at a late hour noted, with exasperation, that Arkwright and Charles Hurt had yet to sign it. Moreover, he thought that not everyone at home seemed to be taking matters as seriously as they might because he wished that 'the gents at Wirksworth would keep their feasting and gluttonising till the Bill is past for I have been told of the burning of the papers more than once. Pray desire Charles Hurt to be

---

[1] DRO, D258/41/32z, J. Gell to P. Gell, 27 Feb. 1789.

[2] DRO, D258/41/32aa, J. Gell to P. Gell, – Feb. 1789.

[3] DRO, D258/50/13m, J. Gell to P. Gell, 14 March 1789.

easy for a little time and then they may riot as they please'.[1]

Arkwright was a frequent source of worry for the Captain who reported that 'The Knight' was seen too often in the company of the Derby party. This concern was shared by Evans, the solicitor, who found it necessary to speak to Arkwright quite severely, saying that 'the whole country' would be alarmed because many had subscribed only when it was known that he was associated with the project.[2] Then, just before a Commons committee was to consider the bill, the Captain wrote again that Arkwright had complained of the inconvenience he would suffer by the canal crossing his lands and that 'the roads would be soiled during construction and the horses used on the canal will eat all the meadows. Between ourselves, he seems to have no regard to consistency of character'.[3] Other disagreements followed at that late stage before Beresford intervened to warn that 'we have reason to believe that Sir Richard Arkwright and Mr Hurt are becoming cool if not adverse to the business'.[4] Arkwright was clearly seen as attempting to gain advantages, knowing that his colleagues were likely to appease him rather than risk dissension at that critical time. Beresford sagely noted, however, that they would have to make a stand sometime because each concession seemed to lead to just another demand. He was correct in his conjecture because Arkwright then alarmed them all further with a new submission to the Cromford party, this time concerning the canal's water supply. He had declared his 'direct opposition to our plans unless a certain alternative can be adopted which Mr Jessop declares is quite impracticable'.[5]

Now although Jessop did not seem to have finalised the position of the water intake to the canal by that time, he probably intended to draw directly from the Cromford sough, a mine drainage adit which flowed into the River Derwent close by the canal terminus. Arkwright's new proposal was to raise the level of the existing weir in the river by his Masson Mill so that water would flow from there, by an open channel, across his land (where he was having a mansion, Willersley Castle, built) and by an aqueduct spanning the river before discharging into the canal on the other side. This idea meant that the existing weir would have to be raised by 16 ft to provide the necessary head. It was a needlessly costly solution with most of the advantages accruing to Arkwright,

---

[1] Ibid.
[2] DRO, D258/50/14u, J. Gell to P. Gell, 21 April 1789.
[3] DRO, D258/50/14t, J. Gell to P. Gell, 20 April 1789.
[4] DRO, D258/50/14i, F. Beresford to P. Gell, 9 April 1789.
[5] DRO, D258/50/14j, F. Beresford to P. Gell, 28 April 1789.

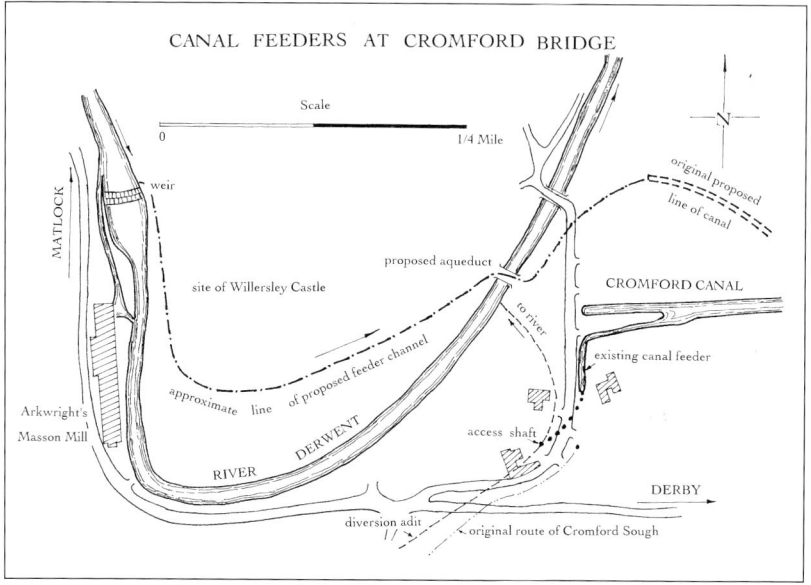

Figure 2 Canal feeders at Cromford Bridge.

whose water storage and power for the mill would thereby be substantially increased. Moreover, the feeder channel and sluices would lie in full view of his house and thus be under his control in any future dispute. And the Cromford sough, which Arkwright diverted frequently for his own use at the mill, would remain untouched for the purposes of the canal (Fig. 2).

Jessop and the attorneys were scandalised by this latest intervention and at first the engineer refused to consider it. Arkwright then began to complain of his expenses; he had retained counsel and several attorneys to oversee his interests and yet felt that all their charges should be met by the company. Small wonder that the unfortunate Captain Gell was 'so taken up in reconciliations'.[1] Finally, for the sake of maintaining peace, he managed to persuade a reluctant Jessop to accept the proposals, although he and his officers secretly had no intention of complying with Arkwright's wishes, once construction work began. The sorely tried Captain wrote to Philip to say that Arkwright at last seemed satisfied and that 'I am his friend. He even smiled today which he has never done before on this subject. What do you think of him now? I think more than

---

[1] DRO, D258/50/14g, J. Gell to P. Gell, 30 April 1789.

I shall say'.[1]

Benjamin Outram had not been so circumspect when expressing his personal opinions on the untrustworthy Arkwright because Captain Gell was to inform his brother that 'he [Arkwright] complains of Outram talking disrespectfully of him, of which he hears in the country, such as 'he ought to have his head off'. I wish Outram would not say any such thing, nor is it proper that he should'.[2] Their young assistant was clearly puzzled when writing later to Philip Gell from London with his reports. In passing, he noted that 'your brother the Captain has been this morning with Mr Beresford and Sir Richard Arkwright, who behaves rather strangely. I do not know what to make of him'.[3] Outram had yet to learn to keep his own counsel because there are always people around only too willing to carry tales. Arkwright, in his turn, said nothing to Outram to show that he was aware of those careless remarks. He had an unforgiving nature and his revenge was duly exacted at a later stage, somewhat to Benjamin's disadvantage.

Outram was working very hard in his father's practice at the time besides assisting on the promotion of the canal, but he obviously relished this abundance of work and his new experiences. Travelling up and down to London from Alfreton was not easy: 'I got to London on Monday morning about six o'clock after a fatiguing journey partly on horseback and partly on the shaking box of the male (sic) coach, no inside place to be had, but numbers waiting for conveyance at every stage'.[4] He wrote several letters to Philip Gell, giving good resumés of progress in the House. Once he referred to a matter which suggested that he and other proprietors (notably Captain Gell) were not averse to using bribery on occasion. One of the Cromford Company's agents was a Mr Bloxam who had been frequently praised for his work on their behalf. His efforts were not entirely unselfish, however, and for several weeks he had been hinting to Outram that he would appreciate the gift of a good gun dog, preferably a young and seasoned pointer. Ostensibly this was because

> Mr Bloxam wishes to have a pointer that has been hunted a season or two which along with a staunch one that he has already got, he means to present to Mr Pitt's private secretary from whom he is

---

[1] DRO, D258/50/14h, J. Gell to P. Gell, 1 May 1789.
[2] DRO, D258/50/14ta, J. Gell to P. Gell, – May 1789.
[3] DRO, D258/50/14sa, B. Outram to P. Gell, – April 1789.
[4] Ibid.

asking favours, and who has hinted that he wants a brace of good pointers. You will do an essential piece of service of sending him a dog immediately—a favour well-timed has a double effect—I beg you to do this directly for he cannot bring forward his request till he has got his dogs'.[1]

Whether or not Philip Gell approved of such measures is unknown but the request was ignored, as were others as late as October that year when the bill was clear of danger and work on the canal was in progress. Pitt's secretary, or more likely Mr Bloxam, never did get that dog.

Jessop had not confined his report to engineering matters alone but had also provided a rudimentary economic review which referred to the commodities to be conveyed by the canal. He stated that 'he had been informed that the quantity of coal taken from Cromford would probably be about 25,000 tons annually' (presumably this was an error: he meant from Pinxton to Cromford) but as regards the export of lime from the locality, he had made a calculation 'to form some Idea of the probable Consumption ... . I have supposed that for 16 miles on the River Erewash and 12 miles down the Trent to Ratcliffe ... as the Expense would increase, the Demand would be less and less till at Newark it might end in a Point'. He estimated that 179,200 acres of agricultural land would be accesssible and if this were manured once in ten years with three tons of lime per acre, then the average annual consumption would be 53,760 tons. Costs of carriage were given as 7s. 6d. per ton, increasing to 10s. if transported as far as Newark. He concluded, rather unwisely as it turned out, that 'I must leave it with those better acquainted with Agriculture than I am to correct this Calculation'.

His opponents readily supplied such experts during the proceedings in Parliament in June 1789. Two lime-burners came from Ticknall for that purpose and they pointed out that Jessop's calculations assumed that lime was not already supplied to the Erewash and Trent valleys. They claimed that the 18,000 or 19,000 tons exported annually from their district was more than enough to meet the needs and that another 53,760 tons would be superfluous.[2] Fortunately for Jessop, they were found wanting on the price for their lime. Crich lime delivered for less than 7s. 6d. per ton would undercut the Ticknall product at 8s. 4d., the latter cost being a result of expensive land carriage to the Trent. This seriously

---

[1] DRO, D258/50/14eb, B. Outram to P. Gell, 3 June 1789.

[2] HLRO, Evidence on Cromford Canal Bill, G. Banton and G. Hutchinson, 22 June 1789.

undermined their case and they were, furthermore, put in a bad light when cross-examination revealed that a fraudulent proposal to cut their price to 7s. per ton had been widely publicised in order to detract from the viability of the Cromford case. That round certainly went in favour of the Cromford party.

The inquiry continued at the bar of the House of Lords on 22 June with the appearance of John Varley, 'Surveyor and Engineer', a witness for the Derby party.[1] Varley, who was to criticise the engineering aspects of the scheme, seemed an unwise choice on several counts. He had been made resident engineer on the Chesterfield Canal on the death of James Brindley in 1772 and, although those works were completed under his direction, he had proved an incompetent contract administrator. After irregularities had been found in the books, he was fortunate to keep his job; yet the same thing happened five years later whilst he was employed as engineer to the Erewash Canal Company. On that occasion his ignorance of paperwork and accounts led to his dismissal,[2] hence it is surprising that this ordinary fellow should have been selected to give evidence in Parliament to debunk the designs of the country's leading civil engineer.

Varley's evidence was negative and lacked substance. His uneducated manner and inability to substantiate his calculations did not inspire the confidence of the committee, at least that was the view of Captain Gell, who gleefully dismissed him as a witness of little account. Varley began by stating that the ground crossed by the navigation was unsuitable 'stony and loose soil' which could not be made to contain water. He made no mention of water-proofing by clay-puddle lining, until he was pressed to admit that this was common practice. Varley then described his experiments made on a 24-mile level pound of the Chesterfield Canal which, he claimed, proved that the water loss due to leakage and evaporation was about 73½ tons per mile of waterway per hour. He applied this figure to the Cromford Canal (assuming 16 miles of waterway) and thus asserted that 28,224 tons of water would be needed each day to make up this loss. He then added an amount required for lockage to give a total water usage of 34,516 tons each day, or 241,612 tons per week, just six times the quantity assessed by Jessop! Varley seems to have got off lightly under cross-examination; few must have been impressed by his overstatements and his calculations were not taken

---

[1] Ibid., J. Varley, 22 June 1789.

[2] C. Hadfield, *The Canals of the East Midlands (including part of London)* (1966), pp. 40, 41.

seriously. Even so, Jessop's design figures were hardly adequate for the efficient operation of the canal. If, as he had estimated, the annual export of quarry products to the Trent valley would be 53,760 tons, then about 23,500 tons of water would be needed each week for this purpose alone. Ostensibly this was well within his assessment of 41,040 tons but it left too little for other exports, for leakage, evaporation and wastage, quite apart from losses through human error. All were inevitable future occurrences.

Jessop's figures were also criticised by the next witness for the opposition, Robert Mylne FRS, the Scottish engineer and architect who was justifiably famous for his engineering accomplishments, which included the Blackfriars Bridge and various harbour schemes.[1] An irascible fellow, he knew Jessop well because both were members of the Society of Civil Engineers. There did not seem to be much love lost between them, although the dislike seemed to emanate from Mylne rather than the amiable Jessop. At the time Mylne was engineer to the Lea Navigation and yet he admitted, under cross-examination, that he had never designed, nor built, a canal or a lock. His style as witness was to find fault with everything; moreover he was pompous and sought to impress everyone with his technical knowledge, which was frequently couched in language difficult to understand. For example, when commenting on the water abstractions proposed and the effects these would have on the mills of the Derwent, he said:

> considering the nature of these works it will be of very considerable detriment indeed. They are various in their purposes and very different in their effects. The prime mover and original power, which sets them all in motion is the quantity of water. Any quantity taken from the water of the Derwent is the cause of deprivation of the principal mover.[2]

Mylne went on interminably in this manner, no doubt confusing and boring both counsel and Members alike with his verbosity. Not surprisingly, he displeased Captain Gell who told his brother that 'being a Scotch man he is very prolix ... Jessop says his calculations are false, he means to refute them'.[3]

Jessop had expected that there would be fierce opposition from

---

[1] C.M. Norrie, *Bridging the Years* (1956), pp. 11, 33.

[2] HLRO, Evidence on the Cromford Canal Bill, R. Mylne, 3 July 1789.

[3] DRO, D258/50/14qb, J. Gell to P. Gell, 4 July 1789.

riparian mill-owners but in spite of this he had been sympathetic to their interests when designing his scheme. His requirements were trivial compared with normal river flow; moreover, he had decided to locate a water intake at Cromford which was well upstream of the several tributaries draining into the river between there and the industries at Belper and Derby. Mylne ignored this significant fact and insisted that abstractions of one-twentieth of the flow at Cromford meant one-twentieth less power for the mills situated anywhere downstream. Conveniently, he implied that the quantity abstracted was one-twentieth of the flow at all times, ignoring the fact that Jessop only intended to draw off this fraction of the lowest flow of recent years. He went on to say that the average velocity of the river was one mile per hour and the full effects of the 'deprivations' would be felt in Derby at least 24 hours after the weekend abstractions were completed. In some ways he seemed bent on misleading the committee but several gentlemen present would not have been taken in, including that eminent authority on canals, Francis Egerton, the Duke of Bridgewater. According to the Captain, he 'shook his head generally when he [Mylne] was giving his opinions'.[1] Mylne was to make a curious, nonsensical remark with which he sought to illustrate his arguments: 'if I may be allowed the expression; a man finding by his trade that a lever of 20 inches long is necessary with his strength to perform the business, if an inch on a foot is taken away, all his business is stood still'. His eminent audience would have made very little of that remark.

Mylne's observations on water usage were at variance with Varley's outrageous computations and these widely differing views could not have been lost on the Members. He reckoned from his own experiments on 20 miles of canal that losses of 5,000 tons of water per mile each week could be expected but in the case of the Cromford Canal: 'I suppose they will take a great deal of pain to make it tight and it will only be 4,000 tons'. He calculated that the losses on the Cromford Canal up to the junction with the Erewash Canal would be 48,854 tons of water per week and to this he added lockage water to allow for the estimated export of 53,700 tons of lime, 25,000 tons of coal and 6,240 tons of other goods. The quantity of water used weekly was thus estimated to be 86,284 tons. This was twice the amount required by Jessop's calculations but, on this occasion at least, Mylne was probably not too far off the correct mark. And although he had debunked every feature of the canal on which his opinion was invited, on one matter he was certainly correct.

---

[1] Ibid.

TABLE 2.1

Principal Mills on the Derwent, 1789

Location	Wheels	Head	Value of buildings	Return p.a.	No. of workers	Type of factory
1. Belper	3	6ft	£26,247	£36,400	800	cotton
Makeney & New Mills (Strutt)	9	8ft	£11,000			cotton
2. Darley (Evans)	7	5ft 8in.	£13,000	£20,800	450	cotton lead paper
3. Derby or St Michael's (Derby Corporation)	6	4ft	£11,000	£22,152	90	silk copper zinc

Source: HLRO, Evidence on Cromford Canal Bill, July 1789.

He pointed out that it would be difficult to ensure that, at all times of the year, no more than 41,040 tons were drawn from the river during the permitted weekend periods. The accurate measurement of river flows is never an easy matter, particularly when levels are constantly changing. He doubted, quite rightly, that it could be done without very close control and, as a result of his comments, a clause was inserted in the Act to ensure that independent inspectors were to be appointed to supervise the weekly event.

An indication of the local importance of the River Dewent in the domestic and industrial communities can be gleaned from the documents prepared by Outram and tabled in the House.[1] These listed mills and works 'from the Tail of the Cromford Mills to the Place where it falls into the Trent near Sawley'. There were upwards of thirty mills and a 'water engine' which supplied householders in Derby, besides 'common cocks' for the poor. There were 53 water wheels in use with a total head

---

[1] HLRO, Evidence on the Cromford Canal Bill, July 1789, List of Mills on the River Derwent.

of 55 ft 3 in., and the value of buildings and machinery was £84,247, with a return per annum of £97,092. Most important, however, was that 1,800 people were employed 'and moreover the Salmon Fisheries produce £500 a year'. A few of the bigger mills and their owners were listed as in Table 2.1.

Mylne was familiar with the characteristics of the water-wheels used in those mills and he stated that many were of the undershot type which rotated by the force of fast-flowing water striking the base of the wheel. He considered, quite rightly, that these were wasteful, outmoded machines, preferring the breast wheels because 'in these modern times it is generally allowed by the best engineers that the water acts to the best advantage by gravity, by bringing it slowly and coolly'. By this quaint comment Mylne showed a glimmer of understanding of the nature of hydraulic power and demonstrated that engineers, even at that early date, realised the significance of the relationship between the quantity of water supplied and the head. This led, in Victorian times, to the design of impulse and pressure turbines which utilised, more efficiently than water-wheels ever could, the range of water resources (in terms of the flow rate and head) encountered in nature.

Mylne mentioned a related factor to his cross-examiners and the assembled House. In his pompous manner he said that 'There is an effect, which is rather delicate in its nature and not easily understood. The taking away any proportional part of a first mover is a proportional loss in the value of the works but the business to be performed by the remainder of the power is not of so good a quality as that which was performed by the whole power'. It is doubtful if any of his listeners knew what he was talking about but he simply meant that if a wheel had been properly designed for a given water supply and head, then its efficiency would fall off if, for any reason, the flow of water was reduced.

It was about that time that Captain Gell sensed a change of mood in the House and he reported home that 'The Lords are certainly tired of us and I am sure we are sufficiently so of them. It is difficult to get attendance so late. Many are going. I am of the opinion that the Chancellor will do for us when the water is cleared up, for many are swayed by his opinion'.[1] The worthy Captain nevertheless continued his lobbying and was forever anxiously counting aristocratic heads, for and against. He had long since realised that success or otherwise would owe much to the whims and fancies of these uninterested peers; small wonder

---

[1] DRO, D258/50/14x, J. Gell to P. Gell, – July 1789.

he wrote that there would be 'no more canals for me'.

The evidence of the Cromford engineers was refreshingly different from that of their opponents, particularly as to the effects that the project would have on industry and the interests of the riparian owners of the Derwent. Both Jessop and Outram proved to be well-informed and the former, who bore the brunt of the attacks on the project, responded convincingly. However, it is noticeable that Jessop's reputation helped him evade some difficult questions, a few of which he answered imperfectly, or even incorrectly. The opposing counsels were not well-versed in the limited knowledge of engineering mechanics which prevailed at the time but, even so, opportunities were lost by their failure to ask commonsense questions about some of the vague assumptions upon which the water supply provision of the canal was based, and which was the principal reason for opposition to the scheme.

In explaining the water supply needs for the canal, Jessop stated that even in a normal season the quantities abstracted would amount to only 1/140th part of the river flow. He observed that the water 'wasted' (i.e. not used for power) on weekdays in such seasons was as much as 50 times or more than he had 'asked for' and thus abstractions for the canal could have no effect on other users. Moreover, out of 60 wheels in use, about 50 used twice as much water as was necessary because they were in such a bad state of repair. 'Most of the other 10 wheels are much better and some of them perhaps as good as can be made, but those mills never work on Sundays and therefore have the water running waste most part of the day'. He also stressed that out of 180 ft of fall down the river, use was made of not more than 110 ft. He calculated that there was four times as much power unused as that already derived from the river by industry.[1]

Jessop was then directed to withdraw and two local worthies, William Kirkland, a miner from Cromford, and William Longden, a stocking-maker from Derby, were called to give evidence on behalf of the Cromford party regarding water flows and levels of the Derwent during dry seasons. Their simplistic evidence, in the absence of any systematically derived hydrological data, was all that could be mustered by the promoters. Both men lived close to the river and claimed to have observed the floods and depth changes which had occurred over the years. They confirmed Jessop's belief that the river was as low as it had ever been when he recorded its flow as 570 tons per minute, late in

---

[1] HLRO, Evidence on Cromford Canal Bill, W. Jessop, 1–7 July 1789. The number of waterwheels was in fact 53 (above, p. 24).

1788. Their methods were hardly scientific, however. Kirkland, for example, said 'There are steps lead (sic) down to the river to fetch water out of it, and by means of these steps and the rising and falling of the water, I judge of the height because [I] regularly fetched water'.[1] He unwisely claimed that, in November 1788, the level was 2 in. lower than in the dry season of 1785, after which counsel set about to confuse this humble witness with questions about the precise levels in different months and years, all of which soon showed up the inadequacy of his memory, unsupported as it was by written memoranda.

Outram was then called and questioned on the capacity of the weir pools along the River Derwent and the operation of the waterwheels.[2] He knew a great deal about each mill, just as the Captain had said, and could quote flow rates and operational data with surprising confidence for a man of 24. His evidence suggested that the mill-owners had little cause for concern because canal abstractions would be trivial compared with the normal river flow. He also stressed that the Derwent received substantial discharges from several tributaries before reaching the industries of Belper and beyond, all of which had been conveniently ignored by Mylne. Outram's description of the Derwent Navigation drew attention to the unsatisfactory character of such obsolescent waterways: 'There are a number of rapid streams and shallows in it, which render it extremely difficult to get a boat up in dry seasons ... it would take as much water to bring up one boat in a dry season from Lord Harrington's Mills as this proposed canal would be empowered to take in a month'. As regards the movement of vessels through the shallows, this was achieved 'by opening the sluices of the mill ponds and by that means making an artificial flash, and by lighters to take out part of the cargo'.

On that day, 6 July 1789, proceedings were drawing to a close but the two engineers continued to display an impressive understanding of the River Derwent's flow characteristics. Jessop followed up Outram with support for the view that the Navigation could not suffer by the water abstractions:

> If it [the abstraction] was taken every day in the week in very dry seasons it could not reduce the depth of water more than one-eighth of an inch over the shallows and being taken only on a Sunday it might for some portion of the Monday reduce it near an inch. This arises, not from want of water in the river but from the

---

[1] Ibid., W. Kirkland, 6 July 1789.
[2] Ibid., B. Outram, 6 July 1789.

very great imperfection of the navigation, for less than one-fivehundredth part of the water should be sufficient to make it a complete navigation ... the amount of injury would be if a boat should happen to pass on the Monday, the boat must discharge two tons more into the lighters than they would have occasion to do...not likely to happen more than five or six times in a year, it may be easier to make compensation for it.[1]

Thus he pointed out that engineering schemes which could possibly be afflicted by the vagaries of the weather should never be built to accommodate every contingency but, instead, should be designed to reasonable criteria and, when the extremes occurred, those suffering damage should be compensated for their losses.

Jessop's concluding evidence ranged widely over the details of his project and included a description of the aqueduct proposed by Arkwright, as well as the effects that the Butterley tunnel might have on springs and the natural water table, and the manner in which streams might be taken across and under the canal. He was questioned closely on the method of measuring the exact amounts of feed-water taken into the canal and perhaps he made lighter of it than he should. This line of questioning continued into the following and final day but Jessop still fenced cleverly with the opposition's counsel without giving precise details of what would have been a difficult measurement problem at that date. At one point, counsel asked Jessop if he really meant to say that the velocity of flowing water at the surface of the aqueduct was the same as that at the bottom. With every confidence the great man replied 'I certainly mean to say so, or I should be more ignorant than a schoolboy'. Questioning ceased with that rejoinder, but present-day knowledge shows Jessop was wrong; water velocities in open channels are by no means uniform.

A significant point raised by counsel concerned the rate at which water would be fed into the canal at the point of intake: 'I wish to know what calculations you have made to prove the water will run in, in twenty-four hours, considering the length of canal and the small fall'. Counsel must have been prompted to ask such a tricky question by someone (probably Mylne) and Jessop was unable to answer except by saying, 'I have not made any calculations but speak merely from my general acquaintance with the subject'. He was allowed to escape with this lame response but, with the state of knowledge of the time, it is

---

[1] Ibid., W. Jessop, 7 July 1789.

difficult to believe that anyone would have known how to determine this. In fact, present-day calculations show that filling could not have exceeded about 60 cu. ft per second without creating surge waves, scour and over-topping of the canal banks. Jessop's design required an inflow of only 17 cu. ft per second, so he was fortunately correct on that occasion.

Victory was close at hand, as the Captain wrote home that

> The Lords are tired of us and many declare in our favour ... the trial is to be for three days to come and Lord Sandwich is in the chair. Lord Scarsdale has never appeared and Upton [Derby's counsel] prevented him he told me today. This youth I really think has more assurance and impudence than most people possess.[1]

Lord Stanhope, who had not been as consistent a supporter as Captain Gell would have liked, came in for a severe lambasting as 'he is for teaching everybody. The Bishops religion and the Chancellor the law, and now Jessop, as engineer. They all say our evidence is clear and steady'.[2] A week later he wrote again to say 'we are all over and passed yesterday and the message sent to the Commons, for I went there to hear it read'.[3]

Benjamin Outram, too, sent his congratulations to Philip Gell, perhaps in the hope that all his past efforts would lead to greater things:

> Lord Rawdon and the Derby people seemed disappointed by the offer of terms of accommodation ... I sincerely wish they may be mortified by seeing in a few years time a most extensive trade carried on upon this canal, such as may exceed even the hopes of its promoters ... the Peak and the Wapentake must ever remember with respect this great debt they owe you and those other gentlemen who have been the active promoters of this measure.[4]

---

[1] DRO, D258/50/14x, J. Gell to P. Gell, – July 1789.
[2] DRO, D258/50/14ya, J. Gell to P. Gell, 7 July 1789.
[3] DRO, D258/50/14ob, J. Gell to P. Gell, 16 July 1789.
[4] DRO, D258/50/14ca, B. Outram to P. Gell, 19 July 1789.

CHAPTER 3

THE CONSTRUCTION OF THE CROMFORD CANAL

Within a month of the passing of the Act, a succession of meetings had arranged the management structure of the new company and the officers were appointed. The organisation compared in style with that commonly adopted by other canal companies: there was a general assembly of shareholders, a management committee to control day-to-day events, and a senior staff of clerk, treasurer and engineer. Francis Beresford might have been expected to accept the office of clerk but his legal practice extended over several counties and such a parochial appointment was not for him. That went instead to Francis Evans, his co-author of the Act.

Chairmanship of the company and its committee was accepted by Sir Richard Arkwright. It was surprising, in view of the difficulties he had created for his colleagues in the recent past, that he had been offered this key appointment but there was some advantage in this. Arkwright was a major industrialist who also owned land and water rights locally and it would have been seen as better to have him working for the company within, rather than allowing him to create mischief from the outside. So once again, his boorish ways would be tolerated but since he owned only ten shares in the business the full weight of the other shareholders could outvote him if circumstances ever required it. Philip Gell would seem to have been a preferred choice from the outset and indeed he was to officiate as chairman on occasion, but his indifferent health seems to have prevented much active participation once the company was formally established.

After settlement of their parliamentary expenses of £2,700 and the negotiation of the agreement and bond with the Erewash Canal Company on water rights, the committee's attention was directed to the selection of engineering staff. The latter would prepare the detailed site surveys, the working drawings and specifications for the canal structures and other engineering features, before tenders were invited from contractors.[1]

Jessop was appointed principal engineer with a salary of £300 per annum plus £50 expenses, provided that he would spend one third of his time on the project. Benjamin Outram was appointed superintendent of

[1] PRO, RAIL 819/1, 24 Aug. 1789.

works at a full-time salary of £200 p.a.[1] Captain Gell implied that these appointments had been the subject of some disagreements because, before they had been confirmed, he had written to his brother Philip that

> I know that he [Arkwright] will oppose Mr Jessop. Where you will find a better engineer or a more able practical man I do not know .... His practice on many canals points him out. B. Outram and he are great friends and he may be his resident executioner. I should think Jessop will be much disappointed if he is not concerned.[2]

Fortunately these fears had not been realised but Arkwright was determined to settle at least one account because the committee (chaired by him on that occasion) resolved that

> Sir Richard Arkwright be requested to recommend a proper person to superintend of that part of the canal which goes through his property under the direction of Mr Jessop the engineer, and at the expense of the company and that Mr Benjamin Outram's office of surveyor shall not extend to such part of the said canal.[3]

It is surprising that the committee allowed this malicious resolution, which was nothing but revenge for Outram's earlier impertinence. Once again Arkwright was appeased and the shareholders were expected to defray the costs as well. But if this petty action gave Arkwright satisfaction, it lost him more of the goodwill of his worthier neighbours, none more so than Captain Gell. It was he who still smarted at the treatment received at the hands of Arkwright at the time of the parliamentary procedures. He told his brother that 'I am convinced he has many designs which the world cannot know of. He has not so much of the milk of human kindness in him as the world would give him credit for'.[4]

However, the Captain derived much satisfaction on one important issue. Jessop, rather against his will, had been obliged to defend Arkwright's ideas for the feed-water intake at the head of the canal but when the issue was later raised in committee, Gell successfully instigated

---

[1] Ibid., 5 Oct. 1789.
[2] DRO, D258/41/32ia, J. Gell to P. Gell, 13 Aug. 1789.
[3] PRO, RAIL 819/1, 14 Jan. 1790.
[4] DRO, D258/50/14da, J. Gell to P. Gell, 18 July 1789.

a move to abandon that costly design and to revert instead to the cheaper option of diverting water from Cromford Sough. Thus the aqueduct was never built, nor was the weir at Masson Mill raised, but this must have displeased Arkwright and may have been the reason why, shortly after the opening of the canal, he saw fit to turn the flow from Cromford Sough into his own weir pool. Not surprisingly, this led to acrimonious litigation with the canal company.[1]

In spite of these machinations, the work of the engineers proceeded apace and an advertisemnt appeared in the press during September 1789, stating that plans and specifications for the canal could be viewed at Outram's home in Alfreton.[2] Three contracts were on offer: for the section between Cromford Bridge to the west of the tunnel, including two aqueducts and various deep cuttings, to be completed in two years; the Butterley tunnel and parts of the cuttings at each end, to be finished in three years; and the canal from the cutting east of the tunnel to Pinxton and the Erewash Canal, to be completed in 18 months.

This programme was ambitious. There were no general contractors in business at that time (that is to say, companies with the finance and wide experience which equipped them to take on major works of this type) and so the new canal companies either had to resort to inefficient direct labour methods, or else invite tenders from 'labour only' contractors. But on this occasion there was no shortage of contractors willing to tender for these substantial works and Outram was soon ordering the digging of trial pits in the rocky ground west of the tunnel in order to assess sub-soil conditions for the benefit of the applicants.

On 5 November 1789 the committee considered the tenders and accepted those submitted by two respected engineers namely, Thomas Sheasby of Tamworth and Thomas Dadford of Corbridge, for all three contracts. The first contract was priced at £8,933 11s. 8d. and the second, which comprised the Butterley tunnel and approach cuttings, at £4 15s. per yard run of tunnel (the canal company providing £1,000 for the provision of pumps) and the cuttings at 5d. per cu. yd. The third contract was priced at £9,284 0s.10d. The total contract sum came to £37,113 3s. 6d. against Jessop's estimate of £42,627.[3]

The estimating methods of that time were crude and speculative and the bids received for the construction of the Cromford Canal were in keeping with such practices; they were based on little more than

---

[1] Private correspondence, author with S.S. Stoker, Cromford.
[2] PRO, RAIL 819/1, 16 Sept. 1789.
[3] Ibid., 5 Nov. 1789.

Plate 1 Interior of Butterley tunnel showing supports to prevent damage from mining subsidence. The west entrance can be seen in the distance.

guesswork. Firstly, there were several unsatisfactory features about the contract documents provided by the company's engineer. In spite of Outram's exploratory excavations, there was a lack of comprehensive information given on sub-soil conditions in the tunnels and in the deep cuttings, and there was very little guidance on the designs of structures and other engineering features. The contractors were expected to fathom out much for themselves and then come to agreements on construction details with the resident engineer only as work progressed. Hence it was impossible to submit accurate tenders before work began.

Moreover, as regards the tenders submitted, contractors were expected to lump several substantial items together and to price them as a single entity (such as the tunnels, which were costed per yard run). Such crude practices can only lead to expensive claims and disputes at a later stage of the works. Unless such items are broken down into their constituent parts, and each part then priced separately, it is difficult for the supervising engineer to make comparisons between different bids—and also to pick out mistakes in the submissions, as well as any omissions, overpriced or underpriced items. It is essential to examine tenders in this manner before their acceptance. Such difficulties were only resolved in more recent years by the introduction of a standard method for the measurement and pricing of civil engineering works,[1] but of course, this was devised only as a result of the mistakes and experiences of generations of engineers since Jessop's day.

The most disquieting fact about the tenders for the Cromford Canal, however, was that Jessop's estimate had been public knowledge for several months. It is noticeable that the successful bids were some 15 per cent below this, suggesting that the contractors had assumed that Jessop's figures were generous assessments of costs, based on accurate measurements. They may then have submitted a lower figure in the hope of a successful outcome. In spite of all the difficulties created by these primitive estimating methods it was, nevertheless, Jessop's task to check whether the contractors were competent, equipped and experienced enough in that class of work to finish on time, and also to ensure that their tender prices were realistic. It was not his job simply to recommend acceptance of the lowest bid to the committee and then leave the parties to the contract to make the best of it as the contract progressed. Nobody benefits by such methods, least of all the clients. Regrettably, Jessop appears to have done just that and, in the event, the bids proved to be grossly underpriced and the company was soon to suffer as a result.

These problems were not anticipated at the time and work soon began on shaft sinking and driving of the tunnel at Butterley. Hand excavation methods necessitated the sinking of 33 shafts, each about 75 yards apart (and up to 57 yards in depth) which would provide a sufficient number of working faces to ensure completion within the contract period specified. One steam engine, with a 40-in. diameter cylinder capable of raising water 50 yards, was used for pumping, as well as several horse gins and other hoists. Much of the heading was driven through the coal measures. Some faulting and thin rock strata were encountered but generally excavation was in soft sub-soils.[2] Twelve or more coal seams, between 12 and 30 in. thick, as well as others of iron ore, were intercepted along the tunnel's length and these finds amply confirmed Jessop's report that there were possibilities for establishing 'considerable ironworks'. The length of completed tunnel, which was lined throughout except for occasional short sections where stable rock was perforated, was 2,978 yards,[3] although railway and roadworks built over the western portal in years later increased the length to about 3,100 yards.

---

[1] Institution of Civil Engineers, *The Civil Engineering Standard Method of Measurement* (1953) and the associated Institution report of 1933.

[2] J. Farey, *A General View of the Agriculture and Minerals of Derbyshire*, iii (1817), pp. 343, 344.

[3] Ibid., p. 344.

News of the discovery of the seams exposed in the tunnel created much interest and Joseph Outram wrote to Philip Gell that 'In the second shaft they sunk 7 yds. then came to a regular bed of ironstone 6 ins. thick and the next two yds. had 4 regular beds of ironstone 6 ins. each together two feet. They have discovered coal in the west end of the tunnel'.[1] Benjamin Outram who, no doubt, was his father's informant, was even then preparing to do something about these promising finds.

It was surprising that the contractors began driving the great tunnel and its approach cuttings, their most difficult tasks, during mid-winter, a period which more prudent engineers would have avoided for such extensive excavations. They should have awaited the drier spring weather, meanwhile concentrating on masonry, drainage works and short, shallow sections of the earthworks which were less dependent on good weather conditions. But, as they elected to go for a spectacular start, high costs were inevitable. These could not have been helped by their subsequent tactics because, by May 1790, they had further stretched their resources by constructing earthworks along the entire waterway.

The measured work was then valued at £7,216, which indicated reasonable progress, but by November the contractors were clearly in financial difficulties. Outram was instructed by the committee to pay the contractors only as 'they appeared to earn' but he was also told to advance them £170 to pay for a timber delivery.[2] Presumably their credit was no longer in good standing and the committee were fully aware of it. Even so, the crash came six weeks later, on 12 January 1791, when Sheasby and Dadford declared their inability to proceed; they then simply walked off the site. At the time they were both involved in the contruction of the Glamorganshire Canal, for which they had been obliged to agree to a bond of £10,000 for the satisfactory performance of their contract.[3] There was no such undertaking with the Cromford Canal Company and so, much to their discredit, off they went to Wales. A financial statement read to a hastily convened assembly of shareholders showed that the contractors had been paid £18,511 but that measurements had revealed over-payments of £427, with a further £904 in dispute over unfinished work.[4] The company threatened to pursue the matter through the courts but, no matter, the contractors could not be persuaded to return and face their obligations. Meanwhile, Outram had

---

[1] DRO, D258/50/15p, J. Outram to P. Gell, 10 Jan. 1790.
[2] PRO, RAIL 819/1, 24 Nov. 1790.
[3] C. Hadfield, *The Canals of South Wales and the Border* (1960), p. 91.
[4] PRO, RAIL 819/1, 12 Jan., 13 Jan. 1791.

to hurry to take over the works and to prevent workmen drifting away from the site.[1]

There were no recriminations from the proprietors regarding these matters; they were content to let their engineering advisers sort out the problems. Outram took over as manager with instructions to complete the works using a direct labour force. He was also allowed to appoint six assistants to help supervise construction. Because of Outram's 'continued and unwearied labour, care and attention', he was 'justly entitled' to a salary increase to £300, plus £100 for expenses, and this award was back-dated to the day when Sheasby and Dadford left.[2] This appointment was a great responsibility because he was still relatively inexperienced and was then but 27 years of age. However, even up to that time Outram had been fully involved with the setting out and management of the works, quite apart from the tasks of negotiating with landowners over land-takes, trespass and various other disputes. He also had to attend all committee meetings and was regularly disbursing very large sums of money for the purchase of land, and to cover wages and material costs. Small wonder an investor could write that 'young Mr Outram seems as likely to acquit himself like a man of real business in his department'.[3]

During May 1791 the committee awakened to the burgeoning costs of their enterprise. The original share capital was £46,000 (even though Jessop's estimate was £42,697) and up to that time calls had been made on shareholders to provide £36,800. Of this latter sum, £32,638 had been expended, although implements and materials were held in stock to a value of £4,000. If they were to add the remaining £9,200 due from shareholders to stock valuations and cash in hand, then a total of £17,362 was available to finish the canal. After discussion, however, they agreed that it would take more than this to complete and it was decided to increase calls so as to raise the share capital to £52,900. At that stage the actual construction costs had been £25,138.

The engineers were present at the meeting and the assessments were presumably of their making. These were readily accepted by a trusting committee and, indeed, both Jessop and Outram were complimented on their conduct and the general progress of the works. This revised estimate was, nevertheless, close to guesswork and certainly did not compare with the eventual costs. It was deplorable that Jessop preferred not to admit that the original estimate was falling far short of the true

---

[1] Ibid., 24 Jan. 1791.
[2] Ibid., 24 May 1791.
[3] DRO, D258/50/15k, J. Wilkinson to P. Gell, 11 March 1789.

costs, of which he must have been aware. By that stage the works had advanced to such an extent that there had been every opportunity for the engineer to compare his original estimate with actual costs. Accordingly, it would have been possible for his staff to prepare accurate forecasts of the time required for completion, as well as for the final costs of the work remaining, but Jessop failed to have this done. The committee also asked him to declare how much longer his services as engineer would be necessary. Jessop replied that construction would be sufficiently advanced by February 1792 for him to relinquish his appointment. But here again, his forecast was to fall short of that mark, mainly because of unforseen circumstances.

There came a serious setback in January 1792 when Jessop reported a failure of the Amber aqueduct.[1] Although the structure was described as an aqueduct, it is actually an earth embankment about 200 yards long and 30 ft high in places, which runs across the river valley. It is still surmounted by masonry walls, which once contained the water channel, and the bank was originally pierced by three small arches, through one of which flows the River Amber. Another arch was for an accommodation underbridge and the third, now demolished, was a curious gothic arch known as the Bull Bridge, through which passed the main highway. The company minutes refer to damage to one of these arches and it is likely that this had resulted from a major earth slip, caused either by inadequate compaction during construction, or the result of using unstable, and perhaps saturated, material. Characteristically, Jessop took the blame upon himself for this mishap and offered to pay for the repairs estimated at £650. The committee declined to accept this offer, yet agreed to his generous gesture to give his services free for the remainder of the construction period. This type of earth failure, which is common enough even in present-day practice, is not always avoidable. The causes are sometimes difficult to detect until after failure occurs, hence it was unreasonable that Jessop should be expected to take the blame. In the months following there were several similar, albeit minor, earth slips at the Amber crossing and in fact, such incidents persisted for several years after the canal was opened.

On 29 May 1792, the annual report of the engineers gave a summary of progress and costs incurred up to that time. The report also revealed the undue optimism of their earlier forecasts. It stated that east of the Butterley tunnel the canal was navigable and complete except for minor works on bridges, locks and toll houses. The great tunnel was 1,354

---

[1] PRO, RAIL 819/1, 5 Jan. 1792.

Plate 2 The Amber aqueduct: an embankment with retaining walls to contain the canal channel.

yards completed and another 533 yards was open but unlined. The shafts for the remaining 1,081 yards were sunk to waterway level and the equipment and materials needed were installed in the workings. West of the tunnel the cuttings and bridges were nearing completion and the three short tunnels (Gregory, Hag and Buckland Hollow) were open, but the two aqueducts required 'considerable expense to finish'.

The remaining work to be done was priced at £7,462 but taking into account the stores and implements to be sold off on completion (said then to be worth £3,000) the final costs were likely to be £59,995, a figure nearly £7,000 more than the estimate made but twelve months before! At the general assembly of 29 May 1793, the proprietors learned that disbursements had reached £67,271, which left the company with a balance in hand of only £273. Nobody seemed concerned that this sum was 50 per cent greater than Jessop's original estimate and the engineers were not troubled to explain these spiralling costs. Possibly this was because the proprietors knew that the canal was already showing a modest return, even in its unfinished state, and the growing mining and industrial developments close by obviously promised a prosperous future.

There was a further hint of trouble in store in August 1793 when the committee instructed Outram to apply to Mr Lister, a Nottingham architect, to inspect the Derwent aqueduct. The reason for this was only clarified at the December meeting when a letter from Jessop was read out.[1] This told of major defects that had developed in this massive structure and, more particularly, in the spandrel walls. Jessop wrote that 'I have reason to believe that the dimensions of the walls would have been sufficient if the material had been good enough but from a misapplied economy I suffered it to be built with Crich lime which has not set at all ...'.

Drawings of the bridge showing Jessop's recommended remedies have not survived, indeed it is unlikely that any were prepared because the busy Jessop probably found it easier to describe the repairs in writing, knowing that the proprietors would visit the site to assess the situation for themselves. From information gleaned from his correspondence, as well as by present-day visual inspection of the bridge, it seems that the trouble was a substantial collapse of the spandrel walls on the northern, or upstream, face, although the western side of the downstream face was affected also. The east wall on the latter had already bulged during construction but an extra buttress, built against it at that time, had arrested further movement. As a result of these outward displacements of the spandrels, the arch had 'shaken' and a long fissure, probably on the Cromford side, had opened up in line with the direction of the bridge. The foundations were, nevertheless, unaffected and remain so to the present day.

Some speculation on Jessop's design is justified, in spite of the limited data and information available. This was an unusually long segmental arch (of 78 ft 10 in. span, a rise of 19 ft 4 in. and arch radius of 50 ft) which was to carry a substantial load. The spandrel walls had to withstand considerable horizontal forces resulting from the earth infill over the arch, much of it waterlogged, as well as the superimposed weight of the canal channel. Jessop probably had no real understanding of the stresses induced in the spandrel walls, which were 35 ft high at the abutments, reducing to 10 ft at the arch crown. The design should have called for very strong retaining walls, braced laterally across the arch to prevent them from sliding outwards, and also from overturning.

---

[1] Ibid., 10 Dec. 1793.

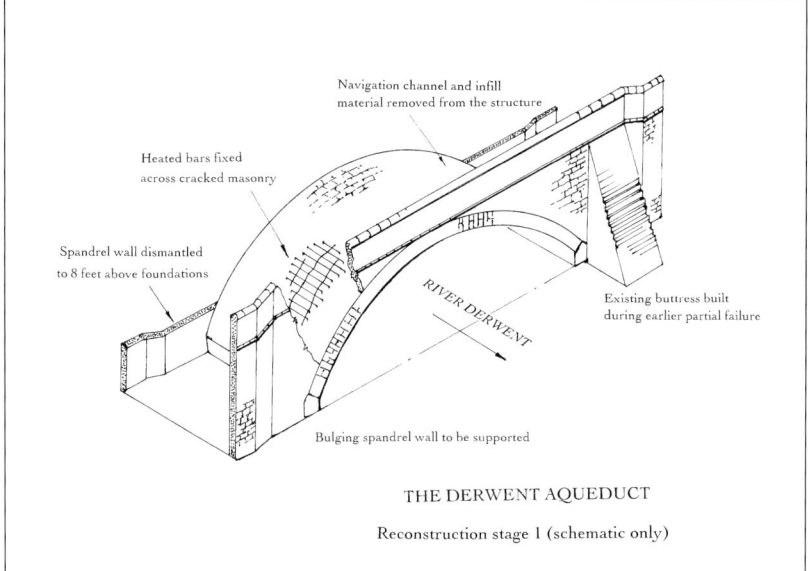

Figure 3 The Derwent aqueduct. Reconstruction Stage 1.

The spandrel walls do not seem to have been built in this manner and were much too slender to withstand the forces imposed. Hence a serious bulging occurred on both faces during backfilling of the structure, and the longitudinal jointing of the arch opened up as a long fissure, as a result of a tranverse stretching action. Failure could not be blamed on unsatisfactory mortar, as claimed by Jessop; the cause was more fundamental than that.

Jessop's recommendations included building a buttress to support the south-west wall, similar to that already strutting the south-east side; stripping the arch of infill material and dismantling the upstream walls to within 8 ft of the abutments; and laying heated iron bars, up to 12 ft in length, across the part of the arch affected by the crack, cramped into position, so that cooling would tend to close the gap (Fig. 3).

Other modifications were more obscure but it seems that he intended to rebuild the northern face with four internal counterforts, each of 12 sq. ft base area; bond similar internal counterforts into the south walls, exactly opposite those on the north face; build a relieving arch 2 ft thick and 8 ft wide, across the fissure and tied into the base of the counterforts (it is difficult to know from Jessop's description whether this was to be built in line with the arch or across it, although the latter seems the more likely); and bolt temporary wooden beams across each pair of counterforts during the time needed to cure (i.e. to set and harden) the mortar.

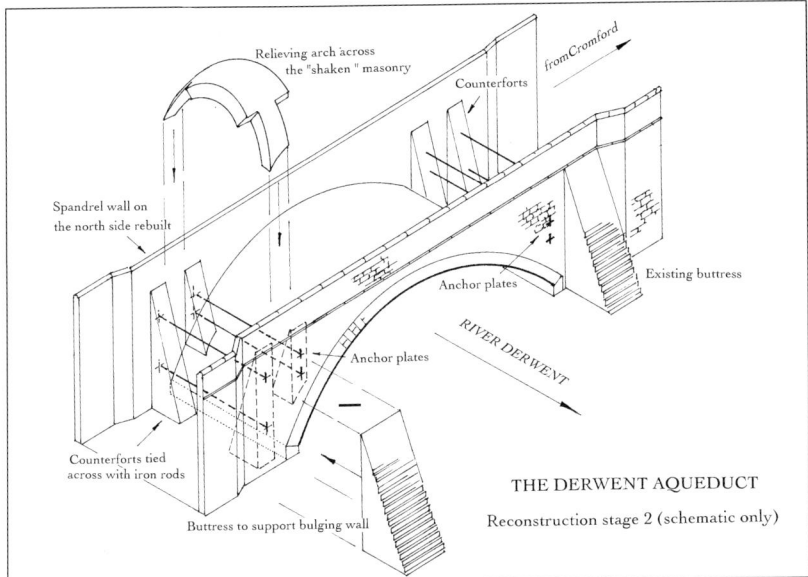

Figure 4 The Derwent aqueduct. Reconstruction Stage 2.

Permanent iron bars were to be fixed between the spandrel walls prior to backfilling with earth (Fig. 4).

Presumably reconstruction followed this pattern but the structure today is still disfigured because the repairs (which were unsatisfactory remedies from today's understanding of structural mechanics) did not cure the basic design defects. A 10-ft long fissure and minor jointing cracks can still be seen over the centre of the arch, showing that further movements of the spandrels have occurred since those days.

In expressing his views on the matter, Jessop told the proprietors that

> I think it common justice that no one ought to suffer for the faults of another. I should be happy if my ability could keep pace with my inclination to make retribution for every loss that has happened or may happen in consequence of my errors ... I shall use every means that I can to have the work effectually re-established and no expense attending it shall be charged to the company.

This extraordinarily generous offer was accepted by the proprietors[1] but how regrettable it was that they saw no cause for sharing the risks of

---

[1] Ibid.

Plate 3 The Derwent aqueduct.

building this massive bridge with their engineer. But of course they had no knowledge of the primitive mechanics then in use by engineers and did not realise that Jessop was coping with the unknown. Structural design practice was very much in its infancy and even engineers of the calibre of Jessop relied to a large extent on experience and empiricism. Just three years earlier, Sheasby and Dadford had walked away from their responsibilities in the face of a difficult task. Jessop preferred to stand his ground, but his honesty and loyalty earned him scant sympathy from the proprietors, several of whom were his friends. Probably as a result of this oft-quoted incident in Jessop's career, it has been said that he was never very good at arch-bridge building,[1] an unfair and unfounded judgment on his remarkable life time of achievement.

The unusual water-supply system had received much thought during the early days of construction and the planned alignment of the canal terminus in Cromford was changed when Arkwright's scheme was wisely abandoned in favour of a direct connection with Cromford Sough. The company was made to pay dearly for this because not only had they to purchase land from the Knight for £1,000 to relocate the terminus,[2] but also they had to agree to landscape, at undisclosed cost, the garden for

---

[1] L.T.C. Rolt, *Navigable Waterways* (1971), p. 64.
[2] PRO, RAIL 819/1, 24 May 1790.

his new home at Willersley Castle.

This change of plan provided a minimum flow from the sough of 74 tons per minute,[1] considerably more than the intake from the river (i.e. 28.5 tons per minute) permitted under the Act. A great advantage was that this increased flow could be drawn off every night, as well as at weekends when not required for Arkwright's Masson Mill. Fortunately, it was still hydraulically possible to discharge this higher rate of flow directly into the canal without overtopping the banks, causing surge waves and endangering the fabric of the waterway. There can be no doubt that this method of supply was used from the earliest days of the canal's existence but to what extent it was legal and how frequently, if ever, the 24-hour weekend filling operation was undertaken remains in doubt. In connection with the latter scheme, Outram had been instructed, when works began, to construct a masonry floor under the arches of Cromford Bridge and to fix a vertical depth gauge such that the river's discharge could be readily determined,[2] flow velocities being recorded by timing the motion of hardwood floats. This order was apparently rescinded, which seems to confirm that the original water-supply method had been abandoned and river flow measurements were, therefore, no longer necessary.

That is how the problems of water supply rested until November 1791, when stealthy negotiations began with the proprietors of a proposed Nottingham Canal which would join the Cromford Canal just above its junction with the Erewash Canal at Langley Mill.[3] This was intended to undermine the latter company's monopoly of transporting coal and other commodities to Nottingham. Its proprietors were allied to the Cromford party and the timing of their proposals was conveniently linked to the construction of the Cromford Canal. More significantly, however, the design of the new line offered a generous and reliable increase of the latter's water supply at no additional cost, because the scheme included the building of new reservoirs on the Cromford Canal's summit level, above the line of the Butterley tunnel. This storage system would be fed from the Codnor Park Brook on the Erewash catchment;[4] thus the discharge into the Nottingham Canal would be via the Cromford Canal's summit level.

---

[1] Ibid., 13 July 1791, Report by B. Outram and P. Roe on the produce of the Cromford Sough etc.

[2] PRO, RAIL 819/1, 24 Aug. 1789.

[3] Ibid., 10 Nov. 1791.

[4] Ibid., 8 Nov. 1791, P. Gell to H. Parker, chairman of the Nottingham Canal Co.

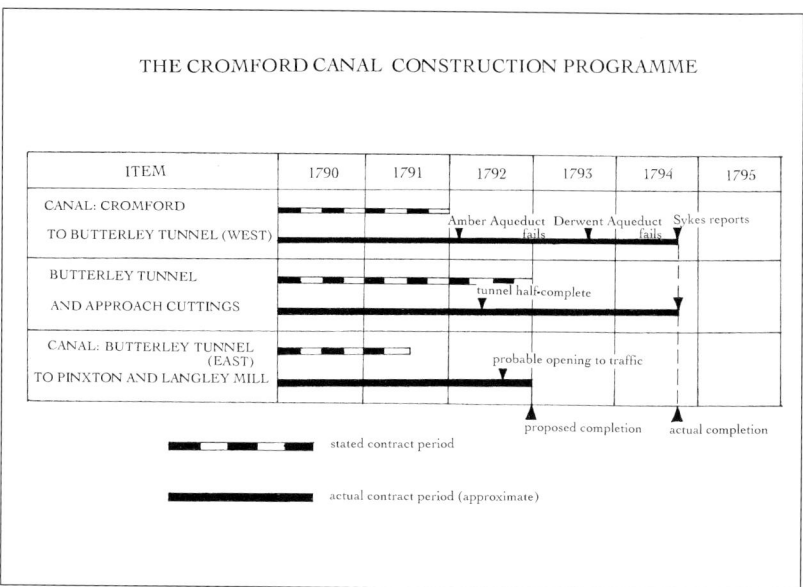

Figure 5 The Cromford Canal construction programme.

By the time of the general assembly of May 1794, costs for the Cromford Canal had escalated to £76,596 but work was drawing to a close and the proprietors turned their attention to inviting an independent engineer to examine and report on the state of their canal. Robert Whitworth was their first choice but he was not available, so Thomas Sykes of Chesterfield, an engineer of small repute, was invited instead. Sykes reported on 29 August 1794. He found no defects in the Derwent aqueduct but recommended strengthening embankments in Lea Wood and also those of the Amber aqueduct. The tunnel had been executed in workmanlike manner and, although he found seepage through parts of the linings, he did not consider this to be unduly serious. Apart from reducing the steepness of the side slopes of the cuttings east of the tunnel, he recommended only a few repairs to eliminate leaks from sections of the navigation channel and a few of the locks (Fig. 5).

The final accounts were presented to the committee on 3 February 1795 when it was noted that costs had risen to £83,055 (95 per cent above Jessop's first estimate), although these figures included work additional to the original scheme.[1] Work had been targetted to be completed by the end of 1792 but that had also proved too ambitious; it

---

[1] PRO, RAIL 819/1, 3 Feb. 1795.

had taken two years longer. In spite of the costs, the proprietors now had an excellent, well-built and potentially highly profitable waterway. The subsequent history of this canal is another story but suffice to say it was always the success that Jessop had expected. By 1802, all debts had been paid off and the returns to shareholders steadily rose from 6 to 28 per cent by 1840. Benjamin Outram's appointment terminated in May 1795, although he was retained to ascertain the discharge volumes of certain watercourses which the company was statutorily obliged to record. For several years, until his death, Outram regularly undertook repairs and improvements to the banks, reservoirs and structures, and at the same time, as managing partner of the Butterley ironworks of Benjamin Outram & Company, he became a major user of the canal he had helped to build.

CHAPTER 4

THE NOTTINGHAM AND NUTBROOK CANALS

Nottingham Canal

During the years the Cromford Canal was being built, schemes for connecting other waterways to it, and to the Erewash Canal, were either being devised, or else construction had already begun. Mention has previously been made of one of these, namely the Nottingham Canal, with which Jessop and Outram were involved together once again.

Certain Nottingham businessmen had rightly anticipated that the collieries along the Pinxton branch of the Cromford Canal would thrive on completion of that waterway, to the disadvantage of those pits closer to Nottingham. Accordingly, it seemed likely that most of the town's supplies would be carried along a circuitous route down the Erewash Canal and along the River Trent and they supposed, reasonably enough, that the Erewash Company might take advantage of their monopoly.

Following a public meeting held in Nottingham on 26 October 1790, Jessop was appointed to survey, estimate and report on the proposed waterway which was to link the Cromford Canal, just upstream of the junction with the Erewash Canal, with the Trent at Nottingham.[1] The scheme would also include a short connection from Lenton to the Trent at Beeston. Jessop stated that he was familiar with the locality and could select a suitable line. Although illness prevented him from surveying the route, he was able to direct James Green of Wollaton, agent to Lord Middleton, in this task but prepared the report and estimate himself.

The investigations were completed by September 1791, the estimate being £36,499 plus £7,000 for reservoirs.[2] Another £2,500 was added later to allow for modifications to the alignment, in order to satisfy Lord Middleton's objections.[3] The line was to be 14¾ miles long, with 20 locks rising from Nottingham to Langley Mill, including a stop-lock at

---

[1] PRO, RAIL 854/1, 27 June 1791.
[2] Ibid., 21 Sept. 1791.
[3] Ibid., 25 Oct. 1791.

the junction with the Cromford Canal.[1] There were no spectacular natural obstacles to overcome and it was designed, over the greater part of its length, for vessels 75 ft long by 14 ft beam.

There were no obvious water sources for this new waterway which did not interfere with rights already held and jealously guarded by the Erewash Canal Company. At this early stage of planning Jessop had no answer to this problem, apart from suggesting that supplies could be obtained from nearby collieries and, if these proved insufficient, then reservoirs could be built in dry valleys (those fed only by 'winter rains and heavy showers') along the route.[2] The committee, in view of these expressed uncertainties, decided that reservoirs would have to be built, but for the time being locations were unspecified.[3] A good plan for a water supply was proposed in November 1791, however, and this was immediately adopted. One or more reservoirs would be built, entirely at the expense of the Nottingham Company, on the summit level of the Cromford Canal and just over the line of the Butterley tunnel.[4]

This idea was likely to have been suggested by Outram, who was then hard at work on the Cromford Canal. Both he and Jessop had probably harboured doubts about the adequacy of that canal's water supply, none more so than Outram who had been charged temporarily with the early commercial operation of the waterway. The new proposal would have been recognised as a remedy for both canals' difficulties. The Erewash Company, however, would have seen through the proposal as nothing but a deliberate attempt (which of course it was) to circumvent the restrictions imposed under the Cromford Canal Act. The reservoirs were to become a major issue in the parliamentary battles for the Nottingham Canal Act.[5]

A committee of the House of Lords was convened in April 1792 to consider the bill.[6] Once again, Robert Mylne represented the Erewash Canal Company on engineering matters, and on the water supplies in particular. On this occasion he was supported by John Smith, 'collector and surveyor' of the Erewash Company. Mylne adopted his usual aggressive tactics of opposing all aspects of Jessop's design as impracti-

---

[1] H.R. de Salis, *Bradshaw's Canals and Navigable Rivers of England and Wales* (1904), pp. 149–51.

[2] PRO, RAIL 854/1, 15 June 1791.

[3] Ibid., 21 Sept. 1791.

[4] PRO, RAIL 854/2, 18 Dec., 24 Dec. 1792.

[5] 32 Geo. III c. c (1792).

[6] HLRO, Evidence on the Nottingham Canal Bill, April–May 1792.

cable but, when giving evidence, both he and Smith unwittingly exposed the makeshift and vulnerable nature of the water-supply system used for the Erewash Canal. The crude and inefficient system adopted years before had never been developed as it should.

Smith described the waterway and the Erewash river valley, which had a substantial catchment and run-off. He stated that reservoirs had never been constructed and instead, over the years, feeders had been cut to connect the canal directly with the River Erewash and its several tributaries. Although there were floods during most years, only minimal efforts had been made to collect and conserve such excess flows and, during the fourteen years of the canal's existence, just a few artificial wides and deeps had been excavated in the river bed for that purpose. It was not surprising that trade in summer was frequently cut, by as much as 30 per cent on occasion, because of such avoidable water shortages.[1]

The first reservoir for the proposed Nottingham Canal was designed for 2,800 locksfull of water (798,000 tons), all to be drawn from the Codnor Park Brook. Jessop, in his evidence to the committee, stated that he intended the entire water of the brook would be retained, except for 'a quantity equal to that which may be measured in the three driest months of the year'. He meant by this that only the discharges exceeding the average dry weather flow rate occurring during these months (unspecified, but presumably the summer months) would be retained throughout the year. The weirs leading into and out of the reservoir would be adjusted accordingly and it followed that, whenever the reservoir was full, the stream would run unimpeded into the River Erewash. This was a fair-minded proposal, comparable with present-day practices, and which would gain approval even with the current concerns for the environment.

Mylne argued, however, that such concessions would deprive the Erewash Canal of a substantial part of its supply, yet he was unable to disguise the fact that there was more than sufficient water for the efficient operation of that canal over what was wanted by the Nottingham Canal Company. He then went on to outline what he considered to be a fairer solution. The Codnor Park Brook should be allowed to run freely for eight or nine months of the year and only excess flood waters discharging in the remaining 'stormy months of the year which are a cause of floods and annoyances to the valley below' should be retained in the reservoir.[2] Mylne stressed that the waters of the new reservoir

---

[1] Ibid., John Smith, 30 April 1792.

[2] Ibid., Robert Mylne, 20 April, 4 May 1792.

could (and most likely would) be used for the benefit of the Cromford Canal as well. His proposals were vague, and yet he was probably correct when he also stated that the Cromford Canal Company were already interfering illegally with the Erewash supplies. He may have been referring to the Codnor Park reservoir, located just below the Cromford Canal summit level. He offered no evidence to support his statement but, no matter, neither of these significant issues was pursued by counsel.

Jessop then stepped forward to defend the scheme in his customary bland and succinct manner, claiming that

> the flood waters of the country, which are more injurious than beneficial, are equal to the supply of ten times the quantity that will be required for the intended canal and that any necessary part of the flood waters may be retained by reservoirs without injury to the Erewash Canal or to any other property whatsoever.

Moreover, he emphasised that there would be advantages accruing to everyone concerned because 'there will be considerable trade on the intended canal from the Cromford Canal and all the waters necessary for that part of its trade will be supplied by the Cromford Canal, which will have none of the Erewash water'.[1] He omitted to mention that water discharged into the summit level of the Cromford Canal from the new reservoir would be of significant benefit to navigation along the upper section of that canal (as already noted by Mylne). But no matter, it seemed obvious that there was ample water for all concerned and so, yet again, that important aspect was never questioned. It may be that their lordships failed to understand the complexities of the hydrology of the Erewash valley as described by the ponderous Mylne and perhaps they found it easier to follow the simpler (but sometimes inaccurate) statements made by Jessop. Once again, Mylne was destined to be on the losing side and the Act for the Nottingham Canal was passed, with only minor amendments, in May 1792.

Benjamin Outram's involvement with the canal began at the first general assembly of June that year when he was appointed engineer for the purpose of gauging the water flows down the Erewash valley, and particularly for the discharge into the new reservoir at Butterley. James Green was made superintendent for construction of the canal at 300 guineas per annum and Jessop was appointed engineer for the design and

---

[1] Ibid., William Jessop, 4 May 1792.

direction of the works at his usual fee of three guineas per day, plus expenses. At the next monthly meeting, payments for advisory and preparatory works were made to each engineer, viz., £262 to Jessop, £346 to Green and £52 to Outram.[1]

By December, both Outram and Green were setting out the controversial Butterley reservoir. A substantial work, it was built over the west side of the tunnel and was some 50 acres in surface area, the average depth being 12 ft. The dam was 200 yards long and 33 feet high at its centre; water was fed into the canal through a pipe and control cock into one of the tunnel shafts.[2] Jessop was very much in demand at this time of the Canal Mania and in October 1794 he was reprimanded by the Nottingham proprietors for his infrequent visits to the site. They recorded that they 'had hoped that Mr Jessop, after undertaking the concern, would have paid greater attention to it'.[3] Clearly, Green was not proving as successful as they had expected because they were dissatisfied by the 'erroneous construction of many works' and because expenditure had, by then, reached £54,489. Jessop, perhaps stung by the criticism, was on the site the following week and accompanied some of the proprietors on an inspection, afterwards stating that the waterway would be finished by March 1795.[4]

Unfortunately, severe frosts, followed by floods, damaged the waterway that winter and this further delayed progress. Outram was called in during April, as conditions improved, in order to view the state of the works and to prepare estimates for its completion.[5] He reported in June that repairs and finishings would cost £3,105, much of this to be expended on earthworks, wastewater weirs, parapet walls on bridges, and the securing of bridge pier foundations from scour.[6] The canal was, by then, probably 90 per cent complete. Green was still retained as resident engineer until the works were finished but from that time onwards Outram took on all of Jessop's responsibilities as the consulting engineer for the Nottingham Canal Company.[7] Besides arranging for minor modifications and maintenance of the canal, he also undertook more

---

[1] PRO, RAIL 854/2, 26 June, 24 July 1792.

[2] J. Farey, *A General View of the Agriculture and Minerals of Derbyshire*, iii (1817), p. 346.

[3] PRO, RAIL 854/2, 8 Oct. 1794.

[4] Ibid., 15 Oct. 1794; a visit had been made on the 10th.

[5] Ibid., 30 April 1795.

[6] Ibid., 17 June 1795.

[7] The appointment is not recorded in the minutes but Outram was frequently consulted after 1795.

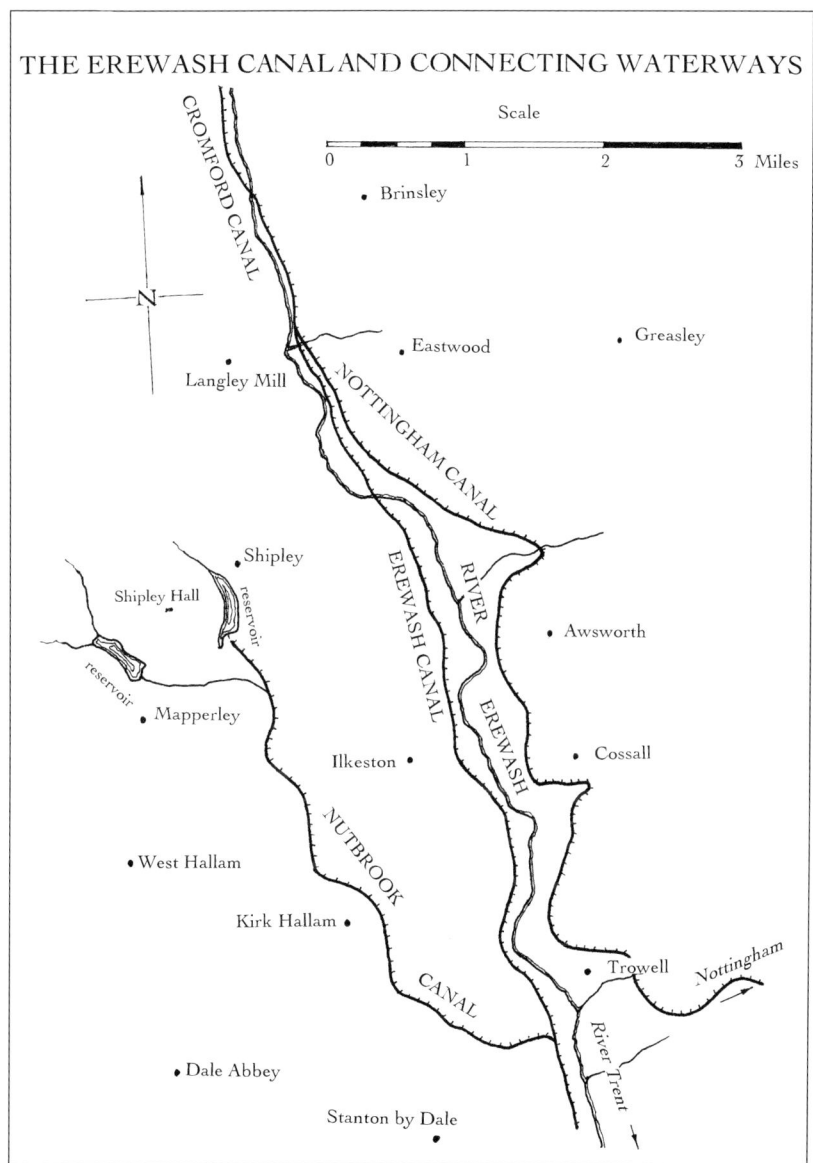

Figure 6 The Erewash Canal and connecting waterways.

important matters such as the design and construction of additional reservoirs to compensate the Erewash Company for waters taken from their catchments; he also negotiated with coal-owners over compensation for flooding caused by the canal and for coal left under the navigable channel to prevent subsidence.

The Nottingham Canal finally opened during April 1796.[1] Completion was, by then, well behind schedule and the final costs, which exceeded £75,000, were nearly double the original estimate.[2] Once more, the inadequacy of contemporary estimating and contracting methods were exposed and found wanting.

Nutbrook Canal

The Erewash Canal was completed by 1779 (Fig. 6). It was surprising that this was not quickly followed by the construction of an interconnecting waterway from the Nutbrook valley because such a measure had been discussed for many years. This was because collieries in the valley, notably those at West Hallam and Shipley, had long been disadvantaged by their comparative isolation from major markets at Nottingham, in Leicestershire and elsewhere because of inadequate transport. Moreover, the Erewash Canal Act allowed for a generous reduction of tolls should coal-owners in the valley invest in such an advantageous venture. In spite of this, it was another twelve years before any action was taken.[3] The only progress made during that period was the building of a short section of canal into the Nutbrook valley from the junction with the Erewash Canal, to connect with a tramroad leading to mineral deposits, leased by Earl Stanhope at Stanton.[4] Also, a primitive wooden tramroad was laid to link the lock at Shipley Gate via a steeply-inclined route to the Shipley Colliery, which was dismantled in 1796.[5]

Eventually, after many negotiations and disagreements between landowners in the valley, an Act was obtained in 1793 'for making and maintaining a navigable canal from collieries at Shipley and West Hallam ... to the Erewash Canal'.[6] Stanhope's short canal was to be built into the new waterway,[7] which was planned by William Jessop from surveys prepared by John Nuttall during 1791 and 1792.

Of Jessop's report and estimate, only the estimate appears to have

---

[1] PRO, RAIL 854/2, 13 Dec. 1796, when it was reported that the canal had been navigable for about eight months.

[2] Ibid., 6 April 1796.

[3] P. Stevenson, *The Nutbrook Canal, Derbyshire* (1970), p. 24; C. Hadfield, *The Canals of the East Midlands (including part of London)* (1966), pp. 49, 72.

[4] Stevenson, *Nutbrook Canal*, p. 14.

[5] Farey, *General View*, iii, p. 288.

[6] 33 Geo. III c. cxi (1793).

[7] DRO, D3808/50, Nutbrook Canal Co. mins., 6 July 1793.

survived.[1] This was a very brief statement of twelve items totalling £12,542, including contingencies. The major items of construction were not broken down into their constituent parts before the quantities were taken off and priced. Instead, for example, Jessop simply wrote '12 Locks of the average Price of £480 and the average rise of 7 feet each ... £5,880'. Similarly, the estimate for the principal reservoir at Shipley, which today would have comprised a detailed list of many separately costed items, was merely noted as 'Making a reservoir Head of 23 feet in height 4 yards wide at top and 36 yards at base. 10,000 cubic yards at 6d. ... £250'.

The Nutbrook Canal was among the least of Jessop's commitments at that time and this might explain the brevity of his estimates and proposals which were submitted to the proprietors. It is unlikely that he had prepared detailed drawings for any of the structures on the canal, or if he had given much consideration to ground levels and sub-soil conditons. His proposals were hurried and sketchy and it is not surprising that the completed works were more extensive than had been anticipated in the original scheme. Once again, Jessop's preliminary planning, design and estimating standards were found wanting but no doubt this was entirely due to the pressures to which he was subjected at that time of the Canal Mania. Many engineering problems were left to be resolved only after work had begun on site and this invariably led to mistakes, disputes and overspending.

The waterway was 4½ miles in length and there were 13 locks (each suitable for Trent barges of 72 ft by 14 ft 3 in.) rising 82 ft to the Shipley reservoir.[2] The latter was fed by the Nut Brook, the original course of which passed down the centre of the lake, but there was another, smaller, feeder reservoir nearby known as Hawley's Pond. Natural obstacles along the selected route down to the Erewash valley were few but five highway and eight accommodation bridges were planned, and each of these types were costed at £85 and £70 respectively. There was also a number of small culverts which carried streams underneath the canal embankments to discharge into the Nut Brook. Thus the navigable channel was not built along the bed of this steam but instead followed its course fairly closely from Shipley, except for the lower 1½ miles where it deviated to the west, by as much as 400 yards, before joining the Erewash Canal. A condition laid down in the Act was that work had to begin at the end nearest to the Erewash Canal junction

---

[1] DRO, D517, Nutbrook Canal MSS.
[2] De Salis, *Bradshaw's Canals*, pp. 287, 288.

and that the three Stanton locks (the lowest of the series) were to be completed before any of the others. That arrangement suited the Stanhope tenants but in any case it made good sense to begin building the canal, at the lower end and work towards the summit. Floods could thus be controlled and prevented from inundating the workings.

The first general assembly of the company was held in Nottingham on 3 July 1793, just before work began on site. The committee and principal officers were appointed, among them Benjamin Outram, who accepted the post as engineer to the undertaking for a fee of £200 'for his trouble and for his expenses'. He was also offered a bonus of £50 if the canal was finished by 1 January 1795.[1] No doubt Jessop had recommended him for the job and it was certainly a convenient arrangement, because Outram was them living in Butterley Hall, only eight miles from the head of the Nutbrook Canal.

Outram was, of course, also superintendent for the Cromford Canal and, although much of that waterway was then navigable and already in commercial use, the Butterley tunnel was but three-quarters complete. That responsibility, in addition to the repairs to the Derwent aqueduct in August 1793 must, at times, have diverted his attention from his duties on the much smaller undertaking of the Nutbrook Canal. These matters, quite apart from the expanding business of the ironworks at Butterley and some serious bouts of illness, from which he suffered during much of 1795 and 1796,[2] must have prevented his close supervision of the works. They may have been partly to blame for the failure to finish on time and why final costs were double the original estimate.

However, by the end of August 1793 the committee members had toured the proposed route of the canal in Outram's company and had agreed on necessary land-take and accommodation works.[3] Outram's next task was to confirm that the value of the work completed on the existing canal through the Stanhope estates at Stanton, by the lessees Thomas Hankey, Alexander Raby and Charles Druce, was worth £300 and the company agreed to take over this section for that sum. They also decided to construct the works by direct labour and one James Pike was appointed 'overseer of the works' (i.e. general foreman) with a wage of one guinea weekly. It seems likely that Outram's assistant, John Hodgkinson, was involved in setting out the works and management.

---

[1] DRO, D3808/50, Nutbrook Canal mins., 3 July 1793.

[2] JRUML, Bagshawe Muniments, 8/4, various letters of J. Outram sen. to J., Bagshawe.

[3] DRO, D3808/50, Nutbrook Canal mins., 27 Aug. 1793.

Work proceeded only slowly, judging by the enginer's annual report which was read to the second general assembly on 4 June 1794 by the Revd Richard Dodsley, the committee chairman. Although the canal's earthworks were well advanced and the reservoirs at Shipley and Hawley's Pond were 'two-thirds done', only four bridges were almost complete and three were 'yet to build'. Of the locks, three were nearly finished, two more in part and eight were not yet begun. Much was made of the fact that 'considerable quantities of lime, sand, timber and about two-hundred thousand bricks were ready for use' and that preparations were in hand to make another 1,200,000 bricks that summer 'which be all that be wanted to complete the works'. Dodsley also announced that spending had risen to £8,424, just two-thirds of Jessop's estimate, and yet the progress he had described did not seem to match that sum. The claim that all the construction materials were then available on site was something of a ploy to allay the fears of and discourage further questioning by the better informed among the shareholders. Fortunately, no one complained or queried either the costs or the progress of the works.

Nevertheless, Outram and Dodsley were no doubt aware that a substantial deficit could be expected and shortly afterwards the engineer set to and worked up a detailed estimate for completing the works.[1] Each of the locks, from the fifth upwards, was carefully and separately costed, as were all the remaining items, even down to the planting of quickthorn hedging. That estimate was £5,391, a sum which suggested that final costs could be nearly £14,000. At the next general assembly in September that year, Dodsley announced that 'it appeared that the sum of £13,000 directed to be first raised ... will not be sufficient to complete'. He then asked the proprietors for an additional £3,900 which was probably reasonable enough because Outram's estimate excluded the purchase of 70 acres of land needed for the canal and reservoirs, as well as the salaries and expenses of officials.[2]

Six months later, just before the committee meeting of 25 March 1795, Outram appears to have prepared two more estimates.[3] The first was for £4,027, made up of £2,315 for the works and £1,712 for outstanding debts for which the company was liable, as well as rentals and damages to lands, and interest due to subscribers. Inexplicably, that

---

[1] DRO, D517, Nutbrook Canal MSS, B. Outram, Estimate of Expence to complete Works on the Nutbrook Canal, Sept. 1794.

[2] DRO, D3808/50, Nutbrook Canal mins., 7 Sept. 1794.

[3] DRO, D517, Nutbrook Canal MSS, B. Outram, Works to do on the Nutbrook Canal, 24 March 1795.

estimate was not divulged to the general assembly.[1] Instead, Dodsley advised the proprietors that another £2,600 would be needed to finish the canal, this being based on the second of Outram's estimates, of £1,740 for the canal works and £860 for debts, damages and interest charges. It can only be assumed that Dodsley was intent on misleading the proprietors and that Outram, somewhat unwisely, had provided the misinformation. If that was the intention, they were certainly successful because once again nobody seemed to query the estimates or complain about the mounting expenditure.

Fortunately, Outram's written report for the June meeting of the committee showed that work was then well advanced; the ninth and tenth locks were nearing completion and he expected the remaining three would be finished within two months. All the gates and ironwork were made and ready for assembly. Bridges were nearing completion and only minor earthworks were unfinished. The only major task remaining was a toll-house, to be located in Kirk Hallam.[2] This building, the only domestic property known to have been designed by Outram, was a substantial but singularly unattractive structure of four storeys. Although it survived closure of the canal in 1895, it was finally demolished by the local authority in 1961. One problem resulting from the positioning of the Shipley reservoir was that the existing coach road and bridge to Shipley Hall had to be replaced and accordingly the committee approved a grant of £60 towards the building of new approach roads and bridges across the head of the reservoir.[3]

Outram's final report of January 1796 was delivered to the committee by John Hodgkinson because of their engineer's continuing illness. This stated that all locks and bridges were built and only minor tasks on the reservoir weirs and canal earthworks remained to be done. The toll-house was still unfinished, probably due to alterations to the original design, but even that was habitable by the late summer.[4] Hence it seems that the canal was not fully operational before the autumn and, even so, miscellaneous minor works continued into 1797. The final costs for the project are not known exactly but were probably a little over £20,000. Outram's £50 bonus was never paid because the time of completion far exceeded the eighteen months originally planned.

[1] DRO, D3808/50, Nutbrook Canal mins., 25 March 1795.

[2] DRO, D517, Nutbrook Canal MSS, B. Outram, Report to the Canal Company, undated but probably June 1795.

[3] DRO, D 3808/50, Nutbrook Canal mins., 14 July 1796.

[4] Stevenson, *Nutbrook Canal*, p. 37.

CHAPTER 5

THE DERBY CANAL

Before the onset of the Canal Mania of the 1790s, the factories of Derby relied on inadequate highways and the archaic Derwent Navigation for the transport of goods. Whilst the river, navigable since medieval times, had been markedly improved during 1720 from the Trent to the Holmes in Derby, by the local engineer George Sorocold, its shortcomings were frequently exposed by floods and droughts. These rendered the river unnavigable for long periods and indeed William Jessop's evidence to the House of Lords Committee considering the Cromford Canal Bill, had clearly drawn attention to these facts.[1]

The River Trent was said to be no better; a report of 1793 stated that

> its imperfections are glaringly flagrant. Not a year has passed ... that near Twenty vessels at a time have not been stopped at the shoals when there has not been more than eighteen inches of water for several days together. Not a year has elapsed, that vessels have not been stranded and their cargoes nearly destroyed ... It is said that the obstruction by *shoals* only happens after floods. Let the fact for a moment be admitted; floods are frequent and shoals and banks of sand and gravel are, and ever will be, the inevitable consequences of them.[2]

Thus the movement of goods along the Derwent was further hindered by long delays on the Trent, below the confluence of these rivers (Fig. 7). Coal, in particular, was expensive and was often in short supply in Derby, in spite of the proximity of the mines of the nearby Bottle Brook valley, which remained undeveloped and unprofitable in the absence of suitable transport. Obviously, Derby had been in need of an efficient transport system for many years and yet James Brindley had advocated a canal as early as 1771, to link his Trent & Mersey and Chesterfield canals via Derby.[3] Unfortunately, opposition by vested interests

---

[1] HLRO, Evidence on the Cromford Canal Bill, W. Jessop, 1–7 July 1789.

[2] DLSL, DCCR, *A Statement of the Case of an Intended Canal from Shardlow to Nottingham and Derby, 7 January 1793.*

[3] C. Hadfield, *The Canals of the East Midlands (including part of London)* (1966), p. 67.

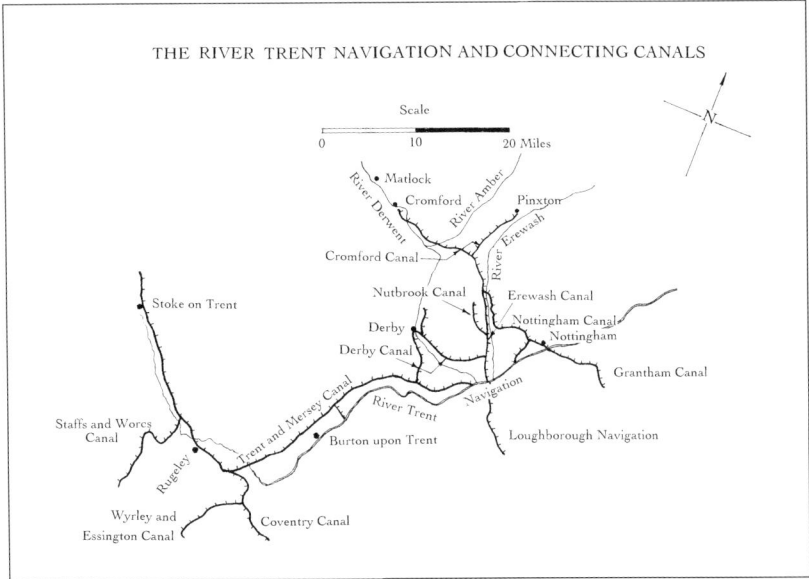

Figure 7 The River Trent Navigation and connecting canals.

prevented its promotion until the matter was raised again in 1791.

In September that year, notices in the *Derby Mercury*[1] referred to two schemes then under consideration by different parties. A meeting had been held to discuss proposals for a canal to join the Trent & Mersey Canal at Swarkestone to Derby. It was reported that the Trent & Mersey and Erewash canal companies were also considering a Trent Canal which would link the Trent & Mersey with Nottingham, with a branch from Shardlow into Derby. Nothing further was heard of either scheme until August 1792, when the former proposal was reviewed by a committee of businessmen and Benjamin Outram was commissioned to survey and estimate for a broad canal. The preliminary work was rapidly completed by Outram, probably aided by the surveyor John Nuttall of Matlock, and he reported to a meeting of the promoters on 8 September 1792.[2]

Outram's original report and estimates have not survived but an article in the *Derby Mercury* of 13 September 1792 stated that the canal would begin 'at or near Swarkestone on the River Trent, passing by Derby to Smithy Houses, with a branch near Derby to the Erewash Canal

---

[1] 15 Sept. 1791.
[2] *Derby Mercury*, 13 Sept. 1792.

near Sandiacre. The expense of executing the waterway would amount to about £60,000'. This estimate was based on the cost of a broad canal from Derby to Denby (£20,231), from Derby to the Erewash Canal (£16,336) and from Derby to the River Trent (£11,280). Added to these costs were those of an aqueduct across the Derwent at Derby (£8,160) and the purchase of the Derwent Navigation (£3,996).

The high costs of the Denby line and the aqueduct caused the promoters concern and Outram, probably a few days later, could only suggest the substitution of a narrow for a broad canal on the Denby line, at a saving of about £5,000. This modification was wisely rejected and more radical changes were then recommended following consultations with William Jessop.

Jessop reviewed his young colleague's proposals and his comments were submitted in a letter to the chairman of the promoters early in November.[1] Generally Jessop approved of most of Outram's plans. He recommended minor improvements to parts of the Swarkestone line but felt that the Denby line merited a complete change of plan:

> The lines from Denby and Smalley Collieries are practicable on either of the Ways that Mr Outram has projected; and if made navigable all the Way, may be amply supplied with Water by a Reservoir: but I am clearly of Opinion, that the most eligible Scheme will be that of a Canal from Derby to Little Eaton; and from thence Railways to the Collieries. If those Railways, which should be of Cast Iron, are substantially laid upon Stone Foundations, and ascend on a regular Acclivity from Eaton to the Collieries, one Horse will easily draw down two Waggons with two Tons on each; and empty, they will as easily be drawn up again, as I understand the Ascent is only one-sixth of an Inch in a Yard in length; those Waggons may be drawn on to Boats and conveyed to Derby, and may be so constructed as to be carried into the Town without unloading. Putting Tonnage out of the Question they cannot be carried so cheap by a Canal the whole Length as by the proposed Railway.

Jessop's opinion was also invited on the alternative proposals for a

---

[1] DLSL, DCCR, W. Jessop to chairman of subscribers of the intended Derby Canal, 3 Nov. 1792.

Trent Canal.[1] This was effectively dismissed by his stating that:

> If the Canals already described were wholly out of the Question this might be the next best Thing for Derby; but for any other Purpose I think it ineligible. I have Reason for believing that few Boats will use the Line parallel to the Trent, if they are not *compelled* to do so; for as the Trent Navigation is generally as good above Nottingham as below, the Boatmen will hardly like to be deprived of the Use of their Sails (which frequently enable them to move with more Expedition and Facility than they can upon any Canal obstructed by Locks and Bridges) and pay a Guinea and a Half per Voyage extraordinary into the Bargain.

Jessop's recommendations were accepted in their entirety both by Outram and the proprietors. A railway would be constructed from Little Eaton to the collieries instead of a canal; adjustments would also be made to the Swarkestone line and its connection with the Trent & Mersey Canal. Additionally, a short length of canal would be built to join with the Trent from the latter, just above Swarkestone Bridge. This was apparently intended to link with a possible canal and railway to quarries and limeworks at Breedon, on the south side of the river, the route for which Jessop had surveyed during 1787.

Accordingly the *Derby Mercury* reported on 15 November that Benjamin Outram, at the 'last meeting of the committee' (probably held a few days before), now proposed a modified scheme which would reduce his original estimate considerably. A broad canal would be built from Derby to Little Eaton, with railways to run from there to Smithy Houses and Smalley Mill. The railways would replace the canal section of six locks which had originally been planned. Also, instead of the canal crossing over the Derwent by an aqueduct, Outram now preferred an unusual, but simpler and cheaper, method. Vessels would lock into and out of the river, the depth of which would be controlled by a weir. This alteration alone reduced the estimate by £4,000.

In his report Jessop had refrained from describing the rolling stock and railway to Denby in anything but general terms, with no mention of design details and dimensions; those matters were left to Outram to decide. The latter was already thinking these out prior to the publication

---

[1] DLSL, DCCR, *The Derby and Swarkestone intended canal. A statement of the advantage and public utility compared with those projected from Derby to Shardlow* (nd, probably Sept. 1792).

of Jessop's report. His letter of 22 October 1792 to John Radford of Smalley, a proprietor, enclosed a paper outlining ideas to be put to the committee and a few of Radford's notes on that document have survived.[1] These refer to the 'Expense of a Railroad, Bed of Stone 15 Inches thick well-broken, upon this Sleepers and oak Rails these covered with Cast Iron Plates at 15 Shillings per Yard, the small Metal about ¾ of a Hundredweight per yard'. Much has been written of Outram's pioneering work as a railway engineer but it seems clear that, at this early stage of his career, he had no experience of that mode of transport. The primitive method described was long-established practice and certainly represented no new advance in railway technology. Although he went on in later years to refine traditional railway construction, there can be little doubt that in this sector of engineering, as in others, he was greatly influenced by Jessop, who had long been familiar with the practice.[2]

The revised designs adopted for the Derby Canal and its railways were a marked improvement on Outram's original plan, with such costly features as the Derwent aqueduct and the steeply inclined Denby branch canal. The petition to Parliament for the scheme (referred to in the *Commons Journal* as the Swarkestone Canal) was ready for presentation on 2 February 1793. The bill's subsequent passage involved but one division as to whether or not it was to be read a second time. It survived by 62 votes to 50, a sure indication of the strength of opposition, but this, nevertheless, spelt the end of the Trent Canal proposals. The petition for the latter, which had been prepared by the engineer John Varley of the Erewash Canal Company, was presented on 20 February but it was abandoned on 29 April following the passing of the Derby Canal Act on 17 April, and the granting of Royal Assent on 7 May 1793.[3]

After acquiring their Act, the first meeting of the committee of proprietors was held on 6 July 1793.[4] Benjamin Outram had been appointed engineer on 11 May and was already at work with George Wootton, directing excavations for the canal and preparing for brick manufacture. The committee approved of their industry and Wootton was made superintendent of works at £150 per annum. Much of the earth-

---

[1] Nottingham University Dept of Manuscripts, Dr/66/13a, B. Outram to J. Radford, 22 Oct. 1792.

[2] C. Hadfield and A.W. Skempton, *William Jessop, Engineer* (1979), pp. 17, 81 169.

[3] 33 Geo. III c. cii (1793).

[4] DLSL, DCCR, Minutes, 6 July 1793.

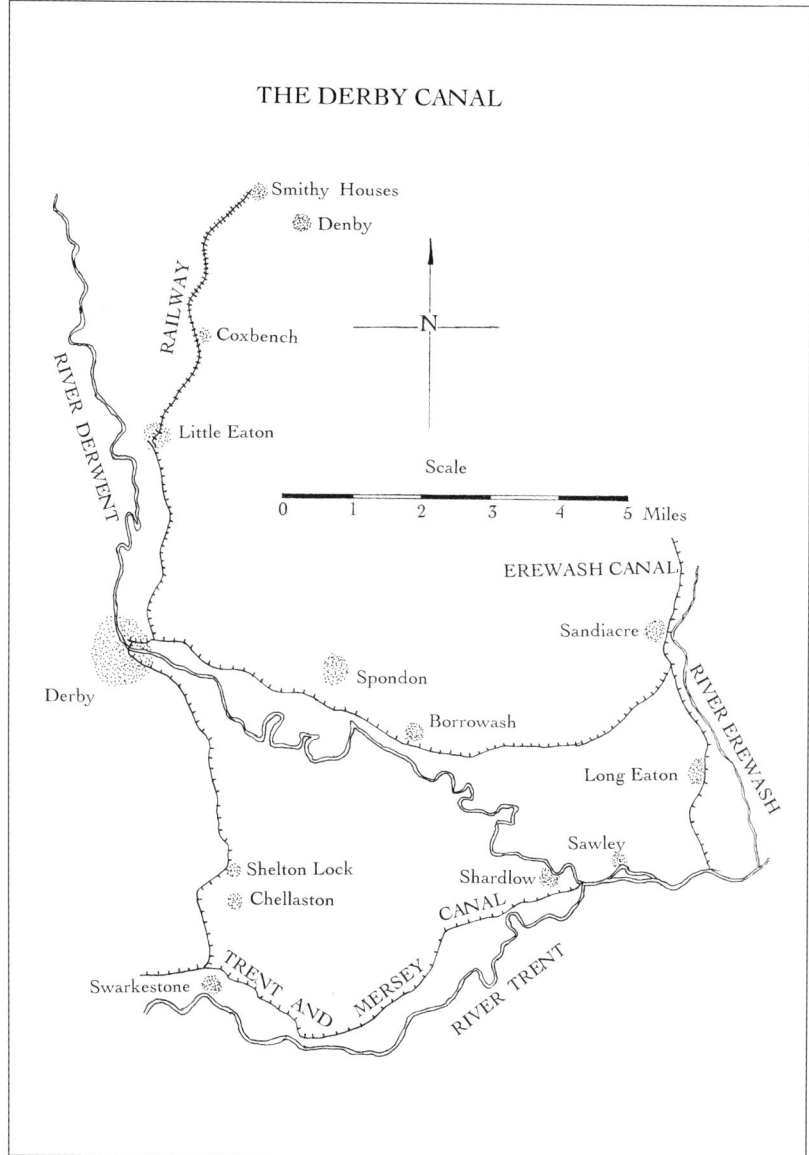

Figure 8 The Derby Canal.

moving and bridge-building seems to have been contracted out on a labour-only basis and Wootton was thus expected to oversee his site supervisors, known as 'surveyors', and to visit all sectors of the works at least twice weekly. He also had to arrange for the provision of materials, to organise transport for the contractors, to control daywork gangs, to keep a close watch on site security and to prevent trespass. His was a daunting task but, wisely, work was planned methodically. It began on the Little Eaton branch and railway, after which it continued on to the Sandiacre line and finally, the crossing of the Derwent and the line to the Trent at Swarkestone. William White was appointed bookkeeper and paid £105 for yet another onerous array of duties which included measuring the excavations and all tradesmen's work, then authorising their payments (much as would a quantity surveyor in present-day practice), and also making up wages for direct labour gangs and keeping accounts for the entire enterprise.

In the final designs for the Derby Canal, two lines of canal radiated from the centre of Derby, where the Derwent was crossed through a pound, 300 ft long, created by the Long Bridge weir. A wooden causeway was built on timber piles adjacent to the latter to serve as a towpath. The Derby to Swarkestone line was 5½ miles long. From the river, it fell through the Pegg's flood lock into a short pound,[1] passing over and beyond a cast-iron aqueduct in the Holmes before leading into Gandy's wharf. The canal then rose through Day's lock into the summit level which was nearly four miles long. The waterway proceeded east of Osmaston, falling through 12 ft at the two Shelton locks, before it joined the Trent & Mersey Canal near Swarkestone Bridge. Beyond this, a short length led down through three locks to the River Trent. In spite of Jessop's earlier intentions, this section soon fell into disuse, its original purpose unfulfilled.

The Derby to Sandiacre branch was just over nine miles in length from the White Bear lock, which lifted the canal from the Long Bridge weir pound to the same level as that of the summit level of the Swarkestone line. A short cut led off the branch, rising through the Phoenix lock into a navigable pound of the River Derwent above St Michael's weir, and downstream of St Mary's bridge. Boats could continue from there along the river to the Darley Mill, about 1¼ miles to the north.[2] The Little Eaton branch turned off from the Sandiacre branch about half a mile from the river crossing, but the main line continued for about eight miles, falling 29 ft through two locks at Borrowash, and by another pair close to the junction with the Erewash Canal. The Little Eaton branch was three miles long and rose 17 ft through four locks, terminating at the

---

[1] Private correspondence with R.V.F. Torrington, who confirmed that the Pegg's lock gates were mitred towards the river; this is also shown on the Ordnance Survey large-scale plans of 1880. The entry in *Bradshaw's Canals*, p. 102, is incorrect.

[2] J. Farey, *A General View of the Agriculture and Minerals of Derbyshire*, iii (1817), p. 356.

wharf in Little Eaton from where the railway (or gangway as this became known locally) began (Fig. 8).

The canal was designed as a broad navigation throughout; the channel was excavated 44 ft wide at the top, reducing to 24 ft at the bottom, the depth being 5 ft, except for the 1¾ miles long summit pound of the Little Eaton branch, which was cut 6 ft deep to act as a reservoir.[1] The locks were 90 ft long and 15 ft wide at the top, battered down to 14 ft 6 in. at the base, although the maximum size of boat normally using the waterway was 72 ft by 14 ft, with a draft between 3 ft and 3 ft 8 in.[2]

Water was obtained from the Derwent and its tributaries, but mill-owners and other industrial users insisted on severe restrictions. The canal levels were designed such that water could be admitted from the river only through the Phoenix lock and this was permitted for a period of 24 hours from 8 o'clock every Saturday evening, when the factories were closed. If this proved insufficient for the weekly operation, then those lock gates could be opened for four hours on Thursday evenings. Similar restraints were placed on the water supplies into the summit level of the Little Eaton branch, which were also derived from the River Derwent or from some of the watercourses flowing into it.

Farey observed that by 1806 the canal was fed through the Little Eaton branch from the Bottle Brook and also by a feeder channel nearly a mile in length and 13 or 14 ft deep in places, which led from an old dam across the Derwent below Duffield Bridge. He stated (but apparently in error) that inflows then were limited to '4 hours on Sundays and Thursday evenings'. He clearly felt that any controls were unreasonable at that time because there 'generally is such a profusion of water' in the river in Derby. Farey made no mention of the flows admitted through Phoenix lock; this must have become the principal feeder for the canal as the demand for water power by industry declined in later years, and the restrictions on the canal company were relaxed (Fig. 9).

Supplies for the Swarkestone branch were provided through a water tight culvert which was built as an inverted siphon through the Long Bridge weir. This connected the Sandiacre line, just above White Bear lock, with the Swarkestone line above Day's lock. The water surfaces were at the same level at these points. The Derby Canal Act referred to a 'proper cast metal pipe' which was to be installed for this purpose, but during repairs to the weir in the 1970s, the conduit was exposed and

---

[1] Ibid., p. 358.

[2] H.R. de Salis, *Bradshaw's Canals and Navigable Rivers of England and Wales* (1904), p. 102.

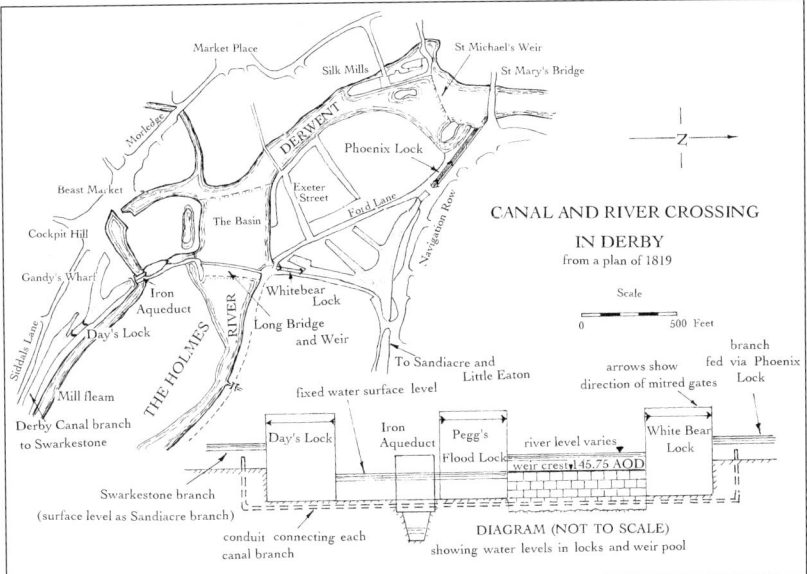

Figure 9 Canal and river crossing in Derby.

found to be a rectangular passage, measuring approximately 3 ft high by 2 ft wide, and formed entirely in the stone of the weir's foundations (it had also been breached in 1958 when water discharged under quite high pressure).[1] The culvert extended on each side of the river and ran along, and under, the towpaths to discharge into the canal, either through sluices or over side-weirs. The total length of the conduit was stated by Farey to be a quarter of a mile.

The cast-iron aqueduct which carried the canal across the western arm of the Derwent at the Holmes is believed to be the second of its kind to be built during the canal era. Until that time, aqueducts were supported by stone arches, the number and geometry of which depended on prevailing circumstances. Thus the semicircular arch could be used to cross a deep valley, whereas a segmental arch of large radius might be preferred where headroom below it was very limited. There is, however, a practical limit for the ratio between the rise of an arch and its span, below which the masonry may become unstable and collapse. In these cases, several flat arches of short span are preferable, although such structures may obstruct flow when used for river crossings, and thus

---

[1] Private correspondence with H.R. Potter, hydrologist with the former Severn & Trent Water Authority.

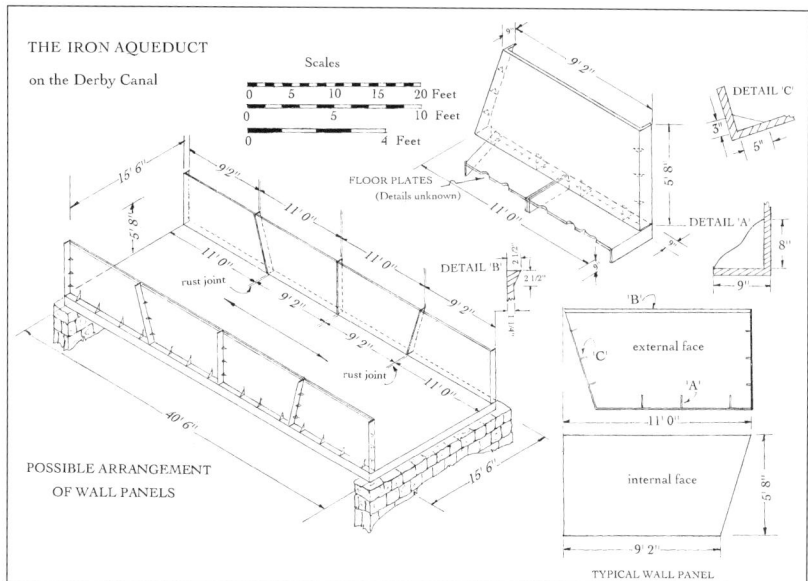

Figure 10 The iron aqueduct on the Derby Canal.

create water pressures on the upstream face which threaten their structural stability.

For situations where headroom below an aqueduct is severely restricted, such as at the Holmes, the best engineering choice is for a structure with a horizontal soffit, and this explains Outram's design of a prefabricated iron trough for this crossing. It seems likely that he heard of this new form of construction from Jessop, who knew of it through his association with Thomas Telford. The latter had erected the first bridge of this type at Longdon upon Tern on the Shropshire Canal, by March 1795. Almost certainly the components for the Holmes aqueduct were cast at the Butterley works during the summer of that year and transported to the site for assembly.[1]

The finished aqueduct was a neat trough, simply supported at each end on low stone abutments. Accurate measurements of the aqueduct do not seem to have been recorded prior to its demolition during town centre developments in 1971, although the present writer measured the structure several years beforehand when it was partially buried in rubbish. The trough was 15 ft 6 in. wide, 5 ft 8 in. deep and approxi-

---

[1] Contract documents have not survived, although this seems a reasonable assumption as the company was in production at the time.

Plate 4 The Holmes iron aqueduct, Derby, during town centre developments.

mately 40 ft 6 in. long. Each wall unit was made up from four flanged and gusseted plates, 1¼ in. thick, bolted together such that they resembled the voussoirs of a stone arch. The units were roughly finished but this is perhaps not surprising, because they must have been among the earliest attempts at the Butterley ironworks to cast large-scale structural units (Fig. 10).

There can be no doubt that Outram had very little understanding of the behaviour of such a structure under loaded conditions, because the aqueduct soon proved to be grossly under-designed. Even when completed in late 1795 and before being filled with water, it must have been seen to flex and vibrate under the lightest of loads. But of course, this was a time when knowledge of metallurgy, structural analysis and design was rudimentary. As like as not, Outram was guided in his design only by personal judgment and a practical feel for the strength of cast-iron.

Not surprisingly, the life span of the aqueduct was short. It failed partially in February 1802 and the force of the escaping water was such that a boat alongside the nearby wharf was dragged from its moorings

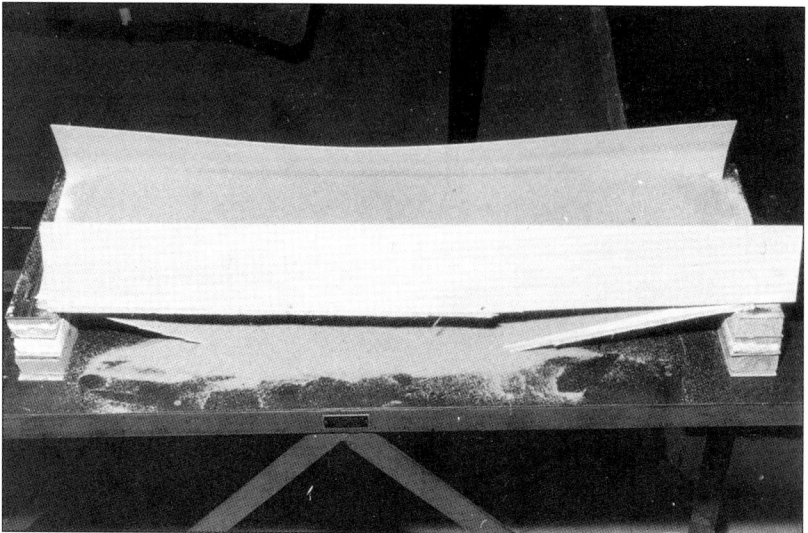

Plate 5 Balsa model to demonstrate mode of failure of Holmes aqueduct.

into the vortex and badly damaged.[1] A contemporary report of the company's engineer stated that 'in consequence of some defect in the metal (as is generally supposed) a short time ago, one of the plates in the bottom of the aqueduct broke and let out the water of the canal'. That brief note provides a clue for the reasons for the collapse; it was probably due to the top flanges of the vertical wall-plates failing in compression, rather than a defect in the metal of one of the base-plates. This would result in a sudden buckling of the top flanges, the effect of which would be to transfer the load to the base of the wall where a longitudinal crack would develop in the thin plates of the floor, close to the junction with the wall. Scale model tests seem to confirm this mode of failure of this weak and lightweight bridge structure.

Remedial works followed this disaster. The methods were not described in the minutes but it is likely that the aperture was closed with timber and puddle clay, in the hope that this would prove sufficient. Nevertheless, the proprietors must have felt some unease about the condition of this important structure; in 1810 they ordered that waste materials lying by the aqueduct had to be removed immediately and instructions were later issued that boats hauled through the aqueduct must not strike the sides of the trough, on pain of a fine of £5.[2]

---

[1] DLSL, DCCR, Minutes, 8 Feb., 8 March 1802.
[2] Ibid., 18 Sept. 1810, 14 May 1811.

In 1812, further failure occurred and it was ordered that 'the iron aqueduct in the Holmes be immediately repaired with deal baulks'.[1] In January the following year, the clerk was instructed to seek estimates 'for building a stone aqueduct and bridge in the Holmes'. Drawings and estimates were provided but an engineer's inspection showed that the remedial works were holding and were not in any immediate danger of giving way. The committee seemed satisfied and in fact the structure survived until October 1930 when the manager reported that the flooring had finally collapsed and had been replaced by a local contractor at a cost of £220.[2] Timber piles were apparently driven at intervals into the bed of the stream and these were bridged with a timber platform sealed with puddle clay. These crude measures seemed effective enough for the remaining years of the life of the canal, makeshift though they appear to have been.

The Derby Canal Railway (described in the Act as a waggonway, or stone road, and known locally as the Little Eaton Gangway) was originally planned to extend as far as Smithy Houses in Denby parish, with a line turning off at Coxbench to Smalley Mill. In spite of the intentions referred to in the Act, it seems that Outram's work at the time was restricted only to the laying of the main line to Smithy Houses. The system, nevertheless, had developed considerably by 1831 when Priestley wrote that

> a railway proceeds by Horsley and Kilbourn to Smithy House, which is 4¾ miles in length. From Smithy House there is a branch of 1¾ miles in length to the collieries at Henmoor, situated 1½ miles east of the town of Belper; another 1½ miles in length, by the potteries, to the extensive coal works near Denby Hall; with a collateral branch out of the last-mentioned branch, ¾ mile in length, to other collieries north of Salterswood.[3]

The design of the original rail track was quite different to that noted by Radford. The idea of using timber rails, reinforced by wearing strips of iron, was abandoned in favour of angle, or flanged, cast-iron rails (which came to be known as plates) similar to those eventually adopted by Outram on most of his railways of later years. The reason for this

---

[1] Ibid., 10 March 1812.

[2] Ibid., 14 Oct. 1930.

[3] J. Priestley, *Historical Account of the Navigable Rivers, Canals, and Railways, throughout Great Britain* (1831), p. 194.

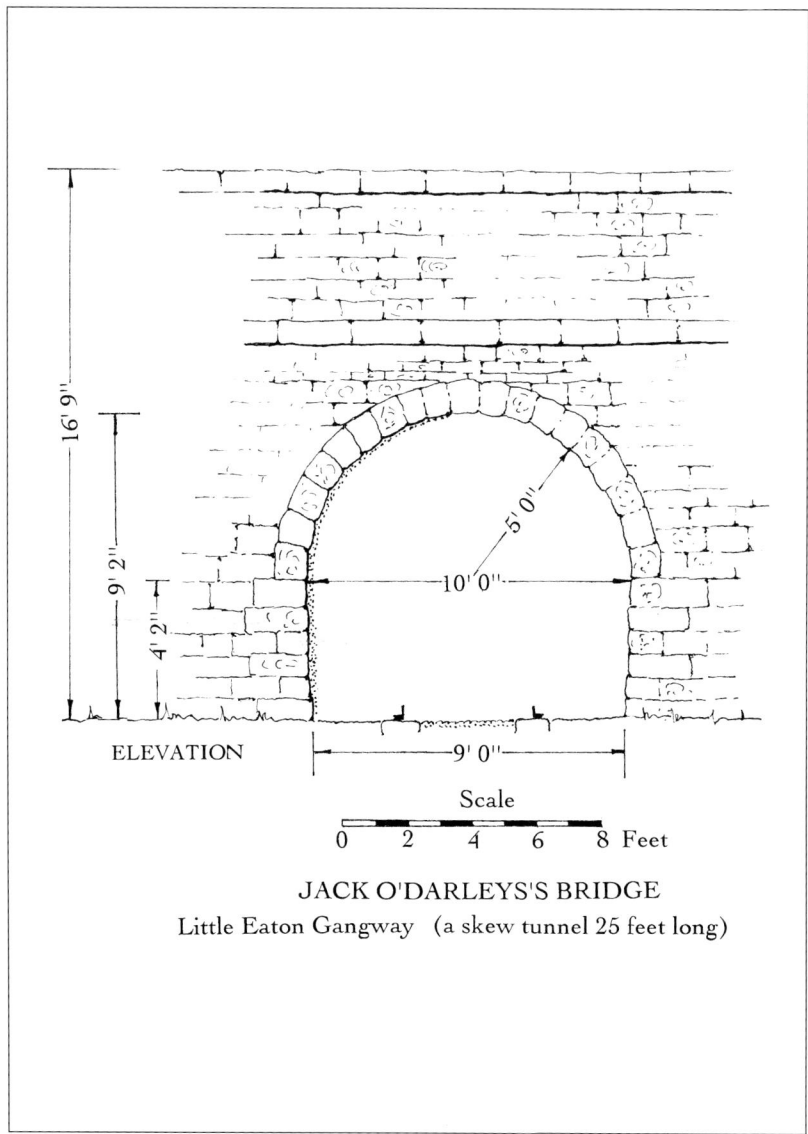

Figure 11 Jack o'Darley's bridge.

change of mind may have been due to the publication by John Smeaton in 1791 of the details of similar railways used during the construction of

Plate 6 Jack o'Darley's bridge.

the Eddystone lighthouse.[1] Jessop, whose father Josias was Smeaton's foreman on those works, was familiar with railway practice and may

---

[1] J. Smeaton, *Narrative of the building of the Eddystone Lighthouse* (1791), pp. 196, 197.

have encouraged Outram to utilise that system. Another significant factor was that Joseph Butler of Wingerworth, the ironmaster who had successfully contracted for the manufacture and supply of rails for the Denby line between 1793 and 1796, is believed to be the first (during 1788) to use flanged rails above ground level.[1] Hence it is difficult to believe that he too did not have some influence on Outram's designs.

One writer has stated that the original rails were 3 ft long and weighed 30 lb, the vertical flange being 4 in. high in the centre, reducing to 2 in. at the ends where the horizontal bed was notched to receive holding-down spikes.[2] To be more precise, it seems more likely that Outram's plates were of the design which he came to use elsewhere: 3 ft long, 3½ in. high in the centre above a 4-in. wide horizontal ledge (the 'sole' on which the wheels of the waggons ran), tapering to 2½ in. at the edges. Each rail weighed 28 lb, although Outram came to recommend that these should never be less than 30 lb and 'on Railways for heavy Burthens, great Use, and long Duration, the Rails should be very stout, weighing 40 lb, or in some Cases nearly half an Hundredweight each'.[3]

Outram (and Jessop too) preferred stone blocks to wooden sleepers. These were specified as being not less than 8 in., nor more than 12 in., deep. Their shapes were immaterial 'and of such breadths (circular, square or triangular) as shall make them 150 lbs. or 200 lbs. weight each'. The underside of a block had to be flat and 'a small portion' of the top was dressed level for the rails to rest upon. A circular hole, 1½ in. in diameter, was drilled 6 in. deep in the centre of each block, into which a 5-in. long octagonal oak plug was firmly driven. The rails had countersunk notches at each end, were butted together and held down by pointed spikes with long heads. These were driven into the oak plugs until the heads fitted neatly into the notches. In later years, many of the original plates were replaced by longer units in which holes were drilled near the ends for the spikes, because experience had shown that the former method of fixing allowed the rails to work loose and part at the joints. The original gauge for the railway was 3 ft 6 in.,[4] which Outram considered appropriate for mineral railways, although a slight reduction was allowed at bends to accommodate the turning wheels of the waggons. At some later, unknown date the gauge was increased to 4 ft

---

[1] Farey, *General View*, iii (1817), p. 288.

[2] D. Ripley, *The Little Eaton Gangway and Derby Canal* (1991), p. 20.

[3] M.F. Outram, *Margaret Outram (1778–1863). Mother of the Bayard of India*, Appendix C.

[4] PRO, RAIL 803/3, 3 Dec. 1799.

6 in.,[1] and thus it remained until closure.

A single line was constructed to Smithy Houses and several passing places were provided along the route, care being taken during construction to ensure an easy and gradual rise from the canal at Little Eaton, by means of a series of shallow cuttings and embankments. Over the distance of 4¼ miles the average gradient of the railway, as built, was 1 in 215, which was very close to that stated in Jessop's report. There were no striking features along what was a nicely engineered route, although the well-known Jack o'Darley's bridge still survives today (Fig. 11; Plate 6), as well as the adjacent twin-arch crossing of the Bottle Brook, built at the same time. The estimate for the railway has not survived but Jessop implied in his report that £12,000 would be required, a sum which may have also included for the canal section connecting with the main Sandiacre line.

The waggons designed for the railway consisted of two components, namely a container for coals which could be lifted from a tram (or carriage) with four cast-iron wheels, each of 2 ft 2 in. diameter and 1½ in. tread. These revolved on fixed axles, there being sufficient float on the latter to accommodate tight curves and any slight variations in gauge. The original waggons were restricted to 37-cwt loads, although standard loads had increased to 2¼ tons by the end of the nineteenth century.[2] Different sizes of waggons were used during the lifetime of the system but a complete replacement of these must have followed the change of gauge previously mentioned. In spite of Jessop's recommendations about transport operations, the eventual practice was to employ teams of horses drawing seven or eight waggons in a gang.

Two factors were emphasised by Outram for the waggons' design. He insisted that the wheel base of each carriage should not be less than 3 ft, such that any rail would not be subjected to a load greater than a quarter of that of the loaded vehicle. He never intended that the railway waggons would be used on public highways, hence on arrival at the destination in Derby, the containers were lifted off the boats by crane and placed on two-wheeled carts for transport to consumers.

The sketchy, inexpertly written committee minutes make scant mention of construction and progress. Work on the Little Eaton branch, and particularly on the preparation of the ground for the railway, proceeded rapidly because Joseph Butler of Wingerworth ironworks was

---

[1] Ripley, *Little Eaton Gangway*, p. 20.

[2] C.E. Stretton, *The Little Eaton Outramway. A paper read at Little Eaton, 29 July 1905*. Stretton, a notoriously unreliable historian, is likely to be correct on this occasion.

awarded the contract for the supply of cast-iron rails (at £10 10s. per ton, equivalent to about 80 rails),[1] the first delivery of which was made in February 1794. The railway was finished as far as Coxbench (about halfway) by the middle of that year and the entire line to Denby by May 1795. The first boat-loads of coal from Drury's pits at Denby were delivered to Derby on 11 May[2] and there distributed for the benefit of the poor. The Sandiacre line was also finished by the end of the month, including the Phoenix lock, through which essential water supplies were admitted into the canal system from the River Derwent.[3]

It was only then, in early June 1795, that work started in earnest on the Swarkestone branch, beginning with the setting out of the section across the Holmes and the location of the Long Bridge weir.[4] Progress was rapid because by 18 February 1796 it was reported that 'the weir across the River Derwent and the cast-iron aqueduct in the Holmes being finished, there remains little to be done 'ere the three branches of the navigation are united'.[5]

At their meeting of 23 October 1795 the committee had authorised payments to Benjamin Outram & Co. of £203 13s. for castings. These components were not specified but almost certainly the payment was for the supply and delivery of the large plates and other items (the total weight of which was about 25 tons) from which the Holmes aqueduct was built. The aqueduct must have been finished before that date because the canal company was often short of funds and it is unlikely that they would have paid out that substantial amount before completion of any contract work. Accordingly, it seems reasonable to assume that the structure was erected by late September 1795. Thus Outram's trough aqueduct became only the second of its type built on a British waterway, although at the time no one seemed to think the event in any sense remarkable.

The entire canal was finished and fully operational by the end of June 1796, the final cost being about £100,000.[6] Under the Act, the canal company was required to purchase the Derwent Navigation, which was duly conveyed for £3,996. This then became redundant but in fact traffic

---

[1] DLSL, DCCR, 23 Dec. 1793.

[2] *Derby Mercury*, 11 May 1795.

[3] DLSL, DCCR, 28 May 1795. Outram was instructed to fix the site of a warehouse on the canal at St Mary's Bridge, implying that the Phoenix arm feeder and the entire Sandicare line were nearing completion.

[4] Ibid., 11 June 1795.

[5] *Derby Mercury*, 18 Feb. 1796.

[6] Hadfield, *Canals of the East Midlands*, p. 68.

river had effectively ceased after 1794.

Outram's services were retained by the company and he was regularly consulted on engineering modifications and management organisation, although this involvement ceased at the end of 1801 with the termination of his practice as a canal engineer. It is nevertheless surprising that he was not called in to advise following the failure of the Holmes aqueduct in 1802. The company's permanent staff then seemed able to deal with all engineering matters, although William Jessop had to be consulted during 1806 to deal with the vexing matter of the water supplies to the Little Eaton branch, a problem which he had satisfactorily resolved by the following year.[1]

---

[1] DLSL, DCCR, 10 Feb. 1806.

CHAPTER 6

THE HUDDERSFIELD NARROW CANAL

The Huddersfield Narrow Canal was one of three waterways which linked Lancashire, across the Pennines, with the West Riding of Yorkshire (Fig. 12). The others were the Leeds & Liverpool Canal and the Rochdale Canal, both broad navigations which followed longer but easier, northerly routes.

The Huddersfield project was formally proposed in May 1793 when there were no completed waterways between Manchester, the Yorkshire navigations and the ports of the east coast. Its promotion was the result of the Canal Mania and the success and high premiums attached to the shares of the Manchester to Ashton under Lyne Canal, the proprietors of which were then seeking extension of their trading links from Manchester. The two canals were to share management structures and the future seemed promising, except from the danger of a possible Rochdale Canal.[1] Had building of the latter begun earlier, enthusiasm for the route to Huddersfield might have been extinguished because of its obviously difficult and costly construction features, besides the operational disadvantages of a narrow canal connecting broad navigations on each side of the Pennines.

Despite these factors, the proposed Huddersfield Narrow Canal gained enthusiastic support at a meeting in Huddersfield on 30 May 1793, when £118,000 was promised of the provisional estimate of £200,000 for the project. Temporary officers appointed included George Worthington, a solicitor of Altrincham, as clerk, and Nicholas Brown of Saddleworth as surveyor. 'Mr. Jessop, if at Liberty', was to be invited to be the engineer.[2] Soon afterwards it was found that he was not available (he was inundated with such requests during those Canal Mania years) but he presumably recommended Benjamin Outram instead.

Outram accepted the appointment and, on 22 October 1793, presented his report in Huddersfield to a crowded meeting of enthusiastic investors. William Pontey, a local businessman and ardent critic of the scheme in

---

[1] C. Hadfield and G. Biddle, *The Canals of North West England* (1970), ii, p. 322.

[2] WYASL, Radcliffe MSS, 319/C/1, Report of a meeting at Huddersfield, 30 May 1793.

HUDDERSFIELD NARROW CANAL

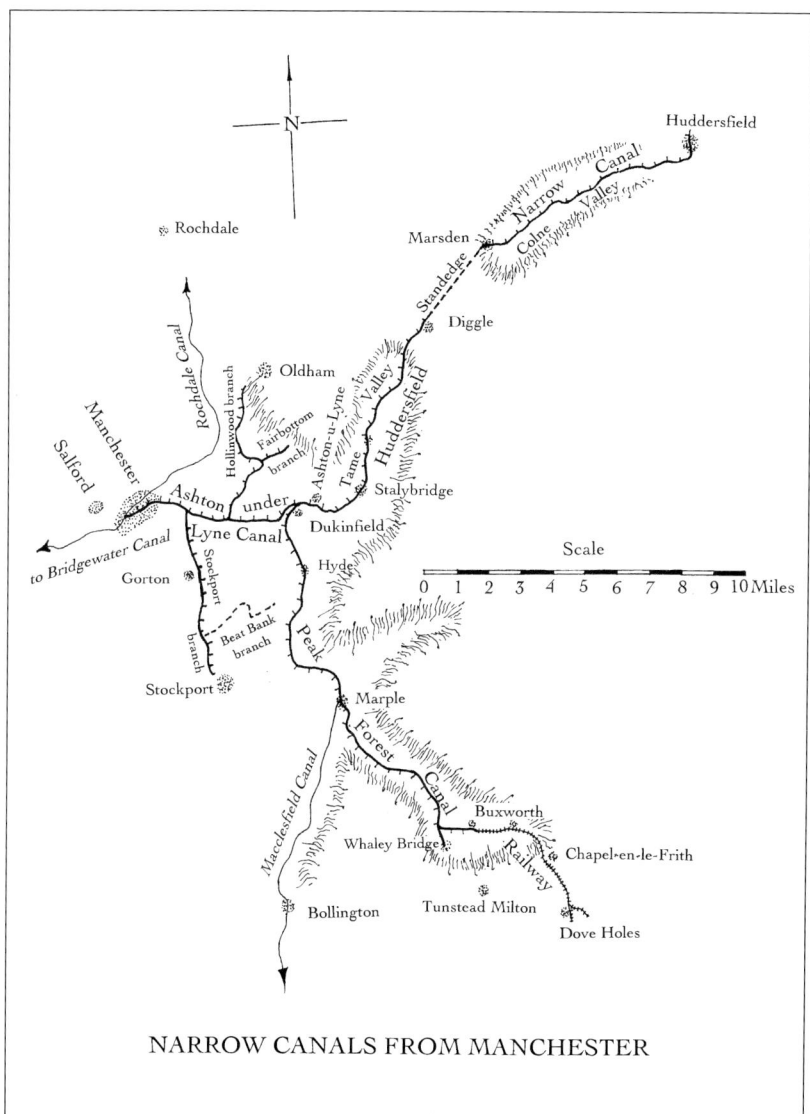

Figure 12 Narrow canals from Manchester.

later years, referred to Outram as 'a very young man [who] like the Bull in Tristram Shandy went through the business with a very grave face',[1] when describing his proposals for the proposed waterway. In that report, Outram recommended a route from Sir John Ramsden's Canal (the Huddersfield Broad Canal which accommodated vessels 58 ft long by 14 ft 2 in. wide) in Huddersfield, to follow the Colne valley to Marsden, climbing 436 ft (438 ft 10 in. as built) to the summit level at 649 ft above sea level. Then it proceeded by a tunnel over three miles in length through Standedge, before emerging in Diggle. The canal ran down the Tame valley on the Lancashire side, descending some 334 ft 8 in. (338 ft 11 in. as built), via Stalybridge to the Ashton Canal, with a tunnel of 200 yards at Scout Mill (220 yards as built).[2] The report did not mention another tunnel near Stalybridge, 198 yards long, which was completed, but opened up in later years. The number of locks was not stated, although the 'lockages' of 436 ft and 334½ ft were listed in the estimate at £65 and £70 per foot respectively. Outram intended to delay a decision on the numbers and locations of locks until the construction stage. 42 locks were eventually built in Yorkshire and 32 on the Lancashire side, the average fall being 10 ft 6 in. per lock. The total length of canal was given as 19¾ miles (19⅞ miles finally).

Outram stressed that the route was

> the shortest Communication yet between Manchester and the eastern Navigations and it will pass through a Country full of Manufacturers ... and by the Vicinity of the proposed Canal to the River the Mills will receive their Articles free from land Carriage to the Canal.

This was true; the numerous mills shown on the survey were located close to the canal because the valley floors were so narrow. Moreover, the engineer was aware of the mill-owners' fears for the loss of their water supplies which would interfere with the 36 waterwheels on which their mills depended for power. He did not intend to take waters from the rivers, however. Instead, reservoirs to feed the canal were to be located in large, deep valleys where 'collected Waters frequently produced Torrents of Floods'.

The major engineering task was the great tunnel at Standedge, for

---

[1] W. Pontey, *A Short Account of the Huddersfield Canal* (c. 1812) (copy in library of Institution of Civil Engineers), p. 3.

[2] Ibid., Report of B. Outram, 22 Oct. 1793.

Plate 7 Marsden entrance to Standedge tunnel.

which Outram envisaged a five-year programme. He claimed that the strata consisted of gritstone and strong shale and did not expect undue difficulties. The estimate provided for 5,380 yards of finished tunnel at £7 per yard, and for the sinking of 18 shafts for £2,767 (31 were actually started; 17 of these were on the drainage adits). Two adjacent shafts were to be sunk on the approximate centre of the tunnel at Red Brook, where steam engines would be erected for drainage and the hoisting of spoil from the workings. Unwisely, Outram did not include any sums for the lining of sections of the tunnel, thereby assuming that even the shale strata would require no support. He was to be proved wrong in that respect. No one then questioned his statement that 'the Hill through which the Tunnel is proposed to be made, appears favourable'. William Pontey was to ponder the question, not unreasonably, in later years as to 'How did he know what would be found in the Tunnel, though about 200 yards below the highest part of the Mountain?'.

Nor did Outram see fit to mention the difficulties, foreseen by some of his critics at the time, of linking his narrow canal with the broad canals at Huddersfield and Manchester.[1] That certainly was a major defect in the design but, had the canal been planned with the generous dimensions of a broad navigation, it is certain that the estimate would have been prohibitive and the project unlikely to have attracted investors. In any case, it is clear that the final estimates were pared down to less than a desirable minimum, in spite of what must have been obvious at the start, that this was to be a difficult project, fraught with imponderables created by its remoteness, uncertain subsoil conditions, and the climate.

In all, the documentation and designs were brief, and in several respects inadequate and misleading for investors. Many complex engineering matters were left to be resolved as works progressed and, although this was an unwise practice, it was not unusual for the time; even the most experienced engineers did no differently. It is not surprising, therefore, that estimates for such proposals were often exceeded and when, inevitably, this happened in the case of the Huddersfield Canal, the critical William Pontey was to write that 'I dismiss it with only one observation ... *Engineers are not infallible*'.

The engineer's original estimate for the construction of the canal, including the Standedge tunnel (which was also costed separately at £55,187) and the reservoirs and all accommodation works, was £178,748. The mill-owners of the Tame and Colne valleys were doubtful about the adequacy of the reservoirs, and so Outram agreed to safeguard their interests by increasing capacity. Accordingly, another £4,000 was added to the estimate.[2] Perhaps because of the euphoria and excitement generated at the meeting, nobody saw fit to ask why the engineer's estimate was so close to the provisional figure quoted during the preliminary meeting held six months earlier. The suspicion remains that Outram had fixed his estimate at a figure just less than that which the proprietors had decided was likely to attract investment. The eventual costs, which were more than twice that figure, lend credence to that view.

There was very little discussion of Outram's report and no opposition to the project before subscriptions were invited. £200,000 in £100 shares was required for the promotion and construction of the canal but 500 shares were reserved for riparian landowners. Pontey described the scene

---

[1] WYASL, Dartmouth MSS, DT 305/3/9, Correspondence of W. Elmsall with the Earl of Dartmouth, 8 July 1793. The canal was to link Sir John Ramsden's Canal in Huddersfield with the Rochdale and Bridgewater canals in Manchester, via the Ashton Canal.

[2] WYASL, Radcliffe MSS 319/C/1, Resolution of mill-owners on the line of the Huddersfield to Ashton Canal, 23 Oct. 1793.

when bidding began:

> Subscription papers were opened at 3 or 4 tables and happy was the wight who could get to write his name, deposit his cash or £1 upon each share subscribed for and then escape from the crowd ... Immediate dealing with shares soon put the premium up to £15 per cent although this soon came down ... The reason I could never ascertain. Probably others like myself were panic struck for very soon after prices began to decline. I sold 3 of the 6 I had hold of in the scramble at a premium of 5 guineas each and very soon after let the same person have the other 3 at par ... .[1]

That hectic scene suggests that, from the very beginning, the company was in the hands of speculators. Although Outram's estimate was to prove too low by far, the company was constantly plagued during the construction years by the failure of investors to pay their calls promptly, as well as by those speculators who evaded payments altogether.

Documentation prepared for the bill was restricted to the minimum required by law.[2] This consisted of a survey and plan of the canal with a book of reference as to land occupations, and also a rudimentary bill of quantities and the estimate.[3] Nicholas Brown was to become an accomplished land surveyor and draughtsman in later years (he and his son practised in Wakefield) but the plan, an early example of his work, was an inaccurate plot, probably based on a compass and chain survey. It was of limited use for engineering purposes but probably served as a useful illustration for potential investors, as well as for those Members of Parliament who gave the project no opposition when the bill was reported, without amendment, on 27 March 1794. The Act received the Royal Assent on 4 April[4] (the Act for the Rochdale Canal was obtained the same day).

The Huddersfield Canal Company was then quickly established and its officers appointed. Worthington was confirmed as clerk, assisted by John Battye to manage day-to-day affairs. Bankers, treasurers and valuers were also appointed. Outram became engineer and he was also nominated, with John Holt, a surveyor, to monitor the discharges and usage

---

[1] Pontey, *Short Account*, p. 6.
[2] Standing Orders of the House of Commons, 14 March 1793.
[3] HLRO, B. Outram, Estimate of the expense of making a navigable canal, 1794 (i.e. the Huddersfield Narrow Canal).
[4] 34 Geo. III c. liii (1794).

of streams by the canal company and riparian owners.[1] Outram's fees were fixed at three guineas per day, plus ten guineas expenses for each journey to the canal.[2] His payments varied between £150 and £175 every six months,[3] which suggests that he would spend four or five days each month on site, followed by attendance at the management committee meetings. Remuneration was not great, considering the long hours he would spend on off-the-job planning, in preparing designs, estimates and reports, and also in visiting companies to negotiate for machinery and materials.

Outram's task was exacting and, as a young man of 29, he was comparatively new to the profession. Hence he did not have an entourage of well-tried and reliable resident engineers and contractors who could be called on to execute works under his control. Instead he had to rely on local appointees, notably Nicholas Brown as resident engineer. The latter was youthful and inexperienced, and was probably selected for the job through the patronage of friends among the proprietors, and not because of any proven ability in supervising engineering works of such magnitude.

Brown was required to devote all his time to the job for a salary of £315, from which he was expected to pay the wages of a book-keeper as well as all personal expenses, except when working for the company more than 15 miles from the canal.[4] He was thus expected to supervise 20 miles of engineering construction across some difficult and remote country, and yet the very conditions of his appointment would discourage him from spending too freely on assistants, or on frequent travel along the line of waterway. The project was understaffed from the outset; it called for at least three resident engineers to assist Brown and, although he was eventually allotted a few overlookers (or clerks of works), initially he had to do all the work himself. Not surprisingly, his management was to prove ineffective. The company minutes are uncritical of Brown but the signs of his inadequate supervision, and the committee's unease, are all too clear. As work progressed there were serious mistakes in setting-out the works. One of these, for example, was the result of misunderstanding Outram's written instructions, for which Brown was required to pay compensation.[5] Another mistake was an

---

[1] WYASW, Quarter Sessions Records, 15 Jan. 1795.
[2] PRO, RAIL 838/2, 11 July 1794.
[3] Ibid., 17 May 1796.
[4] Ibid., 11 July 1794.
[5] Ibid., 23 Nov. 1795.

overflow weir built in the wrong place and then some culverts were constructed too small for their intended purposes.

Moreover, there were frequent claims by riparian owners, proprietors among them, for both real and imagined damages to property, the details of which were rarely recorded in the minute books. Considerable sums were paid out to complainants on the flimsiest of evidence but this was hardly surprising when judgments were made by their friends on the committee. A competent resident engineer would have put a stop to such abuses and avoided disputes by prior consultations with landowners and, above all, by clear written instructions to contractors and direct labour gangs. Brown seemed unable to cope with such matters. Outram must be blamed in part for tolerating the situation but it did not help when he was forced, through his serious illness, to absent himself for long periods during the critical years from 1795 to early in 1797. Perhaps it was no coincidence that Brown was also absent through sickness during the latter part of that period.

Outram was then asked to find another surveyor to help Brown and accordingly the company was to take on William Bailiffe of Marsden, who took charge of the canal west of Standedge tunnel in July 1797 at £100 p.a.[1] Thus Brown's duties were reduced to supervising only the east side of the tunnel. Brown muddled along until at last the committee lost patience with him when an unnecessary culvert was built in August.[2] He was given six months' notice to quit, although this was rescinded the following February when instead he was given a twelve-month extension at half his former salary, with the promise of a bonus if he served with 'faithfulness, ability and to the satisfaction of the committee'. Meanwhile the more successful Bailiffe was given a salary increase of £25.

In September 1797 the committee saw fit to engage the ubiquitous William Pontey 'to attend to and keep memorandum of the payments to the workmen, the letting of small jobs and to superintend the conduct of the surveyor and overlookers on that part of the said canal which extends between Huddersfield and the east end of the tunnel'.[3] Hence the unhappy Brown would have to continue under threat of dismissal with the prospect of Pontey (who previously had been appointed as one of his overlookers) ensuring that he did his job properly! It must have been a difficult time for him and only a weak and inexperienced man such as

---

[1] Ibid., 29 June 1797.
[2] Ibid., 3 Aug. 1797.
[3] Ibid., 15 Sept. 1797.

Brown would have tolerated this situation, which undermined whatever shreds of authority and credibility he still possessed. But matters did not seem to improve and probably even led to the committee appointing a few of their number to set out accommodation and other works.

Not surprisingly, Brown was dismissed after that additional year's service and Bailiffe took over as surveyor and superintendent for all the works on the canal with a salary of £200 p.a. plus expenses, an appointment which continued with fair success until 1801, when he was replaced by John Rooth. Brown, however, continued to prepare surveys for the company on a freelance basis for some time after his dismissal and, indeed, several years after the completion of the canal, he returned briefly as engineer to the company.

When the project began in the late summer of 1794, the first task of the engineers was the setting out of the line of the waterway, the major aqueducts and bridges, as well as the great tunnel at Standedge. Elaborate computational procedures for the precise setting out of works were not practised in those days, hence Brown's chain survey plan, on which the canal appears to have been sketched out in freehand, could only be used as a rough guide. Accordingly, Outram would have set out the canal's centre-line using much personal judgment. Wooden pegs would be driven into the ground, at two- or three-chain intervals, or closer where the canal was built on embankment or in cutting, until the tops of them were level with the proposed water surface.[1] Where embankments were formed, stakes, some probably of considerable length, would be used and in cuttings holes would be excavated before pegs could be driven down into them to the correct level.[2] Alignment between pegs was left to the contractors to decide, a practice which led in places to short meanders. Although unnecessary and avoidable, these are common enough on many canals. Generally, however, the route selected for the Huddersfield Narrow Canal was imaginative and economical.

It is a simple matter to calculate the ground positions of the toes of earth cuttings and banks such that these can be marked by pegs. Raking timber profiles can then be erected so as to define the bank slopes. Such refinements of the basic setting out are always desirable because land-take can then be agreed with landowners before construction begins, and

---

[1] Surveyors formerly used the Gunter chain of 22 yards. It is not a precision instrument and requires frequent calibration because its many links stretch appreciably in use. Brown's stated lengths are thus to be treated with caution.

[2] 'Canals', in *Cyclopaedia, or University Dictionary of the Arts, Sciences and Literature* (ed. A. Rees) (1819), vol. vi.

Plate 8 Interior of Standedge tunnel in the misaligned section, west of Red Brook.

the site properly fenced off. Outram, in common with other engineers of that period, preferred not to do this and always left land assessment and measurement until work was concluded. Not surprisingly, this practice led to disagreements and claims for trespass and damage to property.

More difficult, however, was the setting out and the control of line and level in the Standedge tunnel, which Outram intended to be straight from end to end. The operations which had to be undertaken included an accurate surface alignment between the proposed tunnel mouthings at Diggle and Marsden, in order to fix the position of the intervening construction shafts; the transfer of bench-marks across the mountain to ensure a common basis for levelling;[1] and the transfer of the surface lines and levels down the shafts and into the tunnel workings.

Surviving records fail to describe how these problems were resolved but surface alignment may have been achieved by traversing between the tunnel mouthings using a surveyor's dial (a magnetic compass fitted on to a tripod) and Gunter chain. After plotting the results on the drawing board, the bearing of the tunnel's centre-line could be measured by protractor and then set out on the ground, again using compass and

---

[1] Benchmarks are permanent marks of known level above a datum. They are transferred to different locations using telescopic levels.

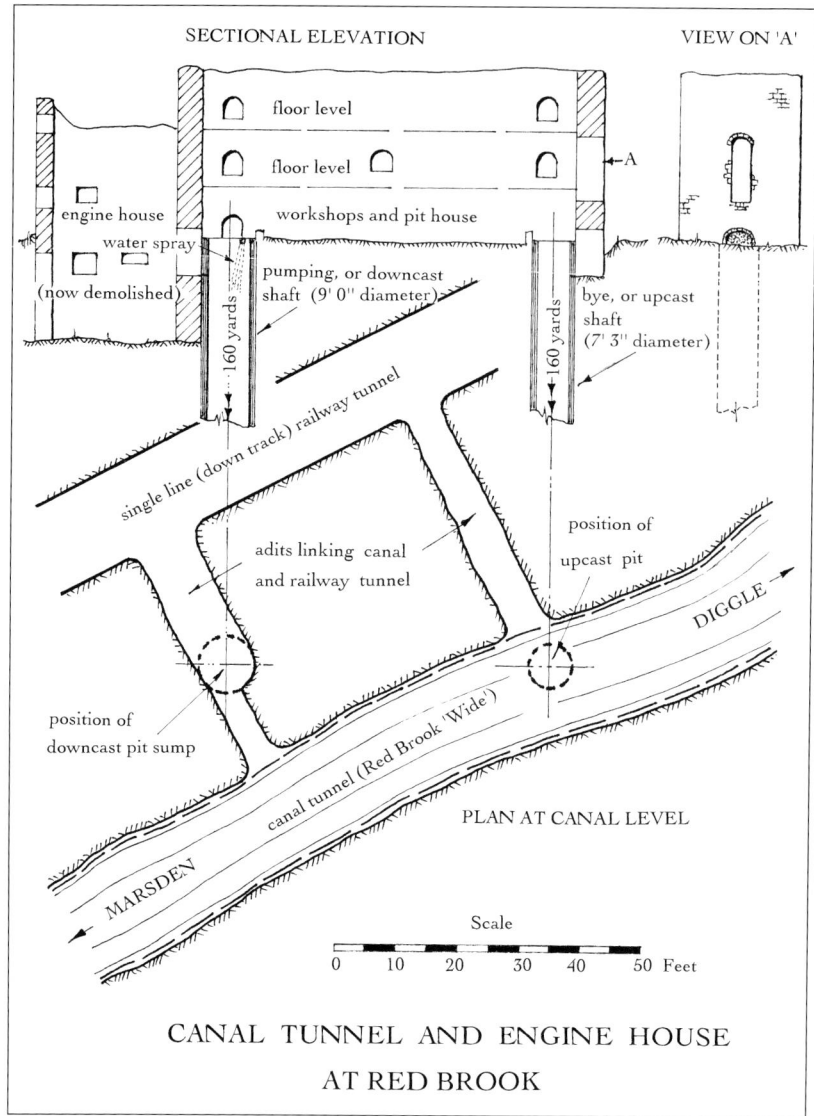

Figure 13 Canal tunnel and engine house at Red Brook.

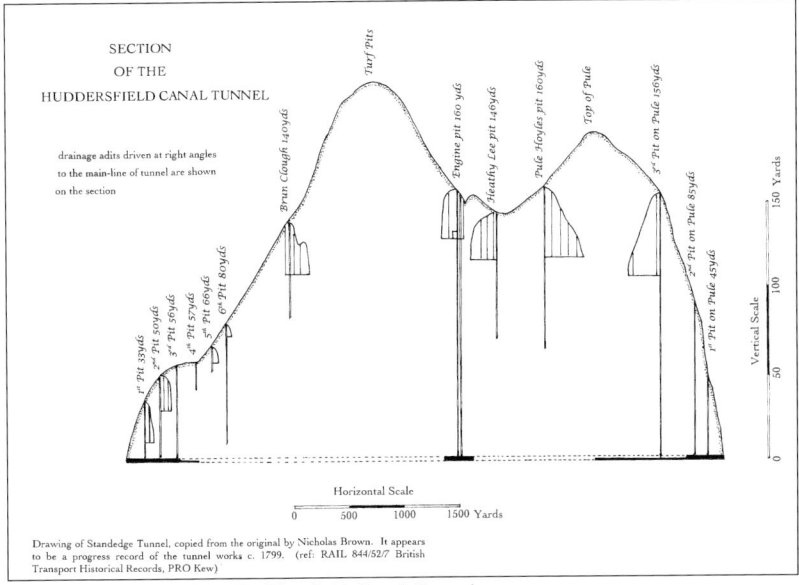

Figure 14 Section of the Huddersfield Canal Tunnel.

chain. Similar problems were overcome on some canals by the method of reciprocal ranging, using very long poles.[1] These were moved laterally, by trial and error, until a good sight-line was obtained across the hillside. Whichever method was used, Outram only managed to fix the surface position of the shafts within 3 and 7 ft laterally of their correct positions. One of the two shafts at Red Brook (the bye pit) was located on the centre-line of the canal as might be expected. The other shaft (the engine, or downcast pit) was deliberately offset from the main tunnel by 26 ft because it contained pipework and fittings for the pumps, which would otherwise interfere with free movement along the tunnel (Fig. 13).

Transfer of the centre-line to the base of the shaft would be accomplished by means of a pair of heavy metal plummets, suspended at pit bottom on cords hung from hooks fixed a short distance apart on the survey line at the surface. Thus a parallel line could be projected into the workings for the guidance of the miners. The correct level of the tunnel was generally obtained by calculating the depth of a shaft and measuring down, as sinking progressed, by chain, from a surface bench-mark.

---

[1] Robert Whitworth lined up Sapperton tunnel on the Thames & Severn Canal by using beacons, and poles as long as 120 ft: H. Household, *The Thames and Severn Canal* (1969), p. 42.

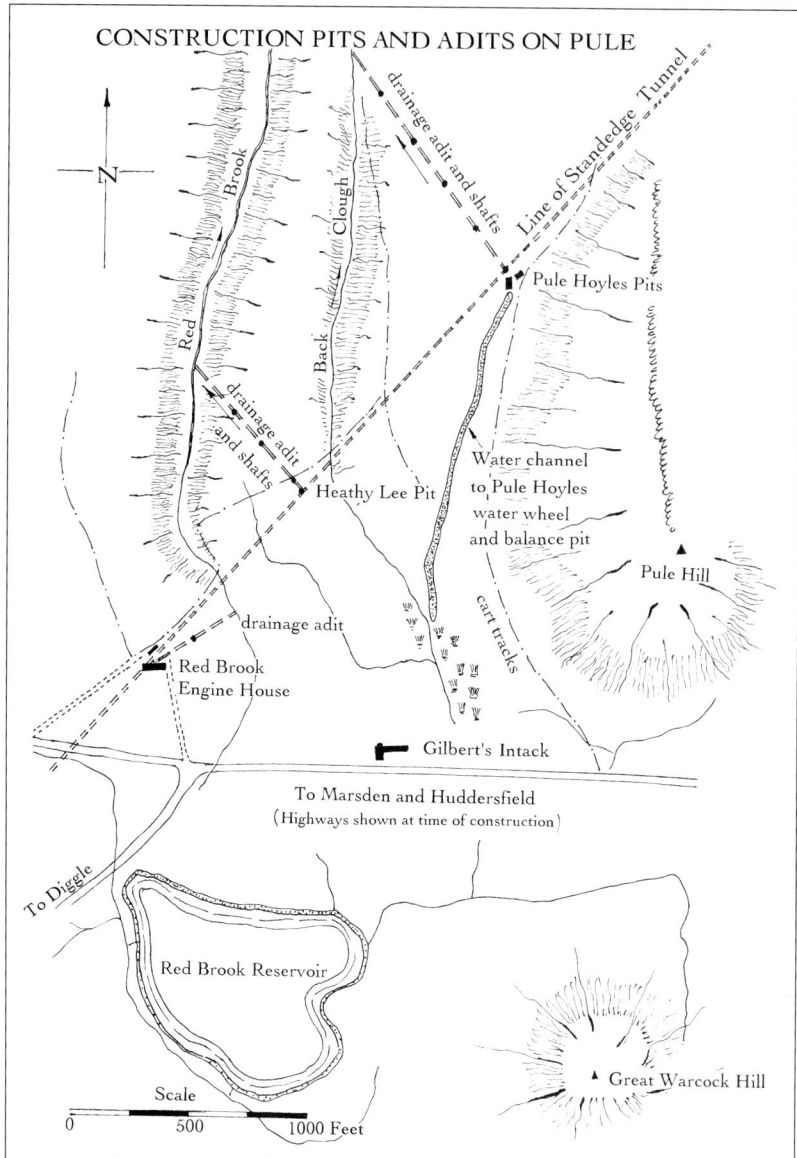

Figure 15 Construction pits and adits on Pule.

Outram's report of June 1796 referred to progress in the tunnel and much can be gleaned from this, and also from a longitudinal section

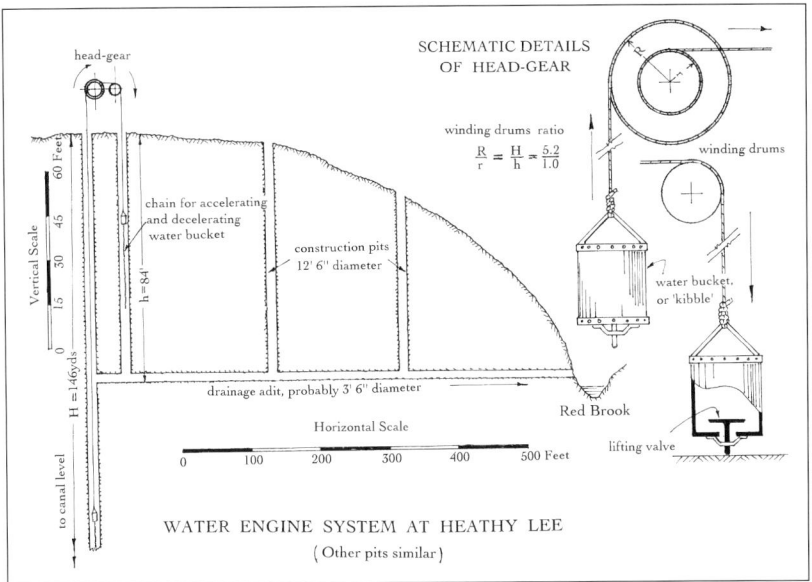

Figure 16 Water engine system at Heathy Lee.

prepared by Brown some time afterwards (Fig. 14).[1] Shafts were initially sunk at 100- to 180-yard intervals from each end and during that year six pits were completed at the Diggle end (their depths varying from 33 to 80 yards) and two at the Marsden end (depths 45 and 85 yards). Pumping costs were very high (the cost of working a steam engine at the sixth pit from Diggle was 11 guineas per week) due to an unexpected ingress of water from the Millstone Grit. As a result, Outram decided to reduce the number of main-line shafts to that at Brun Clough (140 yards deep), two at Red Brook (each 160 yards), and those at Heathy Lee (146 yards), Pule Hoyles (160 yards) and Pule Hill (156 yards).

The tunnel was 9 ft wide, except for a short section mid-way which was widened to 16 ft for the convenience of passing boats. Driving was accomplished by hand-drilling and blasting with black powder. Shaft sinking proceeded in much the same way, although a description by Farey gives a graphic impression of the dangerous conditions of working.[2] A shallow hole would be drilled in the centre of the shaft floor, then thoroughly dried out with oakum before packing with

---

[1] PRO, RAIL 838/2, 7 April 1797.

[2] J. Farey, *A General View of the Agriculture and Minerals of Derbyshire*, i (1811), pp. 325, 326.

gunpowder and a clay seal. On lighting the fuse, the miners would cling, one above the other, to a winding rope and at a signal they were hauled 10 to 15 yards up the shaft, where they remained until the shots were fired. They were then lowered back into the fumes to clear up the broken rock, and to repeat the process. Farey noted that sometimes accidents occurred when the miners were not lifted sufficiently high above the danger zone.

Nine adits were driven through the hillside into the main shafts (approximately at right-angles to the line of the canal tunnel) in an effort to draw off surface water, and also to provide a discharge path for waste from the 'water engines'. These machines were commonly used in the mines of Derbyshire and Nottinghamshire and Outram preferred them to horses, which he stated were 'a great and unavoidable expense on most works of this nature'.[1] Spoil, loaded into a bucket at canal tunnel level, was hoisted to the surface by a water-filled kibble moving down a shallow balance pit. A valve in the base of the kibble lifted automatically when impinging on the bottom of the pit and the water was thus discharged into the drainage adit. Balancing chains then returned the spoil bucket to the canal level once more. This system, which was controlled by head-gears, was installed in only a few shafts initially (Figs. 15, 16). In others, waterwheels provided the motive power but they proved unsuccessful and eventually were abandoned in favour of water engines. Water for the operation of the latter was ducted from nearby streams, or feeder reservoirs, into cisterns adjacent to the balance pits. The remains of some of these temporary works can still be found along the line of the tunnel.

In addition to these economical hoists, four steam engines of the atmospheric (or Newcomen) type were used for pumping out the wettest pits, the largest of these being located at Red Brook. Ebenezer Smith & Co. of Chesterfield had supplied the latter but this had not complied with specification. A 6 h.p. engine had been ordered and yet one of less power had been installed. It was unequal to the task and additionally, certain castings and pipes had broken during service. John Smith was called to attend the committee and he agreed to substitute two cylinders with one of 'equal power to six horses' and also to replace some defective fittings.[2] The engine then seemed to cope with the torrents of water but pumping costs were prohibitive, indeed John Sutcliffe, a Yorkshire engineer, noted that these were as high as £50 per week! By

---

[1] Pontey, *Short Account*, p. 14.
[2] PRO, RAIL 838/2, 10 Feb., 7 April 1797.

1798 Outram tried to improve performance and reduce costs and his efforts led him to apply to Boulton & Watt's northern agent, James Lawson, for advice on the engine's conversion to the Watts' patent.[1] Instead of offering to modify the 70-in. diameter cylinder engine however, James Watt junior suggested its replacement with a 52-in. cylinder, to work at 10 strokes per minute with the existing 14-in. pump of 6 ft stoke. The cost was quoted at £1,170 but this was far more than the company could afford.[2] Outram did not bother to trouble the committee with the idea and after a further enquiry (when Outram appealed to Watt's sense of patriotism for this work of national importance, in an attempt to persuade him to drop his price) the correspondence ceased abruptly. Watt realised that the company was in no position to buy anything but on the cheap and he was having none of that.[3] Smith's engine then continued in use until about 1803.

Good ventilation is essential for tunnelling work and a method, common enough in mining practice of that day, was adopted at Red Brook whereby a duct from the nearby reservoir discharged a small quantity of water as a fine spray into the top of the engine pit.[4] This induced a strong draught of air down into the workings. The same technique was adopted at Brun Clough pit, and both systems efficiently ventilated the tunnel, as well as the adjacent railway tunnels, thereafter.

In the first working season from July 1794, plans were determined by the availablity of specialist contractors able to take on such extensive and difficult tunnelling work. The age of the general contractor had not yet arrived, so that the contracts were let for labour only, the canal company providing pumps and engines, materials for temporary works, powder and other provisions. Fortunately, agreement was reached with one Thomas Lee on a rate, undisclosed in the minutes, for the first 1,000 yards of main-line tunnel at the western, or Diggle, end. Ostensibly this was a straightforward section through alternating strata of shales and Millstone Grit. Drainage was by gravity, thus pumps were unnecessary and the job should have required only a steady and determined effort to complete. Lee's progress was very slow, however, and by September 1797 Outram reported to the committee that the contractor had encountered unforseen difficulties. His losses were considerable and he was

---

[1] BCL, Boulton & Watt MSS, J. Perrins to B. Outram, 26 June 1798.

[2] Ibid., J. Watt to B. Outram, 14 July 1798.

[3] Ibid., B. Outram and J. Watt, 16 Aug. 1798; Watt to Outram, 24 Aug. 1798.

[4] The practice of inducing ventilation by water spray was common in Derbyshire coal mines: A.R. Griffin, *Coalmining* (1971), p. 63.

Plate 9 Red Brook engine-house. Pule Hoyles pits and engine-house are at A; Heathy Lee pit and its adjacent balance pit are at B.

facing ruin. The committee sent a member to the site to investigate and at the November meeting, Lee was offered £300 compensation, with an additional 15s. per yard run for his remaining work, plus a year's extension to his contract. This apparently generous offer was rejected and the company had no alternative but to terminate the contract.[1] He had driven and partly finished about 445 yards by the date he withdrew.

Meanwhile the eastern end was also opened up late in 1794 using direct labour. Driving was through relatively easy blue shale, which had to be arched with masonry, until a fault was crossed into the Millstone Grit between Pule Hill and Pule Hoyles. Although the strata are apparently of the Namurian series,[2] a few thin coal measures were found in the approach to Red Brook and were used to fuel the steam engine there.

It is difficult to unravel the sequence of construction in the Standedge tunnel during the first eight years of its construction. Work should have been carefully planned but instead it proceeded in an haphazard manner. Minds were changed; there were frequent stoppages; tunnel sections and machinery were abandoned for long periods and labour was frequently dismissed, or switched between working faces. The reason for this

---

[1] PRO, RAIL 838/2, 2 Nov. 1797.

[2] Information from the Institute of Geological Sciences, Leeds.

chaotic practice was the constant shortage of capital through the inability, or unwillingness, of shareholders to pay their calls. In spite of this, the management committee persisted in maintaining several working faces with all the associated costs of equipment and labour this entailed, and with only empty promises of capital to come. A better plan would have been to employ small labour forces to drive the tunnel from each end, relying on gravity to draw off the water, and to open up from the Red Brook pits only when funds were available. This seems to have been recommended by Outram but the proprietors were impatient to begin business and misguided enough to believe their methods were appropriate. In fact these did nothing but waste money and manpower.

Outram's first progress report was submitted to the shareholders in 1796.[1] Although there was evidence of some achievement on the canal sections, he could only report limited progress in the 'Great Tunnel'. After two years of effort, a mere 445 yards had been opened up at the west end and 350 yards at the east, and even these sections needed finishing and arching. Much labour had gone into adit construction (1,485 yards) and the sinking of pits. Four of the latter were complete and another ten in 'tolerable forwardness'. The engineer explained that 'the progress of the main tunnel must necessarily be slow until the deep pits are sunk down to the level as, in the meantime, but few miners can be employed at the two ends'. Much was made of the excavation in new quarries and the getting of stone, the building of haul roads, cottages and workshops on the mountain. All these facilities, it was claimed, 'would prevent a repetition of those delays and loss of time which have been already experienced in sinking the deep pits'.

Outram's report was unduly optimistic and attempted to divert attention from his own as well as the company's failings. They were now in difficulties but unwilling to admit it. There was even a nonsensical attempt to turn one of the causes of their misfortunes, 'the very copious supply of water from the tunnel', into a useful asset by claiming that this was 'probably sufficient for the summit of the canal, without the aid of reservoirs'. This disturbing report noted that costs for the tunnel stood at £20,049; in two years they had expended more than one third of the original estimate on less than one seventh of the work in the tunnel, and yet no reference was made to this, nor were revised estimates provided for the shareholders. It is regrettable that Outram lent his name to this misleading document because, as a professional engineer, he was obliged

---

[1] Pontey, *Short Account*, Report by the committee and engineer, 30 June 1796, pp. 12–16.

to write the truth and not allow his judgment to be influenced by clients troubled with financial crises. And yet the company was not entirely to blame because they were constantly plagued by a shortage of funds. For example, by June 1796, when working output along the entire canal should have been rising to a peak, £92,000 of share capital had been called for, of which £22,650 was in arrears, and there was only £889 at the bank.[1]

Meanwhile, by September 1796 a contract had been signed with one George Evans 'for cutting and making the part of the tunnel under Pule Moss and Brun Top'. This was roughly the central section which included several pits and the notoriously wet sixth pit. Evans was to break his agreement in May 1798;[2] the details and reasons are unknown except that Outram was complimented by the committee for the manner in which he cancelled the contract to the company's advantage. Outram then made a surprising offer to contract for completion of the tunnel.[3] This aroused the interest of the committee and he was asked to prepare a quotation 'to finish and complete the tunnel at Brun Clough and Pule Hill on a given time, taking engines etc. provided by the company' and also the remaining short length of cut from the Diggle entrance down to Woolroad near Saddleworth. The estimate was submitted in March 1799 when it was set aside for consideration at a future special general meeting.[4]

The details of that submission have not survived but it is, no doubt, that same estimate mentioned by John Sutcliffe in his book of 1816 which referred to the construction of the Huddersfield Narrow Canal.[5] He stated that Outram had offered to find security and to drive the tunnel in five years, but that some of the committee were of the opinion that it could be completed for a smaller sum. Sutcliffe considered, in view of the time and eventual costs of completing the Standedge tunnel,[6] that this was fortunate for Outram, although he 'had heard his friend Jessop say he might complete it in the time proposed and get money by the contract'. The company should have accepted the offer but before long they were afflicted with a calamity which turned their minds to other

---

[1] Ibid.

[2] PRO, RAIL 838/2, 18 May 1798.

[3] Ibid., 21 Feb. 1799.

[4] Ibid., 20 March 1799.

[5] J. Sutcliffe, *A Treatise on Canals and Reservoirs* (1816), p. 119.

[6] PRO, RAIL 1019/8/1, J. Rooth, Report on the Huddersfield Canal to the General Assembly, 27 June 1811.

Plate 10 Typical bridge at a lock tail on the Huddersfield Narrow Canal.

matters. In August 1799 the canal was badly damaged by severe rains and floods, which were to cost the proprietors thousands of pounds they did not possess. A special appeal had to be made to the shareholders[1] and minds were diverted temporarily from the problems of the tunnel. In retrospect, there seems no reason to doubt that Outram could have taken on the contract because of his experience with Jessop on the Butterley tunnel. Whether he could have made a profit is another matter but, in any case, the opportunity was lost and the matter was never raised again.

Outram had initially estimated for 49 accommodation and highway bridges, 30 culverts and nine aqueducts.[2] In accordance with the practices of the day, there had been no formal negotiations with riparian owners prior to the passing of the Act and eventually ten additional bridge crossings had to be agreed. Most of these bridges were stone arches (costing from £64 to £100 depending on type) built in squared, regular-coursed masonry. A standard design, very similar to that of Outram's bridges for the Peak Forest Canal, was adopted, which comprised a semi-elliptical arch opening, rounded on the shoulders to permit easy access for the horse along the towpath. Stone parapets were built above a string course (a narrow, projecting course of masonry) which defined the surface level of the crossing highway. The wing walls

---

[1] Pontey, *Short Account*, G. Worthington to shareholders, 2 Sept. 1799, pp. 22, 23.
[2] HLRO, B. Outram, Estimate (1794).

Plate 11 Skew bridge at Sparth, built with bed joints parallel to abutments.

curved in plan to terminate in squared, capped pilasters and the spandrel walls also curved in sectional elevation. Built with indigenous Millstone Grit and lime mortar, these simple and inexpensive, yet handsome, bridges were immensely strong and have had to withstand long periods of neglect. Several timber footbridges and swivel bridges were built before 1800 for the sake of economy, but these were mostly replaced by masonry structures before the canal was opened in 1811.

A few accommodation bridges were built oblique to the canal, in line with the existing highway, necessitating skew angles up to 19 degrees. None of these was constructed on the correct principles because the longitudinal joints in all cases were set at right angles to the abutments rather than, more correctly, to the direction of skew. All were built before Outram learnt to apply the more correct technique to the arch of the Store Street aqueduct of the Ashton Canal in 1797 (see Chapter 9). However, none of these bridges shows any signs of instability today, probably because of the light loads they have had to support over the years.

Outram set out seven major major aqueducts and yet none of these compares with his original design proposals, nor is it likely that actual costs bore any resemblance to those provided in the estimates. There is nothing remarkable about their designs; they are simple and functional, carrying few embellishments and, with the exception of Diggle, all incorporate a skene arch (i.e. a circular segmental arch). This type allows

Plate 12 Stakes iron aqueduct at Stalybridge.

some economies in earthworks and lockage, besides providing reasonable clearance below soffit level, although the low pitch ratios (i.e. the ratio of height above arch springing, to the span), coupled with crumbling mortar joints, has resulted in some arch movements at the Uppermill and Scarbottom aqueducts. Some details are given in Table 6.1.

During the great floods of 1799 the aqueducts at Marsden and Stalybridge were completely destroyed. That at Marsden was only replaced in 1806 when funds allowed but the Stakes aqueduct at Stalybridge was located in a section of canal already in use for navigation; immediate action had to be taken to replace it. The River Tame channel at Stalybridge is 55 ft wide, and the headroom below the canal invert was limited to about 5 ft. This was of insufficient height for a single skene arch (the pitch ratio would have been 0.09, which is too low for stability) and so originally Outram had designed a four-span structure instead. During the floods, the narrow openings of this bridge impeded river flow, causing a backing-up of the water to such a depth that the pressures finally overturned the structure. It would have been unwise to rebuild a multi-span arch bridge and so it was replaced by a single-span prefabricated iron trough, similar to that which Outram had already constructed at the Holmes on the Derby Canal.[1] Naturally he elected to

---

[1] Above, Chapter 5.

TABLE 6.1

Aqueducts on the Huddersfield Narrow Canal

Aqueduct	As Planned	Estimated Cost	As Built	Arch pitch ratio	Remarks
Paddock or Longroyd	2/18 ft arches	£298	1/36½ ft skene arch	0.25	Slight skew
Scarbottom or Skyerbottom	Not specified	£260	1/38½ ft skene arch	0.26	Slight skew
Marsden or Waterside	2/8 ft arches	£152	unknown	unknown	Demolished before 1900
Diggle	2/8 ft arches	£160	1/16 ft semi-elliptical arch	0.47	Existing but disused
Uppermill or Mythom Mill	2/18 ft arches	£312	1/35 ft skene arch	0.22	Slight skew
Royal George	3/15 ft arches	£1,075	2/25 ft skene arches	0.22	Costs included earthworks
Stalybridge or Stakes	4/15 ft arches	£1,351	probably as specified	unknown	Replaced by iron trough in 1801

use a much stronger, improved version at Stalybridge, having learned from his earlier experience. Components of the new aqueduct were probably cast at the Butterley ironworks and the high quality of the manufactured units is indicative of the advances made at that foundry in the few years of its existence. The parts were assembled *in situ* and the whole was probably commissioned by May 1800,[1] although Farey stated that it was only finished in 1801 (Fig. 17).[2]

Although Outram must have been aware, shortly after its completion, that the Holmes aqueduct was badly under-designed, he still did not appear to understand, when designing the Stakes aqueduct, how the single-span trough would respond under load. His lack of comprehension is suggested by his reference to it as an 'arch of cast iron' rather than a trough which was simply supported at each end. The floor panels (which had failed in the Holmes aqueduct) were now substantially stiffened, as were the bases of the wall panels. That, however, was not where a possible failure would first develop. The upper flanges of the wall plates

---

[1] PRO, RAIL 832/2, 29 May 1800; the adjacent arch was then under construction for use as the towpath.

[2] Farey, *General View*, iii, p. 373.

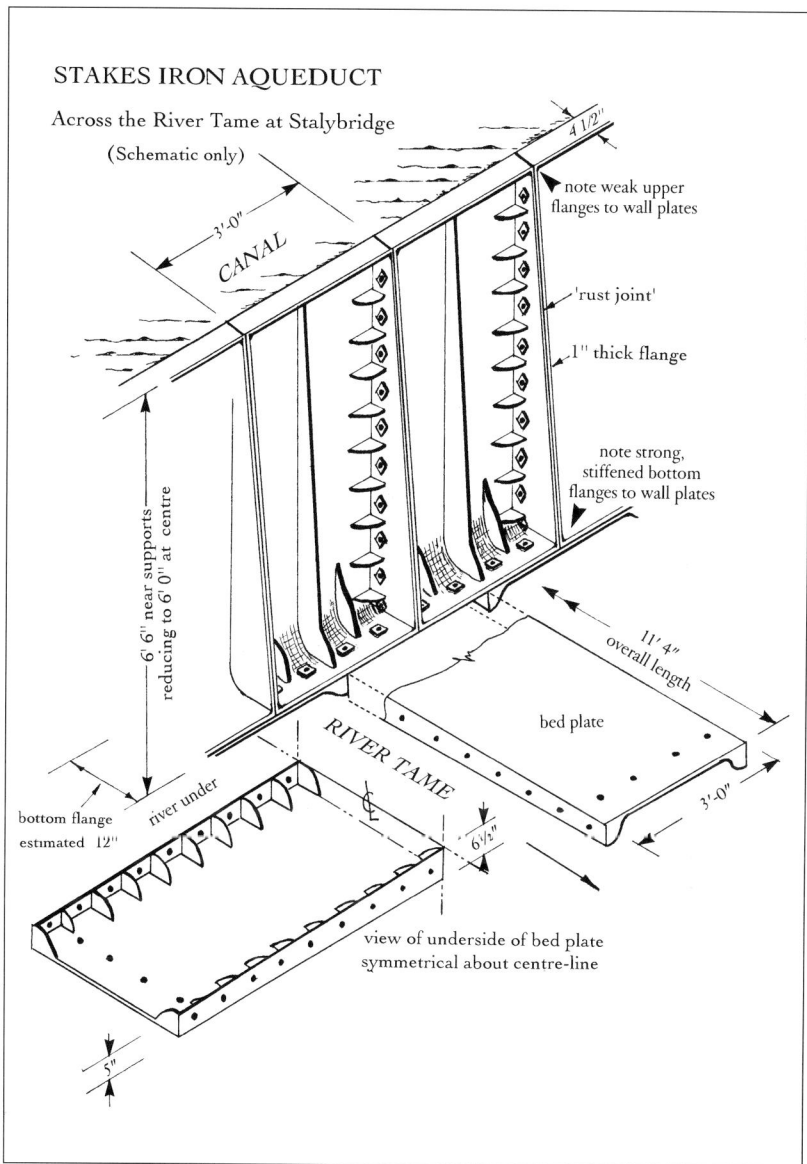

Figure 17 Stakes iron aqueduct.

would always be in compression when the trough was full of water, and they had to be stout enough to withstand buckling under this load. Outram failed to realise the importance of this because the upper flanges were not as strong as the lower flanges. Nevertheless, in spite of this weakness, the aqueduct survived until 1875, when sagging was arrested by a clumsy cross-bracing of 2-in. diameter wrought-iron rods, bolted on to the sides of the structure.[1] The whole was then partially supported by a vertical iron rod hung from the centre of the adjacent stone arch and so it remains to the present day.

Farey inspected the aqueduct in 1808[2] and described its construction, which confirms that the existing trough, apart from its later modifications, still remains in its original form. He stated that it was '... a square trough put together with flanches (sic) which are strengthened by brackets, cast at proper intervals for the rivets between them: the vertical flanches are widest at bottom and have their brackets cast between them'. He also explained that the 'rust joints' between the flanges were made from iron turnings and borax, made into paste by boiling water. Farey felt that the towpath should have been fixed to the trough, instead of which 'a ridiculous stone arch' was built for this purpose on one side. The inaccessible and not very well-known Stakes aqueduct[3] is a sturdy, handsome bridge which is the oldest surviving example of its type still in use for its original purpose. The iron aqueduct at Longdon upon Tern, built by Thomas Telford some five years previously, has long ceased to be used as a canal aqueduct.

Heavily locked routes such as the Huddersfield Narrow Canal were vulnerable in times of drought and this waterway was to become notorious in that respect. Outram, when presenting his report on 22 October 1793, had proposed reservoirs 'to contain 14,900 locks full of water and to supply 100 locks per day for four months altogether, exclusive of ample allowances for evaporation and absorption'. It was fortunate, according to his report, that the countryside thereabouts provided sites in large, deep valleys for up to five times the extent of the reservoirs he required. Hence he could reasonably claim that 'this supply I conceive will be more than adequate to answer the largest trade that can be expected to be carried on the canal', without interfering with local manufacturers' needs. To his credit he recognised that it was necessary to safeguard the mill-owners' interests because, after all, their continuing prosperity and transport requirements were essential for the canal company's survival. And so he could state that 'such a capacious system was needed to discourage the tapping of mill waters in dry seasons'.

In spite of the engineer's reassurances (his optimistic estimates must have been based on intelligent guesswork, rather than sound hydrological

---

[1] The date is marked on the structure.

[2] Farey, *General View*, iii, pp. 373, 374.

[3] The only satisfactory view is from private factory premises.

principles) the sagacious mill-owners were quick to disagree with his calculations and insisted on reservoirs 'for not less than 20,000 locks of water, each lock to contain 180 cubic yards'.[1] Thus Outram, 24 hours after making his confident statement, had to change his mind and accept these proposals which were one-third greater than his own. Moreover, these industrialists, more knowledgeable than Outram of local conditions, insisted that all the streams would flow uninterrupted and that only flood-waters could be taken for the canal's reservoirs. Their cautious reaction was not unreasonable, judging by later experiences, because, after all, the sole source of their power came from the 36 waterwheels[2] fed by the rivers Colne and Tame.

The results of these negotiations on water supplies was that the Act of 1794 required at least 607 million gallons to be stored in reservoirs within three miles of the canal. Withdrawals from watercourses at a distance beyond 2,000 yards were allowed, subject to safeguards for other users. Outram's cost estimates, submitted with the parliamentary depositions, quoted a price of £17,800 for the reservoirs, without stating their positions.[3] They were shown on Brown's survey, however: one location was at Staley Mill, another to the west of Uppermill village, three were in the Wessenden valley above Marsden, and others were to be built at March Haigh and Slaithwaite. All but the first and the last of these were situated on the summit level.

Of these sources of water supply, only March Haigh and Slaithwaite reservoirs were eventually built by the company. The Wessenden chain was constructed by a consortium of mill-owners before 1800 for their own purposes.[4] Waste-water from the system may have been available to the canal company lower down the Colne valley but eventually these three reservoirs were developed by the Huddersfield Corporation many years later for potable water supplies. The amending Huddersfield Canal Act of 1800[5] gave the company additional powers to take flood waters from the rivers Colne and Tame and, seven years later, seven more reservoirs were recommended to augment their supplies. All this, perhaps, serves to show the inadequacy of Outram's earliest proposals.

All the reservoirs were contained by conventional earth dams with

---

[1] Pontey, *Short Account*, p. 6.

[2] The waterwheels are marked on all copies of Brown's survey. Numbers shown vary but 36 is the maximum.

[3] HLRO, Outram, Estimate of the Expence.

[4] WYASL, Dartmouth MSS, D 77/5/2, Minutes of a meeting of mill-owners situate on the River Colne, 4 April 1799.

[5] 40 Geo. III c. xxxix (1800).

clay cores. Although the design and construction methods are unknown, it is likely that spoil was deposited by end-tipping from carts, followed by spreading and compaction, using teams of horses. A period would then be allowed to elapse as the formation weathered and settled naturally.[1] There was very little understanding of the basic engineering principles involved and this, combined with a lack of suitable plant and inadequate supervison, was the probable cause of the leakage and earth-slippage failures which occurred at some of the dam sites for several years following their completion.

Before work began at Slaithwaite dam, a major structure which was to give trouble over the ensuing years, the agent to the Earl of Dartmouth's estates in the Colne valley had written an anxious letter to his master stating that 'there is to be a very large reservoir a little above the town which, if the head should break, would sweep away the greater part of the town but the engineer is perfectly confident there will be no danger'.[2] In fact, Outram had assured him that the slope of the downstream side would be so flat that 'it could be mowed' and that the inside would slope in the same manner (the outer face of the dam was built originally with a slope of 1 in 2½ but the inside face was steeper). Outram's soothing words proved too optimistic because, even as late as 1800, a report noted that the state of all the reservoirs was so bad that 'the whole of the water that could have been collected from the summit would not have been sufficient to support the then little trade upon the canal ... there not being one reservoir out of the five that would retain any water' and moreover, there was a 'fairly well-founded apprehension of great destruction to the property of the country below the Slaithwaite Dam'.[3]

The committee had always seen fit to meddle with the management of their canal and to interfere with business for which they had already appointed staff, hence it was not surprising that they called in the engineer Robert Whitworth during 1797 to report on the progress of the works and 'the propriety of the line as laid down by Mr Outram'.[4] The report by this eminent engineer was widely published and it was damning, to say the least of it, when he reported that the masonry and

---

[1] Formerly, allowances for settling of earthworks could be between one-fifth and one-twelfth of the finished bank height, depending on the soil type: W.J.M. Rankine, *A Manual of Civil Engineering* (1900), p. 339.

[2] WYASL, Dartmouth MSS, DT 303/7/2, W. Elmsall to Earl of Dartmouth, 9 Feb. 1795.

[3] WYASK, Tomlinson MSS, J. Rooth to R. Firth, 24 Feb. 1817.

[4] PRO, RAIL 838/2, 29 June 1797.

the earthworks were the worst executed he had ever seen. Indeed years later, John Sutcliffe (never a friend of Outram), claimed that in a private conversation, Whitworth confided to him that 'the work will be nearly all to do over again, particularly the locks'.[1]

Outram's lock design, although ostensibly economical in excavation costs, was unsuccessful in practice and clearly did not meet with Whitworth's approval. The walls were built with foundations 'very short towards the hill', such that on completion, the back edge at the top of the abutment was 4 ft beyond the foundation. This design was vulnerable as there was an inherent risk of the toe sliding into the lock, unless restrained by the invert masonry. Unfortunately, all the locks suffered from another defect in that the waste-water channels built around them were insufficiently watertight, hence there was constant seepage into the foundations. This caused earth slippage and movement of the masonry and not surprisingly, in times of severe frost, the stonework collapsed 'into little better than a heap of rubbish'. Yet another serious fault was the very short length of the canal pounds set out on the steep inclines, because they gave far too little water-storage capacity for a busy navigation.

In one respect Whitworth was impressed. After examining the partially finished tunnel he wrote that it was 'certainly the most stupendous piece of work of the kind that has ever been projected in the kingdom, both for depth and length'. On the whole this report would have made depressing reading for the committee and yet the minutes make no mention of any admonition of Outram or his assistants. Suffice it to say, however, the defective designs were probably attempts by Outram to save money for his impecunious clients, but the references to indifferent workmanship were sure indications of a lack of supervision. As for the latter, the inexperience of the site staff must be partly to blame, but the difficulties of site access and communications must also have been major contributory factors.

By 1796 the Slaithwaite reservoir was nearing completion because it was needed to supply the lower canal section, then under construction between Huddersfield and Slaithwaite. Work on another reservoir began early in 1798 at Tunnelend,[2] immediately north of the Standedge tunnel entrance, as the canal excavations steadily extended towards Marsden; indeed the committee were urging William Bailiffe to finish this reservoir

---

[1] Sutcliffe, *Treatise on Canals*, p. 118; M.F. Outram, *Margaret Outram (1778–1863). Mother of the Bayard of India* (1932), p. 96.

[2] PRO, RAIL 838/2, 15 Dec. 1797.

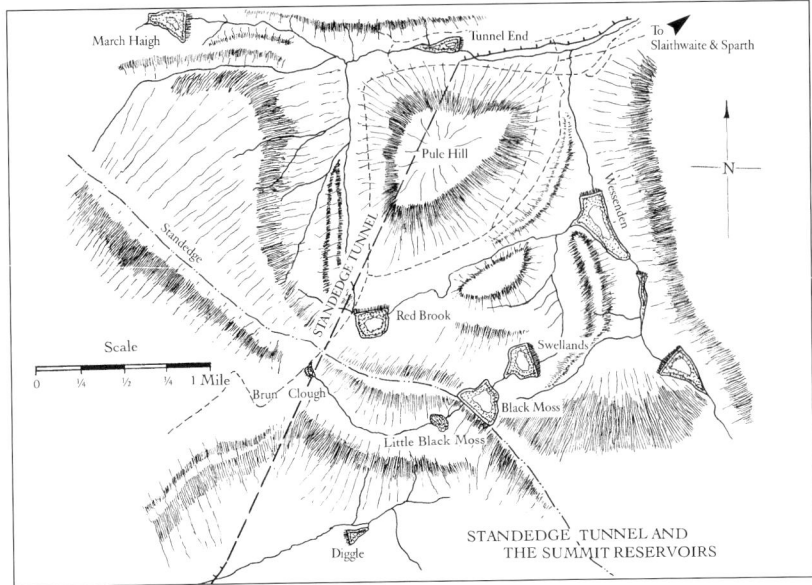

Figure 18 Standedge tunnel and the summit reservoirs.

by November so that navigation could start between Huddersfield and the tunnel (Fig. 18).[1] At the same time, this harassed young man was directed to attend to the Slaithwaite dam which had been leaking badly from the very moment filling began from its feeder streams. On the Lancashire side of the canal, Outram decided to relocate the Staley Mill reservoir at Swinshore Common.[2] Construction of the latter was well advanced by 1796 because its purpose was to feed the lowest canal level leading down to Ashton, which was then nearing completion. Bailiffe was also directed 'to execute the reservoir at Diggle as speedily as may be so as not to exceed in the purchase of lands and execution the sum of £700',[3] the intention being to provide for the higher navigable section then being excavated from Stalybridge to Woolroad. At that time, trade was steadily being attracted to the waterway because the minutes were reporting the purchase of trade boats and the appointment of wharfingers. Moreover, the surveyor was regularly asked to report to the committee on any business attracted to the canal and the tolls collected.

Matters seemed to be going well for the company until disaster struck

---

[1] Ibid., 10 Dec. 1798.
[2] Ibid., 29 July 1795.
[3] Ibid., 21 Feb. 1799.

with the storms and floods of 1799. These severely damaged the immature earthworks along 16 miles of the waterway and also its reservoirs, quite apart from the aqueducts mentioned earlier. Just before the storms occurred, Bailiffe had satisfactorily completed repairs to the dam at Slaithwaite, after which filling began. Now he had to supervise cutting open the earthworks in two places, in a desperate effort to prevent the rising flood waters overtopping the shoulders and completely destroying the structure. Meanwhile the new reservoir at Swinshore Common, which unaccountably had been built without an overflow spillway, had been so badly damaged that it was subsequently abandoned.[1] If that was not enough, the spillway at Tunnelend reservoir had also proved of insufficient capacity and, inevitably, waters washed over the shoulders of the dam, partially destroying it and causing havoc in the village of Marsden below, and the valley beyond. After five years of labour, the proprietors were left only with a wrecked and incomplete waterway, 500 yards of arched and finished tunnel and 1,200 yards of unfinished heading through Standedge.[2]

The proprietors were dismayed at the chaos they found when they walked the line of canal shortly after the floods had subsided. They were close to bankruptcy; small wonder they were close to despair as well. At the meeting which followed it was noted that 'the pecuniary means of the company' were insufficient to repair the damage and to reinstate what little trade they had succeeded in establishing during the previous year.[3] They had no option but to apply once more to Parliament for powers to raise monies,[4] not only to cover the costs of repairs but also to complete substantial parts of their waterway, as well as pay off accumulated debts.

In August 1799 Outram was summoned from Butterley to view the canal and to prepare estimates for the next meeting,[5] but on this visit he also examined the waterway, in company with several proprietors, in order to assess repairs such as breaches and earth slips which had to be put in hand immediately to prevent further damage. At the same time, Bailiffe was told to prepare statements of expenditure made, cash owed to workmen and others, as well as monies owed to the company. He was

---

[1] Information from A.I.B. Moffat, University of Newcastle-upon-Tyne. The engineer, J.F. La Trobe Bateman, while investigating the possibilities of the Swinshore Valley for supplying the Manchester & Salford Waterworks Co. in 1844, came across the remains of the reservoir. He believed its failure was due to the lack of an overflow.

[2] Information from M.D. Stakes, Principal Mining Engineer, British Waterways Board.

[3] PRO, RAIL 838/2, 23 Aug. 1799.

[4] 40 Geo. III c. xxxix (1800).

[5] PRO, RAIL 838/2, 23 Aug. 1799.

even required to value all disposable materials and equipment, so keen were the proprietors to assess the complete financial state of their shaky undertaking. Came the September meeting, Outram was present to discuss his estimates—£3,500 for the repairs, plus a further £3,000 'to prevent further injury by the largest floods'.[1]

While George Worthington was preparing the new bill, Bailiffe was proceeding with the priority work, although he was given only £530 in November to pay for wages and materials. The long-suffering workmen were told that their next pay-day would be one month later. By March 1800, available funds were exhausted yet again and Bailiffe had to discharge the workforce, as well as his two clerical assistants.[2] All work was suspended except for the Stakes aqueduct, and so it remained until after June 1800 when the Act was passed.[3] The company was then permitted to call for a further £20 per share, which would provide £40,000, but, as had happened in the past, only a fraction of this sum was to be paid up by the reluctant shareholders.

Contracts were also let to Edward Banks in June for repairs at each end of the canal and also at several of the reservoir sites.[4] He was required to ensure that the canal was again operational for traffic on each side of Standedge by August 1800. Banks only completed his contract by November but the tunnel still languished, untouched for months past, and building had yet to be started on the section of waterway from the tunnel mouthing at Diggle, down to Woolroad.

Funds were yet again running very low by December 1800 and Bailiffe had to pay workmen in full only if they were owed £30 or less, sums greater than this being paid at 5s. in the £.[5] At this time Outram was invited to address a dispirited committee as to the best way of proceeding with the tunnel from each end and the probable weekly expenditure. He was also asked to investigate the feasibility of building a tramroad across the mountain to link up the canal at Woolroad with that at Marsden.[6] That proposal came from the proprietors of the Peak Forest Canal who were then anxiously pressing for an early completion of the Huddersfield Canal, in order to extend their trade into Yorkshire.

---

[1] Pontey, *Short Account*, G. Worthington to shareholders, 2 Sept. 1799, pp. 22, 23.

[2] PRO, RAIL 838/2, 6 March 1800.

[3] 40 Geo. III c. xxxix; PRO, RAIL 838/2, 9 June 1800.

[4] PRO, RAIL 838/2, 26 June 1800; Banks went on to become Sir Edward Banks of the contracting company of Jolliffe & Banks: H.W. Dickinson, 'Jolliffe and Banks, contractors', *Trans. Newcomen Soc.*, xii (1931–2), pp. 1–8.

[5] PRO, RAIL 838/2, 18 Dec. 1800.

[6] Ibid., 10 Nov. 1800.

This temporary measure (a similar tramroad had just been completed by Outram in place of the Marple locks on the Peak Forest Canal) must have been rejected by the engineer as uneconomical, because nothing more was done about it.

At this time of uncertainty the proprietors were approached by John Rooth, a Manchester entrepreneur, who offered to carry all goods on the canal between Huddersfield and Manchester if commodious warehouses were provided, and if he was also loaned £500, besides a 5 per cent drawback on carriage charges.[1] He was supported by the Ashton Canal Company which was also to provide him with a loan of £1,500 to build a warehouse in Manchester, with a 10 per cent drawback to cover the high costs of ground rents. Rooth's assurances that he would develop trade aggressively on the waterway aroused the enthusiasm of both companies in which he was soon to become an important figure.

William Bailiffe's appointment terminated in February 1801, but thereafter he was retained on three months notice as surveyor and superintendent for the east end of the navigation. A miserly stipend of £75 p.a. suggests that his was but a part-time appointment and that his work was mainly maintenance in character.[2] The proprietors had turned to Rooth in February with an invitation to supervise the works on the west side of the tunnel with the promise that 'a sufficient compensation be made for the trouble he may have in such superintendence'.[3] John Rooth accepted the offer and three months later, in April 1801, became full-time superintendent and manager of the entire works at an annual salary of £300.[4] John Rooth was a self-assured, domineering man who successfully saw the canal through to completion by April 1811, although his engineering abilities were suspect and he made a number of mistakes which were never disclosed at the time. On completion of the works he was retained by the company as agent, until he was dismissed in 1817 in favour of a younger man.[5]

During these managerial changes, several contractors visited the Standedge tunnel with a view to bidding for the remaining work and three of them submitted quotations. The tender of John Varley of Shardlow was accepted in preference to those of Pritchard and Williams,

---

[1] Ibid., 18 Dec. 1800.
[2] Ibid., 4 Jan. 1801.
[3] Ibid., 27 Feb. 1801.
[4] Ibid., 23 April 1801.
[5] WYASK, Tomlinson MSS, J. Rooth to R. Firth, 26 June 1817.

each of whom was paid £21 for his trouble.[1] Unfortunately, Varley had deceived the committee regarding his sureties and the offer was rejected. He persisted, however, and by July 1801 had found two other guarantors, although they informed the committee that the bond of £2,000 proposed, if forfeited, would only be paid six years later! Varley then declined to execute the works on the terms which he had already agreed one month before. Small wonder the committee abruptly rejected these revised proposals and refused to deal further with this unreliable fellow. Rooth was told to prevent him from visiting the works again.

Outram was told of the failure to engage Varley; he was requested to give his opinion as to how the tunnel might be finished, also to advise on the location of additional reservoirs and 'to suggest any other matters which he may think to benefit the interests of this company'.[2] This does not seem to have happened and it was the last occasion that Outram was mentioned in the minutes. It is likely that he terminated his association with the company about this time. Perhaps he was getting tired of the proprietors' self-inflicted problems and the constant cash shortages which must have affected the quality of materials, workmanship and above all, the works programme. In any case, very little work was then in progress, future prospects seemed bleak and Outram had growing commitments at the Butterley ironworks. He probably felt he was wasting time and energy working for the company and decided that this was the time to go. The minutes make no mention of any disagreements and so the break was probably amicable enough. Outram's place was never taken by another engineer, although in 1802 the proprietors made an unsuccessful approach to William Jessop to help them plan completion of the canal.[3] In spite of this, it must have been some consolation to the committee to note that, under Rooth's supervision, there had been a decided improvement in the organisation and progress of their enterprise.

Varley appeared again at the committee meeting of August 1801 and offered once more to contract for work in the tunnel, this time in accordance with the terms he had originally agreed, but which he had withdrawn the previous month. He confirmed that his new sureties would enter into a bond for £2,000,[4] to be paid one year after forfeiture, and

---

[1] PRO, RAIL 838/2, 26 March 1801. Pritchard was of the firm of Pritchard & Hoof, which drove the Harecastle tunnel on the Trent & Mersey Canal in 1827: L.T.C. Rolt, *Navigable Waterways* (1969), p. 106.

[2] PRO, RAIL 838/2, 22 July 1801.

[3] PRO, RAIL 838/3, 23 Dec. 1802.

[4] PRO, RAIL 838/2, 7 Aug. 1801.

this now satisfied the committee. By the end of September, Varley's sureties had reneged once again and so the committee, with some exasperation, recorded that they 'will not enter any agreement with Mr Varley and will not receive any further proposals relative thereto'.[1]

John Booth of Greenacres had already been appointed as inspector to supervise the work of Varley, but now he and Rooth were instructed to begin work immediately at both ends of the tunnel, using direct labour. Driving of the pilot heading at the western end began but at the eastern end work was delayed until the spring of 1803. There was a good reason for this delay; it was suspected that something was wrong with the levels because the workings inside were flooded. This led to a visit in November 1802 of David Whitehead, Outram's former apprentice surveyor on the Peak Forest Canal, to compare the levels at both ends of the tunnel and to 'check whether the line of the tunnel is carrying on the proper direction'.[2]

In transferring the levels across the mountain, Whitehead exposed some serious mistakes.[3] At the Marsden end, the invert at the tunnel entrance was found to be 2 ft deeper than the sill at the first lock. The first 274 yards of the tunnel were complete and finished with masonry lining and a further 922 yards were bottom-headed and unfinished, but the invert of the tunnel dipped 3 ft down from the entrance to the working face. It is difficult to understand how the contractors, quite apart from Outram and Brown, had failed to notice this because water must have accumulated and interfered with their work in what was supposed to be a horizontal tunnel. But that was not all. Of greater concern was the discovery of a difference in level between the two ends of the tunnel. The invert at Diggle was much higher than that at Marsden and although the actual levels are unknown, all the masonry arching, as well as the foundations of the sidewalls of 489 yards of finished tunnel had to be dismantled and rebuilt. Needless to say, nothing of this disastrous state of affairs was reported to the shareholders.

Following Whitehead's visit, further efforts were made to engage a reliable contractor for the central 1,100 yards of tunnel under Red Brook. The committee had calculated that it would be worth paying up to £8,000 extra over the cost of just working the tunnel from both ends, always provided that they could expedite completion by two years. Samuel Marsland, a committee member, then advised his colleagues that

---

[1] Ibid., 2 Oct. 1801.
[2] PRO, RAIL 838/3, 1 Nov. 1802.
[3] WYASK, Tomlinson MSS, J. Rooth to R. Firth, 24 Feb. 1817.

he had located a suitable contractor, Jonathan Woodhouse of Ashby de la Zouch, who was keen to take on the work. Woodhouse was approached and he submitted a tender, but this was exceedingly high by anybody's standards.[1] He had offered to complete in three years for the price of £13 per yard, plus the £8,000 already quoted. He also required the use of all the company's equipment, which was to be retained by him at the end of the contract. Not only that, but he insisted on a down-payment of £5,000 on starting, £2,000 after four months and £2,000 every three months thereafter, plus bonuses. These demands received a guarded welcome and were referred to a special assembly of shareholders two months later. By that time, however, Woodhouse had changed his mind and 'had other proposals to offer'. These were not explored any farther because a motion requesting Woodhouse to bring them forward was defeated.[2] It seemed that there was no alternative but to leave the task to Rooth and Booth, and in time this proved to be a sound decision.

The central section then lay fallow. Meanwhile the direct labour gangs made slow but steady progress from each end of the tunnel. At the eastern end the pilot heading was driven, often through difficult strata of Millstone Grit, at an average rate of 23 yards per month during 1804; thereafter, rates varied from 15 to 18 yards per month. From the western end, driving through shales and Millstone Grits proved marginally easier and the rate of working increased from about 13 yards per month in 1803 to between 30 and 37 yards per month.[3]

In 1805 the company decided to seek its third Act, 'to raise a further sum of money for the discharge of their debts and to finish and complete the Huddersfield Canal' and Rooth was required to provide estimates for the application.[4] That matter concluded, the eminent Scottish engineer Thomas Telford was invited to visit the canal and to cost and programme for the remaining tasks. Arriving on site he found that work had then been accomplished as shown in Table 6.2.

In his report of January 1807, Telford claimed 'a thorough knowledge of the state of the works' because he had 'examined most of them twice and have seen the locks proved as far as could be done by filling and

---

[1] PRO, RAIL 838/3, 23 Dec. 1802.

[2] Ibid., 10 Feb. 1803.

[3] Information from British Waterways drawing D91/9/60R, copied from original at the Waterways Museum, Stoke Bruerne.

[4] 46 Geo. III c. xii (1806).

emptying the water as low as the level of the water of the canal'.[1] His efforts were mainly concentrated on the preparation of a construction programme and estimates for the remaining works in the tunnel but he also planned the finishing of the canal from Diggle to Woolroad. He recommended the construction of more reservoirs and costed minor repairs for the entire canal. Telford noted that the drainage problems in the tunnel were not as acute as formerly:

> the water has either drained off, or now falls in chiefly near the bottom, by which means the state is in general becoming dry and firm and the whole promises to form a perfect tunnel, with only a comparatively small proportion of arching with masonry. At present the heading at the eastern part of the tunnel is carrying on in rock of a very hard millstone grit, at the western part in hard black shale'.

Telford became well known for his careful planning and cost estimates of engineering works and this report to the company was a clear demonstration of his talents in this respect; it was a far cry from the company's practices of former years. He confidently predicted progress for the years ahead as, for example:

West end.

XI. 280 yards, commencing at the termination of the last length and ending at the fourth pit, at the rate of 8 yards per week which will occupy up to the end of November 1807.

XII. 564 yards from the fourth pit to Brun Clough pit, at the rate of 8 yards per week, from the 1st of November 1807 to the 9th of March 1809, being 70½ weeks.

Included with Telford's report was a detailed bill of quantities and estimates for the replacement of major equipment such as pit head-gears, and even minor items such as spoil buckets, ropes and chains. He listed the cleaning out of unfinished pits and the closure and filling of several others, including the partially completed Heathy Lee pit. Nothing was

---

[1] WYASK, T. Telford, *Abstract of the Report relating to the State of the Huddersfield Canal, 29 Jan. 1807*. Telford's MS bills of quantities and drawings of the tunnel are in the library of the Institution of Civil Engineers.

TABLE 6.2

Progress at Standedge tunnel, 1806

	Finished Yardage	Headed Yardage	Unbroken Yardage	Total Yardage
Eastern end	331	1,531	452½	2,314½
Red Brook	107	50	–	157
Western end	510	1,535	934½	2,979½
Totals	948	3,116	1,387	5,451

left to the imagination in this workmanlike document in which he priced the remaining works at £82,498 overall, £45,000 of which was reserved for the tunnel. The company's staff followed the sequence and timing of the plan very carefully over the ensuing years. In the tunnel Red Brook was reached from the east late in 1808 and the pilot tunnel was broken through by the middle of 1809. Finishings and linings were also well advanced by that date. The canal opened in April 1811, just five months later than the engineer had predicted. Although Rooth must have benefited greatly from Telford's directive, it was regrettable that in later years he claimed to have built the canal and its great tunnel 'without the assistance of any engineer'.[1]

Telford had examined the works during the winter of 1806 and he wrote in his report that 'the season was totally unfit for proving the general line of direction over the mountain, and dialling and levelling below, but each end appears very direct in itself'. He was confident that all was satisfactory because he understood that 'they [the survey data] have been tried and found accurate by several properly qualified persons'. Telford was correct at the time because it was not until after his visit that the section from Brun Clough to Red Brook was driven, with a maximum deviation of 26 feet to the west of the true alignment (Fig. 19).[2] That was not all; there were gross errors of 40 and 14 yards in the longitudinal positions of two of the shafts on Brown's original

---

[1] WYASK, Tomlinson MSS, J. Rooth to R. Frith, 24 Feb. 1817.

[2] Author's measurements taken from an LNWR Standedge Tunnels Plan (No 11231A, 22 Jan. 1894), formerly in the possession of British Rail, Leeds.

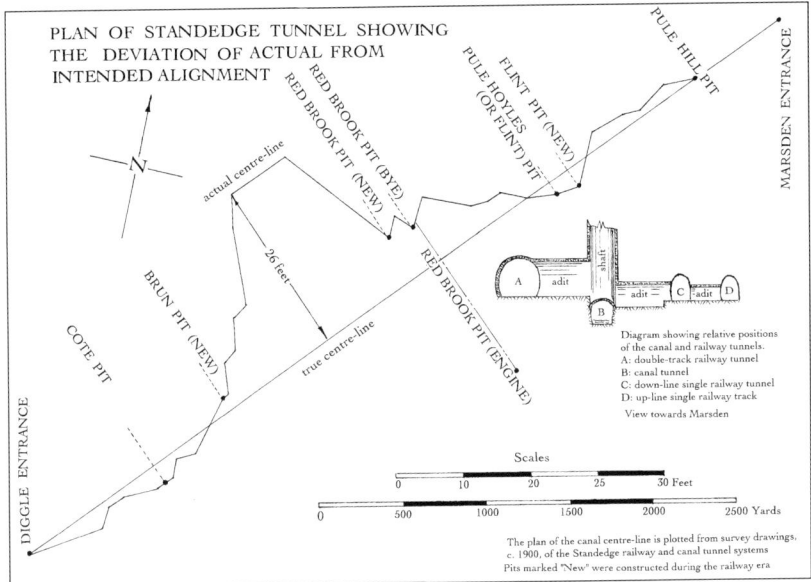

Figure 19 Plan of Standedge tunnel.

survey and these were undiscovered until the mountain was finally pierced in 1809. The corrected length was then stated to be 5,477 yards.[1] With the advent of the railway era, three more tunnels were driven through Standedge, two single-track tunnels completed in 1846 and 1870, and a twin-track tunnel in 1894. All utilised the canal tunnel for the removal of spoil through some 46 cross-adits. The Diggle end of the canal tunnel was realigned and extended to accommodate each of the last two railway tunnels, to the present length of approximately 5,698 yards.[2]

Of the incomplete waterway between Woolroad and Diggle, Telford recommended that three of the eight locks should be built in 1807, four in 1808 and one in 1809, so as to allow time for 'the work to harden' before opening the canal. This section, which included two bridges, culverts and wharves, was priced at £15,000. Another £16,238 was reserved for repairs to the rest of the canal, all of which were to be ready 'for tests during the winter of 1809 so as to allow for defects to be remedied in the summer of 1810 previous to the opening of the tunnel'.

---

[1] Pontey, *Short Account*, J. Rooth, *Report on the Huddersfield Canal*, 29 June 1809, p. 33.

[2] Hadfield and Biddle, *Canals of North West England* (1970), ii, Appendix II.

For all these works and repairs Telford stated that 'Mr Rooth would require to have two respectable assistants; one to superintend the works on the west, and the other those on the east side of the tunnel; and persons who have been bred masons, would best answer this purpose'.

The engineer offered much sound advice regarding water supplies for the canal. He wrote that 'it is too late to express a regret that such large sums of money had been expended on such narrow dingles of small capacity 'as these were liable to rapid silting up by detritus washed down by mountain streams'. Telford considered that the summit level should be furnished with a reserve supply equivalent to 40 locksfull of water per day for three months, each of 28 working days. Assuming each lock to be of 200 cu. yds capacity, this gave 672,000 cu. yds, to which he added a like amount to cover losses due to evaporation and leakage. These quantities could be supplied in part by the combined capacities of reservoirs at Tunnelend, of 8 acres and 30 ft head; Diggle, of 4 acres and 45 ft head; and Diggle Moss (which later became known as Black Moss), then under construction, of 10 acres and 15 ft head. These would contain some 421,013 cu. yds, leaving 922,987 cu. yds still to be provided. Telford recommended another reservoir to make up the total capacity, to be built east of Diggle Moss. He wrote that

> until a regular survey has been made, I cannot speak with any certainty; yet so far as can be judged by inspection, it appears that with a head of about 45 feet, a surface of about 50 acres may be acquired and which would contain about 1,210,000 cubic yards which, added to the the other reservoirs, would give a surplus of above the supposed demand of 287,013 cubic yards, or 1,435 lock fulls, being a provision of 35 days at 40 lock fulls per day.

Even as late as 1807 the Slaithwaite dam was still giving trouble and so Telford allotted £2,000 towards its repairs. He also recommended building a three-acre reservoir at Sparth, in the Colne valley, just below the summit level. Rooth had completed the latter in three months, by the midsummer of 1807.

Although work followed Telford's plan over the next three years, no one was able to predict the calamity which afflicted the countryside late in 1810. John Rooth was to report to the annual general meeting of shareholders the June following that 'Early in the morning of the 29th November last, owing to continued rains, the black earth at the tail of the Diggle Moss Reservoir gave way, and let off the water in such a quantity as to cause the loss of five peoples' lives and do other damage to the amount of near four hundred and fifty pounds'.[1] It seems that there were two dams enclosing the reservoir and that at the tail, or east end, failed, possibly by seepage through the base of the earth structure and the water rushed eastwards, inundating Marsden and the Colne valley beyond to Paddock. Houses and factories were wrecked in that night of the 'Black Flood' and the force of the discharge was such that a 15-ton stone was carried two miles down from the summit, and the mountain side was denuded of soil. The company's finances were such that repairs were delayed until 1818, and it was as late as 1825 before the Swellands reservoir, recommended by Telford to lie below and to the east of Diggle Moss reservoir, was completed.[2] Nevertheless, the canal was finally finished and fully navigable by 26 March 1811, to be followed by a ceremonial opening on 4 April.[3]

Through all those troubled years those who suffered the most were the workforce of miners, tradesmen and labourers who toiled in primitive conditions for small recompense. Of their origins very little is known, although names in the Marsden parish register suggest that they were itinerant workers in the main, and mostly of northern English stock.[4] The company was never unduly concerned about the welfare of its workers and the minutes refer only occasionally to such matters. On one occasion the surveyor was instructed to set up a sick fund, the company subscribing but 5s. each week,[5] and on another, one guinea was granted towards the costs of 'burying a workman who died today on the line of the canal'.[6]

Something of the dangers of the workplace can also be gleaned from the register. For example, Robert Whitehead of Stonefold, mason, was killed in the tunnel on 10 March 1795; John Kell, who was killed in the tunnel, a native of Stanhope in Weardale, was buried on 19 March 1810; and George Spark, 'a respectable member of society', was killed in the tunnel in an instant by the explosion of gunpowder, together with Thomas Whitehead of Puleside. Upwards of one thousand people attended their interment.

It seems likely that over fifty men lost their lives in the tunnel and

---

[1] PRO, RAIL, 1019/8/1, J. Rooth, Report to the General Assembly, 27 June 1811.

[2] G. Brown, 'The Black Flood—what caused it?', *Saddleworth Historical Society Bulletin* (1994), pp. 25–7.

[3] PRO, RAIL 838/3, 4 April 1811.

[4] WYASW, D143, Marsden parish register, 1776–1812.

[5] PRO, RAIL 838/2, 4 May 1797.

[6] Ibid., 29 June 1797.

along the line of canal during the construction years. To this number must be added an unknown number of mothers and children who also died, the latter often within weeks of their birth, in the squalid cottages and shacks that were erected adjacent to the pit-heads on Puleside, Gilbert's Intack, Red Brook and elsewhere. The foundations of these dwellings are still to be found today.[1] Little is known too, of the lifestyles of these navigators and their families but local records suggest that male leisure pursuits included nothing much beyond hard drinking, bare-fist fighting, cock-fights and bull-baiting.[2] It is recorded that on one occasion in 1797, a well-known rocking stone on the Ravenstone Brow at Greenfield was blown up by miners from the canal tunnel during a drunken spree. One of them was killed and several were injured.[3]

The final cost of the canal far exceeded Outram's original estimates. This was stated to be £396,267 in June 1811 but in view of the remedial works and modifications completed in the following year, it is more appropriate to include for this period also. Thus total costs by June 1812 were £402,653,[4] of which £123,804 was for the Standedge tunnel, or £22 12s. per yard run compared with the original estimate of £10 5s., taking into account costs of pit shafts and ancillary works.

Not surprisingly, critics such as John Sutcliffe and William Pontey publicly denounced the proprietors and their engineers for their supposed profligacy and incompetence. Sutcliffe, for example, gave some forthright views in his book on the manner in which canal companies were established and their capital raised, and how much the public had been misled as to the 'supposed profits that arise from the making of them'. He criticised speculators who were 'little better than swindlers' and engineers for their 'fictitious estimates', and showed scant sympathy for managing committees which embarked on these schemes with very little understanding of their commitments.[5]

As regards engineering aspects, Outram must be criticised for his sketchy proposals and unsatisfactory estimates. Too many important details of the scheme were changed after work commenced, thus showing that preliminary planning owed more to guesswork than to the lengthy, careful investigations which this complex project merited. But Outram

---

[1] G. Keevil, *Standedge Guide: An Industrial Landscape of Roads, Canals and Railways* (1986), pp. 43–9.

[2] A. Wrigley, *The Wind among the Heather* (1916), pp. 7, 8.

[3] A. Wrigley, *Saddleworth Chronological Notes from 1200 to 1900* (1940), p. 32.

[4] PRO, RAIL 1019/9/1, J. Rooth, Report, 27 June 1811.

[5] Sutcliffe, *Treatise on Canals*, pp. 124, 227; Pontey, *Short Account*, pp. 2, 3.

cannot be blamed for everything. The company's inept management committee, the constant shortage of funds and the inexperience or inadequacy of the persons involved, can be blamed for many of the shortcomings and the confusion of the sixteen years of construction.

But all this is a harsh, retrospective judgement which overlooks the pioneering nature of a project which was exacting, even by today's standards. It is remarkable that a group of local businessmen comprising weavers, drysalters, farmers and gentlemen, with no previous experience or knowledge of engineering, should set up a company, appoint a handful of professionals and work at this massive task using the most primitive equipment and methods. That they should complete the project even in 16½ years is a tribute to their courage and faith in the future of their country. Perhaps the canal never brought prosperity to the shareholders and the speculators among them, but it did much for the common people who inhabited those bleak moors and valleys, bringing employment and communications to an impoverished and remote part of England. For all the failings, the canal was successful. It was a splendid engineering concept and its completion a proud achievement.

CHAPTER 7

THE ASHTON AND PEAK FOREST CANALS:
DESIGN, PROMOTION AND RELATED ISSUES

Most inland waterways in the north-west of England were built as broad navigations. These accommodated a range of vessels with widths exceeding 14 ft and of differing lengths and drafts. The exceptions were the three interlinked canals originating in Manchester which were of the lesser specification required for narrow boats. The first of these narrow canals was the Ashton under Lyne Canal (originally known as the Manchester to Ashton under Lyne & Oldham Canal). This was authorised by an Act of 1792[1] and, when construction began, interest was soon stimulated in schemes for the Huddersfield Narrow and the Peak Forest canals.

The choice of narrow gauge for the Ashton Canal seems to have resulted from the plans of the previous year for the Rochdale Canal, with which it was expected to connect at Piccadilly in Manchester. The earliest proposals for a Rochdale Canal dated from 1766 and were for a narrow waterway which crossed the Pennines to link the Duke of Bridgewater's Canal at Castlefields in Manchester with the Calder & Hebble Navigation at Sowerby Bridge.[2] That project foundered but more determined efforts to secure an Act for a resuscitated scheme began in 1791 and were finally successful, at the third attempt, in April 1794,[3] in spite of much opposition from rival parties. The promoters of this project had also intended to build a narrow canal but they were prevented by the parliamentary committee considering their bill: they were required to build broad locks 15 ft in width, such that vessels from the Mersey could pass through via the Bridgewater Canal. The Rochdale proprietors then 'humbly intreated the House to disagree' with their committee. They wished to build a narrow canal because it was expected that traffic would chiefly be 'with the Midland and Western Canals by the Staffordshire Canal' (i.e. the Trent & Mersey Canal) and then on London by the 'Braunston Canal (the Grand Junction) without any transhipment of

---

[1] 32 Geo. III c. lxxxiv (1792).

[2] C. Hadfield and G. Biddle, *The Canals of North West England* (1970), ii, p. 263, citing *The Todmorden and Hebden Bridge Historical Almanac* (1874).

[3] 34 Geo. III c. lxxviii (1794).

goods'.[1] But their earnest appeal was to no avail and the Rochdale was to be built to the same broad standards as the Bridgewater Canal. The proprietors of the Ashton Canal Company had, logically enough, selected the same specifications as those of their intended neighbour, and no doubt they were more than disappointed by these changes in plan. Unfortunately, by that time the construction of their narrow waterway was well advanced but at least the new Act released them from one uncertainty: it empowered them to join their waterway to the Rochdale Canal at Piccadilly.

When the Ashton Canal was extended by the Huddersfield Narrow Canal and the Peak Forest Canal, all three companies became closely allied because several eminent local businessmen with cotton, land and coal-mining interests took substantial shareholdings in each enterprise. The appointment of the Altrincham solicitor, George Worthington, as clerk to all three also did much to ensure common management policies.

The first Ashton Canal Act authorised the construction of a waterway from Manchester to Fairfield, where it divided. One branch continued via Waterhouses, at which point a short arm crossed the River Medlock by an aqueduct. From there the main-line route continued close by the river, crossing it in two places, between which was a 250-yard-long tunnel. The line then passed through Fairbottom to terminate at New Mill, near Oldham. This latter section was subsequently changed in favour of a shorter route north of the river, which eliminated the need for tunneling and costly earthworks. The other branch ran from Fairfield via Dukinfield (where a short spur crossed the River Tame by an aqueduct, which later became the connection with the Peak Forest Canal) and continued a further half-mile to a terminus in Ashton, where the Huddersfield Narrow Canal began. The company was empowered to raise £60,000 for the project and a further £30,000 could be subscribed by the proprietors should that sum prove insufficient.

The Ashton Canal survey plan of 1792 by William Bennet[2] is but a sketchy outline of the intended routes which, with a book of reference indicating the land-takes, was the minimum required by law. Distances and levels are shown, in addition to highway and river crossings, but no information is given as to locks, accommodation works and water supplies, although an existing 'pumping engine' shown at Bradford may have had some purpose in the latter respect. An engineer's report and estimate have not survived but it seems unlikely that these were ever

---

[1] WYASL, Radcliffe MSS 319/C/1, Petition by Rochdale Canal Proprietors.
[2] Lancashire Record Office, PDC/2.

prepared for what seems to have been a hastily assembled and incomplete scheme. The cost estimates are likely to have been inaccurate and deliberately set on the low side to attract investors. This view is seemingly confirmed by the company's frequent attempts in later years to raise additional capital for the completion of the waterway.

Another Act[1] was sought the following year, just five months after the first had received Royal Assent. William Bennet once more prepared the survey plan,[2] which shows substantial extensions of the original scheme. A branch now continued from the main line at Clayton, passing through Gorton to terminate in Heaton Norris, close by Stockport. From Taylor's Barn on this new branch, another line ran out to the coal-mining area of Beat Bank at Denton. Yet again, the plan is devoid of much detail apart from land-takes, overall distances, levels, and riparian topography, although a reservoir, some 1,100 yards in length, was tentatively sketched in about half-way along the Beat Bank branch. That reservoir was never built and in fact the entire branch was abandoned during construction when it was found that the extensive earthworks (which included a long tunnel) were proving too ambitious for the company's limited finances.

The second Act also authorised yet another line 'from the intended aqueduct bridge at or near a place called Waterhouses ... to Stoke Leach at Hollinwood' but a survey plan of this extension, if ever made, has not survived. A water-supply feeder was to discharge into the head of the Hollinwood branch and eventually this was incorporated in the short, privately built Werneth Canal which led to the Old Lane Colliery, near Oldham.[3] The company was permitted to raise another £30,000 for all these additional projects but, probably as a result of their piecemeal, muddled and ill-planned schemes, they were never free from debt during the early trading years. Further Acts were obtained in 1798, 1800 and 1805, ostensibly to allow the construction of various minor works.[4] In fact such newly available funds were mostly needed for the completion of the principal works on the canal.

The confusion of those early years was hardly helped by the failure to appoint qualified staff and in particular an engineer; indeed the company first advertised for the latter only in September 1793.[5] Until

---

[1] 33 Geo. III c. xxi (1793).

[2] Lancashire Record Office, PDC/4.

[3] Hadfield and Biddle, *North West England*, ii, p. 295.

[4] 38 Geo. III c. xxxii (1798); 40 Geo. III c. xxiv (1800); 45 Geo. III c. xi (1805).

[5] *Derby Mercury*, 26 Sept. 1793.

then it may be that William Bennet had taken on this duty before leaving in 1794 for his appointment as surveyor and engineer for the Dorset & Somerset Canal. The company was thereafter unsuccessful in attracting a suitably experienced man and this inevitably led to mistakes of design and construction. In the absence of any company minutes[1] it has generally been assumed that James Meadows, who was later agent to the company, acted as engineer during most of the construction period from 1793 to 1798. This seems possible because Meadows generally described himself as a surveyor. An example of his survey work dating from that period is proof of his competence.[2]

Nevertheless, it was recorded in the company minutes five years later in 1798, that 'it being the decided opinion of this committee that the works of the said canal have in many instances been improperly managed for want of the assistance of a proper engineer and thereby the interests of said company ... have suffered significantly'. The proprietors then resolved that 'Mr Outram ... is hereby requested by this committee to accept the appointment to the office of engineer ... and to devote as much of his time as his other engagements will permit'.[3] Benjamin Outram accepted the appointment; it suited his purpose to do so because he was then busily engaged on land developments nearby and was also beginning important contract work on the Piccadilly terminus of the Ashton Canal. Probably the proprietors had invited him to act as engineer for the same reasons.

Presumably Outram had not applied for the earlier appointment, even though he must have been aware of it, because he was engaged at the time on surveys and designs for both the Huddersfield Narrow and Peak Forest canals, and the job was also advertised in the *Derby Mercury*. Just a few months earlier, a Peak Forest Canal scheme had been under consideration by the Ashton Canal Company as a possible branch of their own waterway, hence it is more than likely that he had been approached directly by the proprietors and offered the job, even before it was advertised. Outram's failure to seek the prestigious appointment as engineer for all three canals simultaneously remains an unsolved puzzle. It may be that he had refused an invitation because he was busily engaged on the Derbyshire canals, as well as with his increasing commitments at the Butterley ironworks. Perhaps too, he was reluctant

---

[1] Neither the minute book of Ashton Company for 1792–8, nor parts of the second volume, have survived.

[2] *Lords' Journals*, xlv, p. 583; HLRO, Petition of Francis Outram, with plans by James Meadows.

[3] PRO, RAIL 804/1, 8 Aug. 1798.

to accept because of the unsatisfactory state of the works of the Ashton Canal, as well as the impecunious state of the company. These factors would hardly encourage anyone to take over at that stage.

Meanwhile, at some time late in 1793, Outram began his association with the proposed Peak Forest Canal Company, the leading promoters of which included Samuel Oldknow of Marple and Captain George Hyde Clarke of Hyde. Thomas Brown, a local surveyor from Disley, had completed the surveys[1] that year and may also have selected the canal route and its reservoir sites before Outram was engaged as engineer. On 14 March 1794 several proprietors, accompanied by George Worthington and Brown, met a Select Committee of the House of Lords.[2] Because Outram was unavoidably absent, Worthington presented their bill and proved the deposited plans and other documentation, including Outram's estimate of £115,794. John Bowyer Jodrell, a leading proprietor, 'proved the public utility of the undertaking'. The bill was soon reported without amendment and the Act[3] received Royal Assent on 28 March.

Brown's plan, dated 1793, showed the Peak Forest Canal (then described as the Derbyshire Canal) beginning at the Ashton Canal at the Dukinfield aqueduct and continuing for seven miles on the level through Hyde and Romiley, before traversing the deep valley of the river Etherow by the Marple aqueduct. It then rose 212 ft over the next two miles by a flight of locks, after which it ran through Marple and Disley, on the level, for a further six miles to Bugsworth. A half-mile long branch led from Bottoms Hall, near Bugsworth, to Whaley Bridge, although the main line continued for a further mile beyond Bugsworth to Whitehough, where a second flight of locks lifted the canal 129 ft into a pound, one mile in length, which terminated in Chapel Milton. A railway was to be laid from this point to the Loads Knowle and Dove Holes limestone quarries, rising about 270 ft over a distance of three miles.

Outram stated that his estimate was 'to make a Canal and Railway from Dukinfield to Whaley Bridge and Loads Knowl' and it differed from Brown's survey in some significant respects. For example, an item relating to the main line through Bugsworth was vaguely listed as 'To cutting the canal from the Goyt aqueduct towards Whitehough one mile and an half'. No mention was made of an extension to Chapel Milton,

---

[1] Cheshire Record Office, T. Brown, Plan of the Intended Derbyshire Canal from the Aqueduct Bridge at Dukinfield to the Limestone Rocks at Loads Knowl and a Branch to Whaley Bridge (1793).

[2] HLRO, Procs. at Committees on Bills, Peak Forest Navigation, 14 March 1794.

[3] 34 Geo. III c. xxvi (1794).

as proposed on Brown's survey, which would require extra lockage and an additional feeder reservoir. Clearly, Outram had not then decided on the terminus for the canal and was considering construction of a railway, longer than originally planned, for the steeply inclined route up to the intended quarries.

Outram's estimate comprised just 34 items, one of which was for the Marple aqueduct, the major structure on the waterway, which was costed at £12,229 10s.[1] The much smaller, single-arched aqueduct across the Goyt near Whaley Bridge was priced at £5,801 but no mention was made of the Tame aqueduct at Dukinfield, even though work on it does not seem to have begun until the Peak Forest Canal was under construction. It was described in the company minutes as the 'intended aqueduct',[2] hence it seems to have been part of the new project, even though it was shown on the deposited plan of the Ashton Canal. It may have been paid for by that company.

Earthworks were substantial; Outram calculated that some 350,000 cu. yards would be moved to form embankments and cuttings, in addition to the excavations for the navigable channel and some considerable ancillary works. He also expected some trouble from 'sliding ground', which required 'soughing and piling'. This meant that timber piling would be necessary in several places to reinforce steep slopes against the possibility of land slips. Culverts, or soughs, laid under and across embankments were to prevent the ponding of surface water, particularly where the canal skirted sidelong ground. Examples of the latter can be seen near Disley, where blockages of culverts have, on occasion, led to major collapses of the embankments.

The engineer suggested that puddling of the waterway would only be necessary above Marple, 'when the strata is open'. It seems likely, judging from a visual inspection of the ground, that suitable clay for this purpose was found and excavated in thin layers from riparian land along this section. The channel from Dukinfield to Marple was cut 30 ft wide at the top, reducing to 15 feet at the bottom, the depth being 5½ ft. Outram copied the practice adopted for the Cromford and Derby canals by making the upper pound 7 ft deep so that it could act as a reservoir. This necessitated a wider cross-section for the upper canal and also resulted in the greater spans of the overbridges, compared with those of similar design erected on the Huddersfield Narrow Canal.

---

[1] HLRO, B. Outram, Estimate of Expence to make a Canal and Railway from Dukinfield Aqueduct to Whaley Bridge and Loads Knowl (1794).

[2] PRO, RAIL 856/1, 5 June 1794.

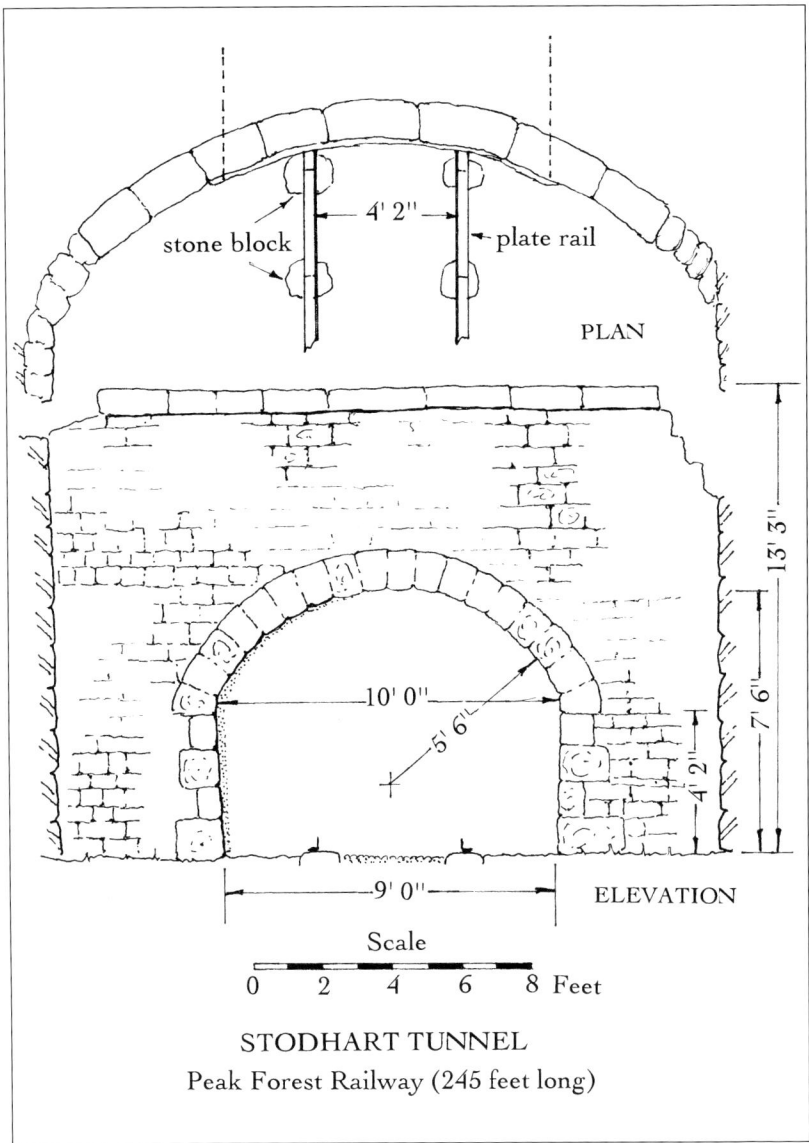

Figure 20 Stodhart tunnel on the Peak Forest railway.

The plan showed a tunnel at Hyde Bank where Outram's estimates provided for 'a deep cutting and tunnel 600 yards @ £6 per yard'. In the event, that tunnel was 308 yards in length on completion but others were also driven at Butterhouse Green (176 yards) and Rosehill (about 100 yards). The latter, which was opened up late in the nineteenth century,

Plate 13 Stodhart tunnel, c. 1929.

was located at the western approach to the Marple aqueduct, its position now indicated by retaining walls built above the former arch springings.

Outram simply quoted £5,600 for 3½ miles of railway track leading to the quarries, a figure which included for all earthworks, bridges and ancillaries. Although not mentioned in the estimate, a tunnel 245 ft long, sometimes known as Stodhart tunnel, was built for the railway above Chapel Milton (Fig. 20).[1] Similarly, he included feeder reservoirs 'to contain 12 thousand locks of water', which were costed at £8,691 14s. There was no mention of the locations of these most important elements of the system, even though Brown had marked three reservoirs on his plans at Todds Brook, Combs, and Hockham Brook.

The uncertainty regarding the location of the canal's terminus and the route of the railway persisted for some months after construction began. In March 1795, in what seems to have been a radical review, the committee was seeking a valuation of lands 'on the line of the railway from Bottoms Hall near Whaley to the termination of the railway in Peak Forest'[2] and the following month the purpose of this move became clear because Brown was instructed to prepare a list of landowners affected if

---

[1] B. Baxter, *Stone Blocks and Iron Rails. Tramroads* (1966); the length is incorrectly given as 100 yards (p. 66) and, more accurately, 85 yards (p. 174).

[2] PRO, RAIL 856/1, 4 March 1795.

the railway began at Whaley, to skirt Tunstead Milton before continuing by way of Marsh Hall to Loads Knowle. The surveyor was also requested to make an 'eye sketch' of the route and the clerks were to furnish a memorandum for the signatures of consenting landowners.[1]

These measures were intended to avoid the difficult, original route of the railway to the quarries which would necessitate the construction of an inclined plane, but the new plans were to no avail; the various landowners were not persuaded by the appeals of two visiting committee members, Frith and Gaskell, to consent to the new alignment.[2] This scheme was thus abandoned because any further legal measures to enforce these changes would mean 'that many difficulties will arise and much loss of time in the execution of the canal and railway will be occasioned in obtaining consent … therefore the idea of such deviation be abandoned'.

As part of this review, Outram and Brown had again surveyed and levelled along the original alignment because in July 1795 the committee resolved that, 'from the calculations made by the engineer and the best information they have been able to obtain, it will be most for the interest of the company to make the canal as far towards Chapel Milton as it can be extended on the level of the canal at Disley and that a railway be made from thence to the limestone'.[3] And so it was decided to terminate the canal at Bugsworth. Nevertheless, that alternative route remained in the minds of the proprietors long after the railway had been completed and was fully operational because, in 1799, they were devising proposals for a new Act which, among other things, would empower the company to make a new railway to avoid the inclined plane at Chapel.[4] Outram and Brown were again instructed to survey a suitable route and to prepare plans. By the end of the year, ambitions had grown such that proposals were to include canal extensions from Marple to collieries at Poynton and Norbury as well as a railway from Loads Knowle to or near to Buxton, in addition to the earlier plan to build another railway from Whaley Bridge to the summit of the inclined plane. In spite of their enthusiasm, all these proposals were abandoned in February 1800, no doubt for financial reasons, and Brown was then instructed not to obtain the landowners' consents to any of the extensions.[5]

---

[1] Ibid., 23 April 1795.
[2] Ibid., 18 May 1795.
[3] Ibid., 8 July 1795.
[4] Ibid., 5 July 1799.
[5] Ibid., 4 Feb. 1800.

Another scheme in which the proprietors of the Peak Forest Canal showed interest, as well as some anxiety, was a projected canal linking collieries at Poynton with Stockport, of which they heard first late in 1795.[1] Two members were appointed to meet the agents for the scheme and to report back, and this was duly done at the December meeting.[2] It seemed that the proposals were more complex than they had first thought; a Stockport Canal was under consideration, as was another, more favoured, canal which began at the Peak Forest Canal and passed by Poynton and Macclesfield to join the Trent & Mersey Canal near the Red Bull at Lawton, and the Caldon Canal, near Leek. Outram was eventually invited to survey and level for this proposed Macclesfield Canal and he reported to a meeting at Macclesfield on 11 March 1796.[3]

He recommended a line from the summit level of the Peak Forest Canal at Marple which passed via the collieries at Norbury and Poynton and extended from there to Rudyard Vale, with a branch to Leek and a descent to Endon on the Caldon Canal. The length would be 29 miles, of which more than 27 miles would be a level pound. He estimated construction costs at £90,000 and annual revenue £10,175. Outram also stated that a communication could be made with Stockport and, although he had not then decided on the route, the meeting resolved that 'a canal or railway from Poynton and Norbury, into the town of Stockport would be of very great advantage'. A further meeting was held on 22 April when a subscription was opened for raising £100,000; Outram was then appointed engineer with Brown the surveyor.[4] Enthusiasm for the scheme seemed wanting, however; funds were not forthcoming and although another meeting was arranged for 7 June, the scheme was quietly abandoned shortly afterwards. A Macclesfield Canal was finally built by 1831, to the design of Thomas Telford, to link the Peak Forest Canal at Marple with the Hall Green branch of the Trent & Mersey Canal.[5]

During 1810, fresh interest was aroused in Brown's original scheme for a canal extension locking up to Chapel Milton because this would allow a connection to be made from there with a proposed High Peak

---

[1] Ibid., 9 Oct. 1795.
[2] Ibid., 21 Dec. 1795.
[3] *Cowdroy's Manchester Gazette*, 2 April 1796.
[4] Ibid., 28 May 1796.
[5] C. Hadfield, *The Canals of the West Midlands* (1966), p. 211.

Junction Canal to Sheffield.[1] The water supply to the pound above the necessary locks at Whitehough would have been provided by an open channel which followed a long, circuitous route from Combs reservoir,[2] but the scheme was soon abandoned on cost grounds and Outram's system continued to operate until the railway was closed in 1920–1.

Although Brown's original survey showed three reservoir sites, only that at Combs was constructed initially. This remains unchanged to the present day but it had been designed such that its surface area of 45 acres could be doubled by raising the height of the dam.[3] That was never considered necessary because increasing traffic resulted instead in the construction of Todds Brook reservoir in 1831. Water from Combs was directed down through Whaley Bridge where it discharged into the canal.

The engineer, once more, may be criticised because his estimate was but a sketchy and inaccurate outline which omitted several significant and costly items. On this occasion, however, the proprietors must take much of the blame because they were undecided on several major issues which were unwisely left until construction began. Small wonder that costs eventually escalated to £177,000,[4] nearly 50 per cent greater than the original estimate.

[1] J. Farey, *A General View of the Agriculture and Minerals of Derbyshire*, iii (1817), p. 409.

[2] Sheffield Archives, Plan of a proposed extension of the Peak Forest Canal to Chapel Milton with a feeder from Combs Reservoir (unsigned, undated, probably *c.* 1810).

[3] Farey, *General View*, iii, p. 408.

[4] Hadfield and Biddle, *Canals of North West England*, ii, p. 312.

CHAPTER 8

THE CONSTRUCTION OF THE PEAK FOREST CANAL

The first general assembly of the Peak Forest Canal Company was held on 5 June 1794 but it was not until the following month that Thomas Brown was appointed 'surveyor, superintendent and book-keeper', with a salary of £315 per annum, from which he was expected to provide a clerk as well as all personal expenses whilst he was about the canal.[1] He engaged David Whitehead as his apprentice surveyor, a young man who became accomplished in his profession and in later years acted for Outram on several occasions. Brown was only required to work four days a week for the company (he had interests in mining and other businesses locally) but he proved to be an able engineer and manager, which must have pleased the hard-pressed Outram. The latter's appointment was never stated in the company minutes but seemed to date from 1 June 1794; his annual fees were £350.[2]

For several days before that first assembly, however, both Outram and Brown were already at work staking the line of waterway and setting out for the Tame and Marple aqueducts. The committee wished to commence some trading and to generate income at an early date, and so their engineers were urged to concentrate on the section from Ashton to Hyde Hall where a proprietor, Captain Clarke, was opening new coal-pits on his estate. It was hoped that the section would be completed by the time the Ashton Canal was finished to Manchester.[3]

Committee members were asked to seek out suitable contractors for the deep cuttings in the Werneth sector and it seems that Samuel Marsland, a wealthy mill-owner, recommended Robert Fulton, a 29-year-old American then living in Manchester, for the task.[4] Fulton was a curious choice because he was not an engineer and had no previous experience as a contractor. By profession he was an artist specialising in miniature portraits, who had enjoyed modest success since arriving in

---

[1] PRO, RAIL 865/1, 10 July 1794.

[2] Outram's first half-yearly salary was due on 1 Dec. 1794; hence his appointment probably dated from 1 June that year.

[3] PRO, RAIL 856/1, 10 July 1794.

[4] R. Owen, *Autobiography. The Life of Robert Owen* (1857), i, p. 65. Samuel Marsland, one of the proprietors, met Fulton through his association with Owen.

England in 1787.[1] He nevertheless accepted this new opportunity with alacrity and quoted successfully for the work in partnership with Charles McNiven, a well-known local businessman.

Fulton had been enthused by science and its engineering applications as a youth. The Canal Mania, then reaching its peak in England during 1794, appears to have re-kindled his early interests such that he temporarily abandoned his artistic career in order to develop and promote his innovative ideas on canal construction and ship propulsion. He was a tall, handsome, and socially gifted man who, in spite of his often impecunious state, was never too abashed to further his personal interests among the genteel classes, and the aristocracy in particular. Through various influences he was introduced to Earl Stanhope and also the Duke of Bridgewater, whose canal works and steamboat experiments at Worsley probably explained Fulton's appearance in Manchester. Whilst living there he also made the acquaintance of Robert Owen, later to become one of Britain's leading industrialists and social reformers, who was then well-established among the city's intelligentsia.[2] The engaging Fulton was introduced and warmly welcomed into this circle, which included the eminent scientist John Dalton and the poet Samuel Coleridge, besides influential businessmen such as Marsland and Clarke.

Fulton's first venture into contracting was not an immediate success because, although the earthmoving contract was signed in October 1794, Outram was to report to the committee the following March that work on the sector had not even begun.[3] The reasons for this delay are unclear but it seems that Fulton, instead of devoting his complete attention to the task in hand, was, among other things, preoccupied with the design of a digging machine with which he hoped to win further contract work on the Gloucester & Berkeley Canal. His situation was hardly helped by the fact that he was also in financial difficulties. This led to threats of legal actions by his partner McNiven from whom he had borrowed money, a situation retrieved only by further borrowing from Robert Owen. A month later, on 23 April 1795, Fulton advised the committee that McNiven was terminating their partnership but he wished to continue to fulfil the contract himself. The proprietors agreed to this and yet surprisingly the minutes recorded no anxiety on their part regarding Fulton's failure to begin work on that critical section of their undertaking.

---

[1] H.W. Dickinson, *Robert Fulton. Engineer and Artist* (1913).
[2] Owen, *Autobiography*, i, p. 64.
[3] PRO, RAIL 856/1, 4 March 1795.

CONSTRUCTION OF THE PEAK FOREST CANAL 131

Those hard-headed local businessmen must have been spellbound by the articulate young American because only the day before he had met them to present his ideas for re-designing the Marple aqueduct, his proposal being to substitute a cast-iron trough for the stone arches already designed by Outram. Even though a contract for building the aqueduct had been awarded three months earlier to the partnership of William Broadhead, Bethel Furness and William Anderson, the committee accepted the idea immediately and Fulton and Outram were requested to prepare advertisements 'for procuring offers of terms for furnishing cast-iron for making the same', and also to arrange for the manufacture of a model of this structure.[1] Outram could hardly have been pleased by this interference with his brief as engineer for the project, and even more so by the proprietors' indulgence towards this inexperienced interloper who had achieved very little for the canal's progress up to that time.

But that was not all. Fulton must have also suggested that the locks at Marple could be dispensed with in favour of an inclined plane and that small tub-boats were preferable to narrow boats. As a result John Lees, a proprietor, was requested to visit Coalbrookdale, in company with Outram, in order to examine the tub-boat system and the inclined planes of the Shropshire Canal. They reported at the next committee meeting on 18 May 1795 that the adoption of such a system was feasible and could save £7,000 over the construction of conventional locks at Marple. The committee hesitated and resolved that 'such determinations shall be referred to future consideration'. The reason for their sudden lack of interest was almost certainly due to Outram who, in spite of his apparent support for Fulton, must have privately expressed doubts about the suitability and profitability of a tub-boat system for handling bulk cargoes of limestone on the scale expected.

Nevertheless, Fulton was placated by an invitation to prepare a paper on the comparative advantages of '20-ton boats and 5-ton boats with 2-ton boats and railways' to help the committee in their deliberations. Meanwhile, Outram was no doubt steadfastly opposing any change of his plans for a narrow canal in favour of an untried system and so in the end Fulton's innovative ideas did not find favour. The committee were, as ever, kindly disposed to Fulton and, in July, he was granted 100 guineas for his prodigious efforts to alter radically the design of the canal and for the provision of several drawings to illustrate his grand plan. Outram and Brown, probably much relieved, then continued with the works as originally intended. There was no mention of the earlier plans for a cast-

---

[1] Ibid., 22 April 1795.

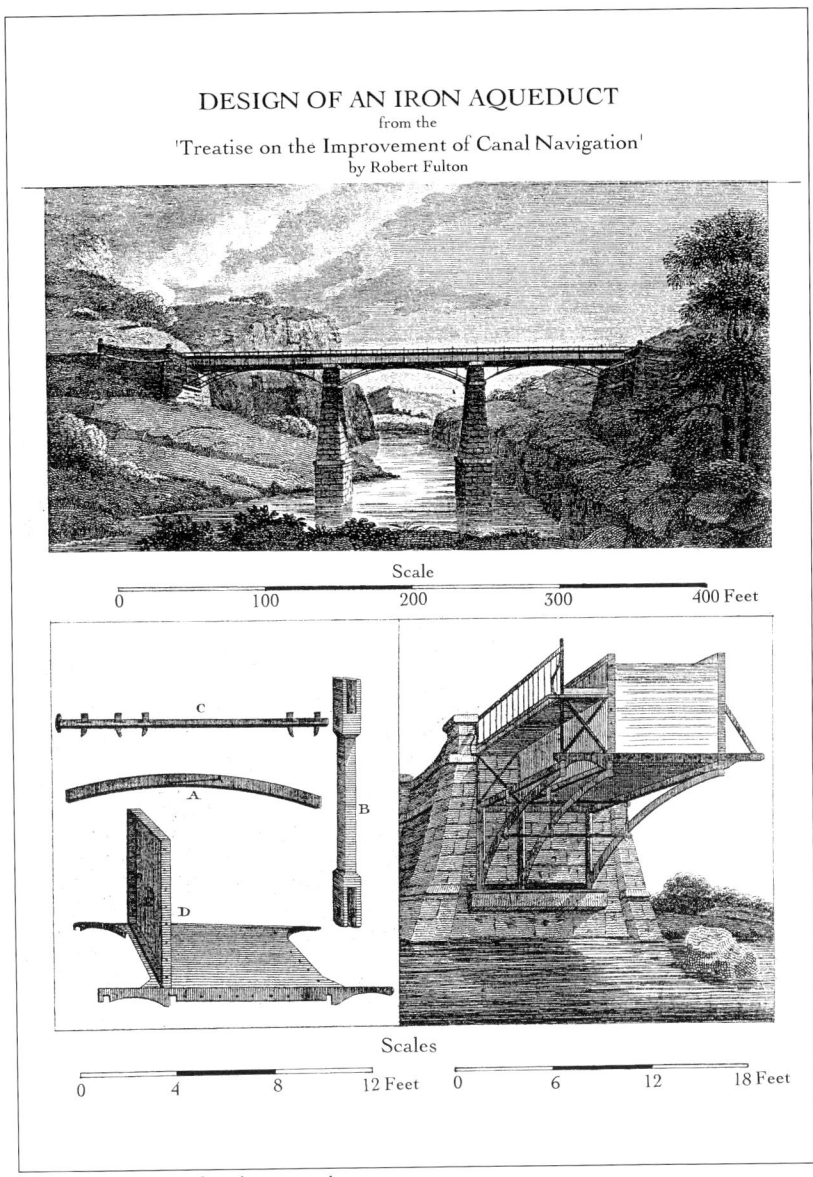

Figure 21 Design of an iron aqueduct.

iron aqueduct and this novel idea, with all its attendant risks, was quietly dropped.

The association with Fulton was not without advantages for Outram. For example, his design for the Holmes aqueduct on the Derby Canal may have stemmed from that time because the manufacture of the cast-iron units for that structure began during the summer of 1795. The original drawings for Fulton's designs for the Peak Forest Canal have not survived, but the American published an article in the *Star* newspaper of 30 July that year on the utility of small canals; no doubt this was much the same as the paper submitted to the committee a month earlier. The article contained an open letter to several eminent engineers, including Jessop and Outram, requesting them to state their objections to the system. Fulton thus implied that Outram had indeed opposed his designs for the Peak Forest Canal.

Subsequently, in May 1796, Fulton, then describing himself as 'civil engineer', published his *Treatise on the Improvement of Canal Navigation*. The work describes, among other things, his thoughts on a small canal system and the use of inclined planes, as well as designs for a cast-iron aqueduct. It is commonly believed that his drawings of the latter are based on the River Etherow crossing at Marple and whilst this supposition is likely to be correct, the dimensions indicated bear little relationship to those of the present masonry aqueduct, the piers of which were under construction at that time.

Fulton's drawings, handsome and artistic though they may be, leave much to the imagination. That of an iron aqueduct shows a central clear span of about 85 ft and outer spans of 100 ft, compared with the three clear spans of the present Marple aqueduct, each of which is approximately 56 ft. The trough was to be fabricated from cast-iron 'at least one inch thick', with vertical side plates 6 ft deep and base plates 6 ft wide. Each span was supported by three cast-iron arched girders, which rose at the centre by one-sixth of the span and every girder measured 8 in. deep by 4 in. in cross-section and was cast in three parts. Several vertical members, which transferred the load of the trough downwards, were bolted on to the girders to form a truss (Fig. 21). It is almost certain that this flimsy aqueduct would have failed under fully loaded conditions and Outram, who no doubt had read Fulton's book, must have realised this because by that date he knew that his recently completed Holmes aqueduct was already proving unsatisfactory. The prospect of erecting Fulton's shaky structure, some 300 ft in overall length, across the Etherow valley was indeed daunting and Outram could only have been relieved when he was instructed to continue building the masonry structure as planned.

Most of Fulton's ideas were not entirely original, being an amalgam of designs derived from several sources besides his own. They attracted some critical response from professional engineers of the day, notably from William Chapman, the civil engineer, who published an entire book in reply.[1] His comments seem to confirm that Fulton's design proposals were widely rejected by the profession. He wrote, tellingly, that 'Those who adopt any favourite system on practical subjects, without the aid of experience to guide them are liable to be carried away by the warmth of their imagination; and are led to apprehend they have attained a something of universal application'. In spite of this scepticism, however, it would not be unreasonable to state that Fulton's publications must have helped to stimulate some interest in the application of more adventurous engineering designs. There are, for example, distinct similarities in style and details, but not in structural strength and rigidity, between his cast-iron aqueduct and the Pontcysyllte aqueduct on the Ellesmere Canal, designed by Jessop and Telford and completed in 1805.

Fulton never succeeded in breaking into the coterie of British civil engineers with his novel ideas but at least he completed his earthmoving contract on the Werneth cuttings. In so doing, he successfully persuaded a still indulgent committee to allow an increase in his rate from 6d. to 6½d. per cu. yd, because of increases in labour rates, before finally departing from the Peak Forest Canal.[2] Soon afterwards he turned his ingenious mind to inventions for naval warfare and the design of steamboats; he returned to America in 1807.[3]

The great aqueduct at Marple was described, shortly after its completion, as being 'among the most considerable of the works of this kind in the kingdom',[4] and it still remains today as a fine example of its genre. Its overall length, including the wing walls, is 368 ft; the central arch is approximately 80 ft above river-bed level and the overall height to the towpath is 95 ft. The pier centres, measured from the Marple end, are respectively 73 ft 6 in., 72 ft 5 in. and 72 ft, although the clear spans are all close to 56 ft. These discrepancies are probably the results of difficulties in setting out measurements for the pier foundations down the steep river banks, and also to slight deviations during erection of the piers. The latter, battered at about 1 in 50, rise some 48 ft to the springing level of the arches. The aqueduct, though massive in outline, projects a light and airy appearance as a result of several architectural embellishments. The river is confined to the central arch and the piers are constructed to arch springing level in rough red sandstone, which

---

[1] W. Chapman, *Observations on the various systems of Canal Navigation* (1795).

[2] PRO, RAIL 856/1, 11 Sept. 1795.

[3] W.S. Hutcheon jun., *Robert Fulton. Pioneer of Undersea Warfare* (1981).

[4] J. Farey, *A General View of the Agriculture and Minerals of Derbyshire*, iii (1817), p. 407.

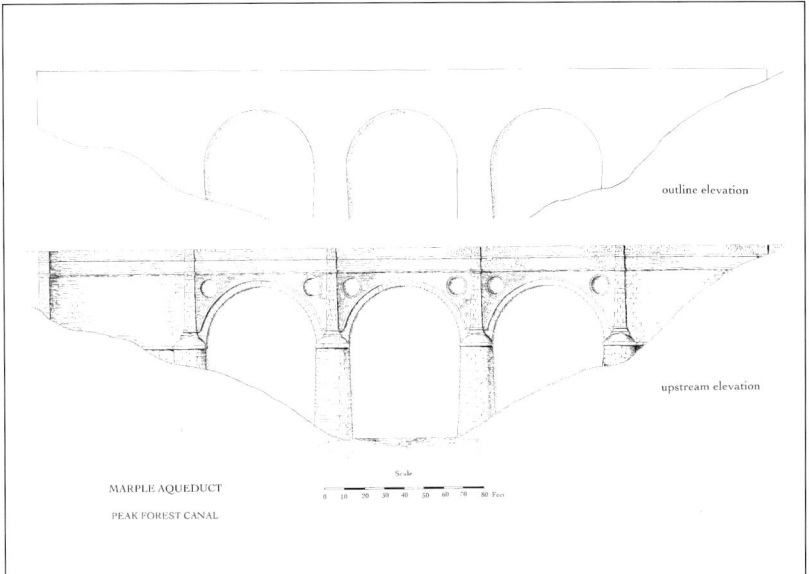

Figure 22 Marple aqueduct.

Outram found in the Hyde Bank quarries,[1] as are the spandrel walls, each of which is pierced by two circular orifices measuring 9 ft in diameter. The pier heads are capped with ashlar hemispheres of yellow gritstone and graceful ashlar pilasters rise from pedestals, sculpted from the latter, up to parapet level. Other features include two distinctive horizontal string courses set above the level of the great orifices, and another around the extrados of the ashlar voussoirs of each of the three arches (Fig. 22).

Marple aqueduct is rightly attributed to Outram. This was his first opportunity to design and construct such a major structure and its engineering would present him with few difficulties, but it is questionable whether he also designed the elegant features that transform an otherwise ordinary, workaday structure into the graceful edifice that still survives today. These may well have been decided in conjunction with the masons and the resident engineer, Thomas Brown, as work proceeded. Local tradition maintains that seven men were killed during the construction of the aqueduct[2] but such claims are to be doubted because

---

[1] PRO, RAIL 856/1, 10 July 1794; Farey, *General View*, iii, p. 407.

[2] J. Wainwright, *A Lecture. Reminiscences of a Life-time in Marple* (1882), p. 7 (copy in Manchester Central Library).

Plate 14 Marple aqueduct.

no accidents are recorded in the minutes, as might be expected of a company which was unusually benevolent to its workmen. Farey stated that the structure was finished by 1797 but in fact the company's financial difficulties prevented completion until two years later.

Planning the progress of the canal's construction was not always left to the discretion of Outram; too often this depended on the whims and convenience of the principal investors. Thus by October 1795 Outram was again being urged to concentrate on the section from Hyde to the Ashton Canal 'on the representation of George Hyde Clarke Esq.'. At the same time he was instructed to stake out for work on the upper level from Bugsworth to Marple[1] to suit the convenience of Samuel Oldknow, who was anxious to sink coal-pits and build limekilns just above the proposed top lock. In effect, the proprietors were impatient to see all parts of the canal progressing simultaneously but they seemed unaware that this could also increase their costs. Only the month previously the treasurer had warned that the company was £1,100 overspent, a trend which persisted throughout the construction period. In the main, however, these cash-flow problems were due to the inability of many shareholders to pay their calls on demand.[2]

More mind-changing took place in May 1796 when the work on the lower level was suspended because the sinking of Clarke's mines had been delayed.[3] In that same month the treasurer complained of more overspending to the tune of £4,000, and so it must have been with some relief that the railway, and much of the upper canal down to Marple, were opened on 31 August and production began in the quarries. On that day the proprietors gathered at Chapel en le Frith 'early in the morning' to view operations 'with several thousands of spectators', when the first six loaded waggons with 'upwards of ten tons of limestone' were drawn by one horse to the head of the canal at Bugsworth, from where two loaded boats navigated along the canal towards Marple. The ease with which heavy loads were passed along the railway and inclined plane 'exceeded all expectations'.[4] The proprietors had earlier doubted the efficacy of the great inclined plane at Chapel but now its performance had clearly dispelled their unease. The potential of the quarries was considerable and the future profitability was obvious to all, yet, the distribution of products was restricted to the locality of the upper canal and would remain so until there was access to the lower canal and the markets beyond.

When originally built, the railway comprised a single track of 4 ft 2 in. gauge, the rails having been supplied by Benjamin Outram & Co. under a contract dated February 1795.[5] In all respects it was similar to the Derby Canal railway. Increasing traffic necessitated a doubling of the line as early as 1803,[6] except for the sections through Stodhart tunnel and a 17-yard-long bridge under the Buxton highway, at the bottom of

---

[1] PRO, RAIL 856/1, 9 Oct. 1795.

[2] Ibid., 11 Sept. 1795.

[3] Ibid., 5 May 1796.

[4] *Cowdroy's Manchester Gazette*, 3 Sept. 1796.

[5] PRO, RAIL 856/1, 10 Dec. 1794, 5 Feb. 1795.

[6] Farey, *General View*, iii, p. 405; PRO, RAIL 856/2, 6 Oct. 1800.

Plate 15 The inclined plane at Chapel en le Frith, c. 1905.

the inclined plane, both of which were 9 ft wide. Passing places were situated adjacent to each of these structures. Farey described the railway as it was in July 1808. From Bugsworth wharf it ran for one mile to Whitehough, rising 129 ft, and from there it continued on the level for a further mile to Chapel Milton. The railway then rose steadily, through 56½ ft in the next $1^{1}/_{8}$ miles, to the foot of the great inclined plane. The latter measured 512 yards and rose 192 ft to the top. There was a further rise of 24 ft in the next three-quarters of a mile to Barmoor Clough (or Loads Knowle) and another three-quarters of a mile led, with an initial small rise, to Dove Hole quarries. This was followed by a level section through the quarry and finally there was a rapid descent into Dove Holes (or Daff-hole) Mouth quarry.

 The plane was designed to be self-acting and accordingly there were two lines of track such that gangs of seven waggons, empty or partly filled with coal and other goods for the quarries, interlinked with short chains, were drawn up the plane by gangs of waggons loaded with limestone, descending under gravity. Both gangs were connected by a long haulage chain which passed around a large tilted wheel, furnished with a controlling brake, which was located in a shallow pit at the top of the incline. The chain was supported by blocks of wood, laid across the plane at 10-yard intervals and a 'double scissors' crossing was installed at the base of the incline. Gangs of waggons could thus be raised alternately left and right, and lowered alternately right and left.

Plate 16 The inclined plane in 1975.

Farey had been told, probably by James Meadows who was the company agent at the time, that a thick rope had been tried for haulage initially but this soon failed (it was also cut by vandals in December 1798),[1] as did its replacement, a patent twisted chain, which was probably purchased from Jonathan Woodhouse of Ashby de la Zouch in 1801.[2] Finally, a chain with 5-in. links, manufactured in Birmingham and costing £500, proved more successful in coping with the rapidly increasing output of the quarries. It had been in use for five years when Farey visited the works and it still 'appeared none the worse for wear'.[3]

The inclined plane formed a concave curve in longitudinal section as a means of aiding the movement of the gangs of waggons. The slope increased gradually for each $1/32$nd part (or 48 ft) of its length, from 1 in 12 at the base to 1 in 6 at the top. These gradients helped to overcome the friction of the rails and chain and aided initial acceleration when the loaded trams began their descent, as well as deceleration as they approached the bottom of the incline. There was another inclined plane, 33 yards long, which led into the Daff-hole Mouth quarry and this also had a double railway. Waggons were lashed to a continuous ropeway

---

[1] PRO, RAIL 856/1, 13 Dec. 1798.
[2] PRO, RAIL 856/2, 29 April 1801.
[3] Farey, *General View*, iii, p. 405.

which was operated by a horse-gin at the top of the bank. By this means, full waggons were simultaneously drawn up, as empty ones were lowered.

The machinery and operating method of the great plane do not seem to have altered much by the beginning of the twentieth century and the technique for controlling the speed of the waggons then was probably similar to that originally devised. A hut, elevated on piles, was erected at the top of the incline for the brakesman and this provided a good view of the gangs of waggons moving below. A 2-in. diameter steel wire rope was then in use and this passed underground at the top of the incline and turned one and a half times around the tilted brake wheel, 14 ft in diameter, contained in a pit with substantial masonry walls. The deep groove in which the rope was turned was lined with blocks of wood, fitted with the grain outwards to provide a good grip as the wheel rotated. The rim of the wheel above the groove was flat, about 5 in. deep, around which was wrapped a steel brake-strap of the same width; this was also fitted with wooden brake blocks, with the grain towards the wheel's rim. One end of the strap was shackled to the masonry at the back of the pit, whilst the front was cranked and shackled above and below the fulcrum of a long lever which passed through the floor of the brakesman's hut situated above. Sufficient purchase on the lever was facilitated by a block and tackle arrangement.[1]

The waggons were originally of similar design to those in use at Derby, although a different style was later adopted. The bodies were then made of rolled plate-iron; each carried 45 cwt and a door at the back of the tram opened up to shoot the contents of limestone on to the boats at Bugsworth wharf by means of a tippler machine fixed to a turntable.[2] Below the great inclined plane, a team of four horses was able to bring down a gang of twenty interlinked waggons loaded with 45 tons of limestone. One horse could take 2 tons of coal or other goods, contained in two waggons, back up the line.

It must have been frustrating for the proprietors to know that insufficient investors' funds were available in late 1796 to permit the rapid completion of the partly built aqueduct and the proposed locks at Marple, because this prevented them from profiting from the potentially enormous output of the quarries. Consideration had been given at the July meeting to cheap temporary solutions. Outram was requested to

---

[1] Baxter, *Stone Blocks*, p. 65, citing G. Taylor, 'The Peak Forest Tramway and Quarries', *Great Central Railway Magazine*, i (1905), pp. 122, 123, 148, 149.

[2] B. Baxter, 'The Peak Forest Tramway', *The Locomotive*, 15 June 1929, pp. 189, 190.

prepare a statement as to the best method for sending boats from the upper to the lower levels at Marple and whether this should be by locks, inclined planes, caissons or other means.[1] Presumably he did this but there was still much unfinished work remaining on the lower canal and it was not until October 1797 that he was again requested to prepare a scheme and estimate for a temporary railway from the top of the proposed locks down to and over the aqueduct.[2]

Outram was asked to price two schemes; one for the laying of a railway across 'turned arches' (i.e. finished arches), and another, assuming that the arches were incomplete. The specification for the railway compared closely with that for the Peak Forest railway above Bugsworth. A quotation for £2,720 was submitted to the November meeting and this included a railway, wharves, waggons, cranes and other appurtenances but excluded the expense of turning the arches and 'for making a proper bed' across the aqueduct.[3] The quotation listed 2,500 yards of main-line railway and 600 yards of lines in passing places, at 10s. per yard run of railway. The cast-iron rails were unusually heavy, weighing 84 lb each. They had to be capable of bearing waggons of 5 tons, although they were most commmonly used with loads from 35 to 45 cwt.

These proposals were greeted with some enthusiasm. The proprietors reckoned that £14,000 could be saved by suspending work on the locks and the channel across the aqueduct; a railway would also advance trading by twelve months. Not surprisingly, Benjamin Outram & Co. were awarded the contract for the railway the following January, for completion by the end of May 1798.[4] The committee wisely decided not to interfere with the construction of the aqueduct, however, and so the railway commenced at the 'Brick Bridge' above the proposed top lock and terminated at the Marple side of the aqueduct.

William Broadhead and his partners had contracted to complete the aqueduct by September 1797 but they were then granted a twelve-month extension without penalty, because the delay was due largely to the canal company's financial constraints.[5] Further extensions had to be granted to the contractors until May 1799, because the arches were only keyed

---

[1] PRO, RAIL 856/1, 8 July 1796.

[2] Ibid., 11 Oct. 1797.

[3] Ibid., 9 Nov. 1797.

[4] Ibid., 2 March 1798; DRO, D503/12/1, Specification for railway at Marple (undated, probably Jan. 1798).

[5] PRO, RAIL 856/1, 16 Feb. 1797.

in during the previous November; Brown was then instructed to give the workmen 'a handsome treat' to celebrate the event. Outram settled the accounts with Anderson and his partners in November 1799,[1] and so presumably the aqueduct and its water channel were finished during the late autumn of that year.

During 1798 work was in progress over the entire length of the waterway and railways, particularly on the lower canal where substantial earthworks, walling and 'arching' (a term used to describe bridging) were under construction. Driving Hyde Bank tunnel through tough rock strata was proving difficult and accordingly Outram brought in Joseph Booth from the Huddersfield Canal to oversee and ensure that the quality of the work was maintained because the gangs were working night shifts.[2]

The impecunious state of the company was not helped by some of the landowners, who were also proprietors, refusing to pay their calls on shares until they were paid for their land taken for the canal.[3] Naturally the committee rejected these illegal demands and warned those shareholders accordingly. One of them, John Thorniley of Romiley, remained displeased and went on to vent his anger on the apprentice David Whitehead, who was observed innocently surveying on his land; the unfortunate young man was thoroughly beaten as a result. The company would not tolerate such behaviour: the clerk was instructed to apply for a warrant to have Thorniley bound over until the next Quarter Sessions[4] but Thorniley met the committee shortly afterwards regarding the valuation of his land, and it seemed that matters were amicably concluded. Nothing more was said or done about the threat of prosecution; this was conveniently set aside and the hapless young Whitehead was presumably expected to suffer in silence, without promise of redress.

The proprietors were tireless in their efforts to develop their quarries and to encourage the use of their products for building, highway construction and agriculture. Accordingly there was some cause for satisfaction because, in spite of the obstruction of the unfinished aqueduct, by June 1798 trade was steadily increasing. Thomas Brown was then instructed to employ sufficient workmen in the quarries to ensure that 20,000 tons of limestone were taken during the year.[5] Independent boat-builders, carriers and lime-burners along the canal were

---

[1] Ibid., 6 Nov. 1799.
[2] Ibid., 10 Aug. 1798.
[3] Ibid., 13 April 1797.
[4] Ibid., 18 Aug. 1797.
[5] Ibid., 6 June 1798.

encouraged to trade by means of generous grants, loans and toll reductions and this further helped increase production in the quarries. Outram was also asked to approach his future father-in-law, Dr James Anderson, the Scottish economist and agricultural expert, to invite him to write a treatise on the nature and effects of lime as a manure. The company then arranged for its publication and the sale of 1,000 copies at 1s. each.[1]

Brown was also being urged to complete all remaining works on the canal. He was assisted in that task when the proprietors appointed German Wheatcroft in July to take over as toll-collector, wharfinger and wharfhouse-keeper at Bugsworth and Whaley Bridge, as well as superintendent of all works on the railway up to the inclined plane.[2] He proved to be an able and loyal manager who applied himself diligently to his several onerous tasks at the meagre salary of £75 p.a.; small wonder that he gave notice to quit twelve months later unless he was allotted another £30 to keep a horse 'to better attend the concerns of the company'.[3] The proprietors agreed to his terms, knowing full well that business was rapidly increasing, as would the duties of the industrious Wheatcroft also.

Long delays in finishing the waterway below the aqueduct meant that the Marple railway languished, virtually unused, until August 1800 when all the works of the canal were finally completed. This justified further developments at the quarries and so the committee resolved to increase the workforce to 200 men. Brown was then told to obtain quotations for another 200 waggons, for which Benjamin Outram & Co. were to supply cast-iron wheels and wrought-iron axles. The decision to double the railways was taken and Outram was no doubt gratified to have his quotation accepted for rails at £11 15s. per ton at the foundry, albeit it with a generous three-year guarantee against breakages, as well as good discount terms. The committee's decision also applied to the newly opened line at Marple because traffic on the latter was already heavily congested, having to cope with the movement of 1,000 tons of limestone each day.[4] That quantity was soon expected to double in order to meet the insatiable demand for both stone and lime. Already the committee were discussing the possibility of commencing the building of the Marple flight of locks, as these were the only effective means of

---

[1] Ibid., 1 March 1799.
[2] Ibid., 4 July 1798.
[3] Ibid., 22 Aug. 1799.
[4] Ibid., 1 Aug., 1 Sept. 1800.

resolving the increasing congestion.[1]

This remarkable expansion of business brought with it acute labour difficulties. Most of the quarrymen engaged were contract 'stone-getters' on piecework, but they proved to be in short supply in what was then a remote corner of Derbyshire. The company offered various incentives: cottages were purchased or rented and rudely furnished as dormitories, and a dinner house provided; a shop was opened by the worthy Wheatcroft where goods were purchased 'at prime cost'.[2] And finally there was surely a measure of desperation in the newspaper advertisement which stated that a good flannel waistcoat and trousers, plus a daily jug of ale, would be provided for each pieceworker employed.[3]

By April 1801 the Marple railway was proving to be a serious bottleneck and the committee was urging Brown to finish the task of doubling the lines; indeed he was 'directed to let nothing prevent him from carrying [these] resolutions into full effect with all possible despatch'.[4] He was even offered a £100 banknote if, within a month from 15 May, he could pass 800 tons of stone down the Marple railway in 18 hours.[5] The astute Brown won the wager easily and 1,170 tons were sent down in the time, although it was a further twelve months before the committee paid his reward.

Construction was drawing to a close and Outram resigned as engineer from May 1801, the same date as his resignation from the Huddersfield Canal Company. It was an amicable conclusion to a successful appointment and the committee 'gratefully accepted his offer to afford them his advice and assistance when professionally requested'.[6] Presumably this was not unexpected; no doubt it was related to the appointment of Thomas Brown, just one month earlier, as full-time engineer to the company at a salary of £500.[7] Two major tasks then faced Brown: the provision of a water supply for the canal from the Combs reservoir and the building of Marple locks, for which he was to receive a bonus of £200 on completion.

For several years past plans for a water supply had received scant attention, but the absence of locks at Marple meant that water levels in

---

[1] PRO, RAIL 856/2, 26 Dec. 1800.
[2] Ibid., 12 Jan. 1801.
[3] *Derby Mercury*, 27 Nov. 1800.
[4] PRO, RAIL 856/2, 29 April 1801.
[5] Ibid., 15 May 1801.
[6] Ibid., 29 April 1801.
[7] Ibid., 20 March 1801.

the upper level could be maintained by minimal discharges from streams, as well as from collieries along the waterway. The situation changed in 1798 when a permanent supply for the lower canal was becoming a necessity. A tentative suggestion was made in committee that a reservoir should be built at Hyde Bank for the purpose but nothing more was done until Outram recommended the cutting of a temporary conduit alongside the Marple railway to collect overflows from the upper level. A temporary duct was then laid across the aqueduct to feed the lower pound.[1] This action must have led to substantially increased flows into the upper level and it was probably the cause of complaints from mill-owners and farmers. Accordingly, next summer Wheatcroft was told not to abstract water from rivers on the line of the canal in times of low flow. At other times abstraction was restricted to 24 hours from 5 o'clock on Saturday evening and then only with the agreement of affected parties.[2]

The obvious long-term solution was to locate a single large reservoir on an upland site and eventually that at Combs, which had been so designated by Brown in his original survey, was selected. Curiously, Outram had seemed unsure of the suitability of Combs when requested by the committee late in 1795 to choose a site and to provide plans for a reservoir. Even twelve months later, neither he nor Brown was 'ready to give an opinion on reservoirs'[3] and it was not until February 1797 that Outram finally produced his scheme. Even so the proprietors, for all their obvious concern, then decided to shelve the matter because 'it cannot be made use of yet'. They also considered that work could be delayed until after 'the next corn harvest' when local labour could be recruited.[4]

Thus the scheme languished until May 1801 when the committee, perhaps excited by the completion and opening of their canal, declared that work on Combs reservoir, and also on the Marple locks, should begin immediately. Brown was told to contact Outram for guidance on the design of both these major and costly ventures but by August of that year the mood had changed when the committee realised that 'the company are not able to raise money to finish the relevant locks and other works'. They then proposed to recommend to the general assembly that an application should be made to Parliament for powers to raise

---

[1] PRO, RAIL 856/1, 2 March 1798.
[2] Ibid., 6 June 1798.
[3] Ibid., 14 Nov. 1796.
[4] Ibid., 16 Feb. 1797.

funds to complete these ancillary works 'with the greatest expedition'.[1] Nevertheless, in the meantime some preparatory works had commenced at Combs. Outram had sent design details to Brown for the dam and 'waste weir', or spillway, and there were sufficient funds in October to begin building because Brown was then negotiating the purchase of the land needed. He continued to supervise site clearance and the building of the earth bank during the next two and a half years.

Combs reservoir remains today much as it was in 1803, although severe storms in January 1976 caused a massive earth slip in the centre of the dam, which necessitated an immediate reduction in water levels until repairs were implemented. The waters are retained by a clay-cored earth dam or 'head', measuring over 1,000 ft in length and 52 ft maximum height. The top of this substantial bank is 10 ft wide and both shoulders slope at 1 (vertical) to 2 (horizontal).[2] In the early days, deep cuts along each side of the lake diverted all normal flows for the use of the mills in the valley below and a long weir at the western end ensured that only flood waters from the Combs Brook were permitted to enter the reservoir. That jealously guarded practice has long since been abandoned. Water for the canal was let off through a large cock, to flow two miles down the valley along an open channel, separate from the Combs Brook, to Whaley Bridge. Nowadays the stream is partly culverted as it passes through the town, before crossing the River Goyt in a large diameter cast-iron pipe to discharge into the canal.

After 1801 the proprietors were much preoccupied with the business of raising funds, mainly for the completion of the Marple locks. They eventually took the unusual step of borrowing the monies privately, rather than petition Parliament, with all the expense which that entailed. The lenders were Richard Arkwright and Samuel Oldknow of Marple, although the latter was unable to provide the necessary funds and soon withdrew, leaving the company to negotiate only with Arkwright.[3] After much dithering, the company decided that a loan of £24,000 would suffice and an agreement for that amount was concluded by August 1803. The loan was paid back in instalments in addition to generous abatements of tolls; it proved to be a good arrangement for the company although Arkwright did very well out of the deal.

Brown then took charge of the lock construction immediately,

---

[1] PRO, RAIL 856/2, 10 Aug. 1801.

[2] G.M. Binnie, 'The evolution of British dams', *Trans. Newcomen Soc.*, xlvii (1974–6), p. 223.

[3] PRO, RAIL 856/2, 19 Jan. 1803.

adopting Outram's plans for sixteen locks, each of 13 ft 1 in. rise.[1] Progress was rapid and the flight was fully operational by October 1805. In the event, final borrowings amounted to £27,000, most of which was expended on the locks and warehousing

Local legend has it that Samuel Oldknow was determined to ensure that one of his narrow boats, named the *Perseverance*, laden with lime from his kilns, would be the first to lock down to the lower level. Oldknow's limekilns were constructed with an elegant Gothic façade, which says something for his concern for the rural environment, and were built into the bank of the upper canal, just beyond the top lock. They were charged with limestone and coal at that level and burnt lime was discharged into narrow boats below, floating in a short arm of the canal, which passed under Posset Bridge to join the main line. The bridge was so named because Oldknow urged his masons on to complete it in time for the opening by providing them generously with ale posset for breakfast.[2]

A feature of the locking system at Marple was a novel lifting gear which operated the paddles in each lock by means of air, compressed by the head of water available in the pound above. The company minutes suggest that both Outram and William Jessop were involved but this device was not a success; the apparatus frequently broke down and had to be replaced by conventional hand-operated gear after only twelve months. Unfortunately the design details of the equipment have not survived and the minutes, composed by a clerk of limited technical understanding, give but a tantalising glimpse of this innovative attempt to mechanise and accelerate the locking process. It was stated that

> in consequence of the lifting gear, the paddles of the locks at Marple frequently breaking by reason of the air being confined in the pipes and the rods not having sufficient space to work in and the air pipes also frequently breaking, the trade of the canal is greatly impeded and the almost daily loss of one or two pounds of water which are obliged to be drawn off in consequence of such accidents … .[3]

Thomas Brown left the company shortly after completing his final assignments on the canal. He returned to a career in mining and other

---

[1] Ibid., 17 July 1801.
[2] Wainwright, *Lecture*, p. 7.
[3] PRO, RAIL 856/2, 21 March 1806.

businesses and his duties were handed over to the new company agent, James Meadows, in December 1805. The company were generous to Brown in their praise and remuneration;[1] they, as well as Benjamin Outram, could not have employed a better, nor more able and loyal resident engineer. The high quality of Brown's engineering management can still be seen today in the fine waterway, bridges, locks, buildings and ancillary works, many of which still survive in their original state.

---

[1] Ibid., 4 Dec. 1805.

CHAPTER 9

THE COMPLETION OF THE ASHTON CANAL

The Ashton Canal was an isolated waterway for the first few years of its existence. By 1796, the sections from the aqueduct bridge crossing Ancoats Lane (now Great Ancoats Street) to Ashton under Lyne and Hollinwood was finished. The Heaton Norris line was completed in early 1797 and that to Fairbottom was operational later in the year. The canal from Ancoats to its junction with the Rochdale Canal in Ducie Street, however, remained untouched until a contract for the design and construction of these remaining works was awarded to Benjamin Outram. He began work on the section during 1797.[1]

In August 1798 the proprietors were hoping that their canal would be finished and navigable by November, but they 'observed with regret that no exertions have of late been made to complete the Rochdale Canal below Piccadilly'.[2] In fact the Rochdale Canal Company failed to collaborate and to connect up the all-important section of their canal, from Ducie Street to the Bridgewater Canal, until May 1800. Moreover, the Peak Forest Canal was only opened in August that year and, although the Huddersfield Narrow Canal was finished up to Woolroad, near Saddleworth, in February 1798, it was not until 1811 that this route was joined to the Yorkshire navigations and the eastern ports. These irritating delays were beyond the control of the Ashton Canal Company and limited the use of their waterway to local movements of coal, textiles and other goods. In spite of its isolation, water supplies appeared to be adequate and, as a temporary measure until its neighbours' works were completed, the company discharged the surplus flows at its Manchester terminus into the Bridgewater Canal (as required under the Act of 1792), via Shooter's Brook and the River Medlock.

During 1798 Outram became deeply involved with the Ashton Canal Company. He accepted an invitation in August that year to be consulting engineer for the entire waterway and, of course, simultaneously he was designer and contractor for the uncompleted canal and bridge works in

---

[1] PRO, RAIL 803/1, 20 Oct. 1798. The minutes refer to Outram's contract for finishing the canal to Piccadilly. The contract date is unknown but was likely to be in late 1797.

[2] Ibid., 8 Aug. 1798.

Manchester. Moreover, quite independently of these appointments, he was the developer of lands, adjacent to the Piccadilly canal terminus, in which he had purchased an interest some time previously.[1] Inevitably, there would have been some conflict of interests arising from these closely related roles and, although the proprietors must have been fully aware of them, Outram's conduct was neither doubted nor questioned.

Outram's contract was for the Piccadilly basin and wharves, and about 750 yards of main-line canal from the junction with the Rochdale Canal up to the Ancoats Lane bridge. This included a major skew aqueduct across Shooter's Brook, known today as the Store Street aqueduct, and a culvert for the brook. The contract was to take about twelve months to complete. The costs of these works is unknown but Outram seems to have been paid £300 to £400 each month 'on account of his contract with the said company for finishing the canal to Piccadilly'. Besides his contract work on the terminal section of the Ashton Canal, Outram was responsible for the designs, specifications and supervision of the company warehouses. These substantial warehouses, at Piccadilly and Heaton Norris, priced at £1,600,[2] were completed by a local builder, David Bellhouse, early in 1799.

One of the problems unresolved by the earliest British canal engineers was the building of masonry arches oblique to a canal, so as to continue the alignment of a crossing highway. Such skew bridges were either avoided altogether, or built incorrectly with no understanding of the principles involved. For example, James Brindley, the most able of the early canal engineers, never worked out a solution for such oblique alignments and his highway overbridges were always set out at right angles to the waterway, much to the inconvenience of road-users of later generations.

The loads on an arch bridge, caused by crossing traffic and the self-weight of the structure, are transmitted down to its foundations by forces which pass from the crown, outwards to both abutments, through the voussoirs of the arch. In the case of an arch built square (i.e. at right angles) to the abutments, the lines of action of these forces are parallel, in plan, to the faces of the arch. The voussoir courses, and their bed joints, are laid perpendicular to these forces, from the crown down to the abutments.

Before 1798, Outram, in common with other engineers of the period,

---

[1] *Lords' Journals*, xlv, p. 583; HLRO, Petition of Francis Outram, with plans by James Meadows (1806).

[2] PRO, RAIL 803/1, 8 and 23 Aug. 1798.

Plate 17 Store Street aqueduct, Manchester.

had sometimes constructed bridges on his waterways with small angles of skew. All of them were built with arch courses parallel to the abutments as described above and, as a result, the forces would follow paths into, and square to, the abutments. Inevitably this led to weaknesses along the lines 'X–X' (Fig. 23). This was because the triangular-shaped masonry on the obtuse angle of the skew was cantilevered out, and thus received no support from the opposing abutments. This practice could lead to partial, or complete, failure if the skew angle exceeded about 15 degrees.

An entirely new technique to overcome these defects was devised during 1787 by the Northumbrian civil engineer William Chapman, whilst he was engaged on the Kildare Canal, a branch of the Grand Canal of Ireland. He was requested by the proprietors to maintain the alignment of the existing roads, certain of which crossed the canal at acute angles. Chapman knew that the usual methods of arch construction would not suffice and, after some experimentation, he discovered an approximate solution which proved to be both stable and simple to construct. He later applied the technique successfully when bridging various drainage channels in the East Riding of Yorkshire.[1]

Chapman realised that the lines of action of the forces in a skew arch should be parallel to the direction of the faces of the arch and, to achieve this, each course of the voussoirs would have to run square to the face of the arch. When applying the technique in practice, however, the forces and bed joints of the voussoir courses are found to be at right angles only in the vicinity of the crown. As the courses (known as spiral, or winding, courses) approach the abutments they become more oblique to the direction of the forces, the nearer they are to the springings. It is there that the slanting courses of the arch meet the skew-backs, of stone or brick, which connect with the horizontal courses of the abutments.

To erect such an arch, timber centres were first erected and placed parallel to the faces of the arch, and diagonal to the abutments. They were covered with plank-sheeting laid parallel to the abutments and then carefully levelled up to the correct curvilinear shape of the intrados of the arch. The position of the courses in the vicinity of the crown would be set out on the sheeting with pencil. The bed joints of all the remaining courses across the structure were then marked out, from the faces of the arch down to the abutments, using long and flexible wooden straight-edges.[2] Masons could then begin laying the masonry or brick courses from the skew backs on the abutments, until they met at the crown of the arch. Work could be very rapid and an arch might be completely turned in a few days.

Chapman was fully aware that his technique produced only an approximate solution to the problem of the skew bridge and he was accordingly cautious in offering advice on the maximum angle of obliquity permitted in practice. The first bridge built on his principles was the Finlay bridge near the town of Naas, where the acute angle between the bridge and canal centre-line was 39 degrees. Chapman examined the bridge frequently during the years following its completion and 'never observed any crack in it'.

Modifications were made to Chapman's basic method during the railway era and numerous bridges were built according to what became known as the English, or Helicoidal, method of skew-arch construction. An exact solution to the problem was also devised during that period which ensured that the bed joints of the courses were everywhere per-

---

[1] 'Oblique Bridges', in *Cyclopaedia, or Universal Dictionary of the Arts, Science and Literature* (ed. A. Rees) (1819). The article was written by William Chapman.

[2] GRO, D2159, Statement by Thomas Telford on the setting-out of oblique bridge arches (nd, c. 1820).

COMPLETION OF THE ASHTON CANAL

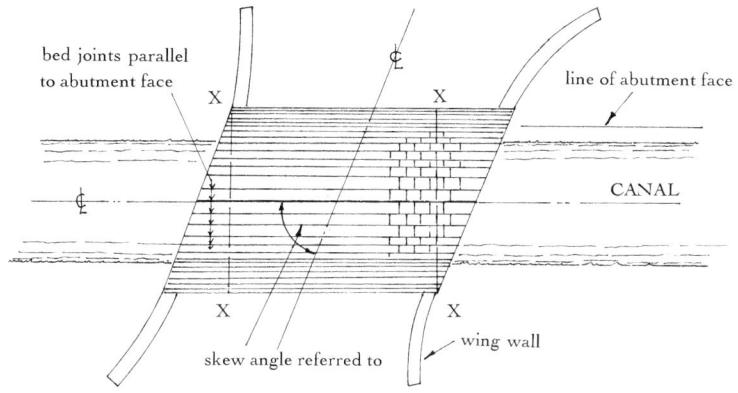
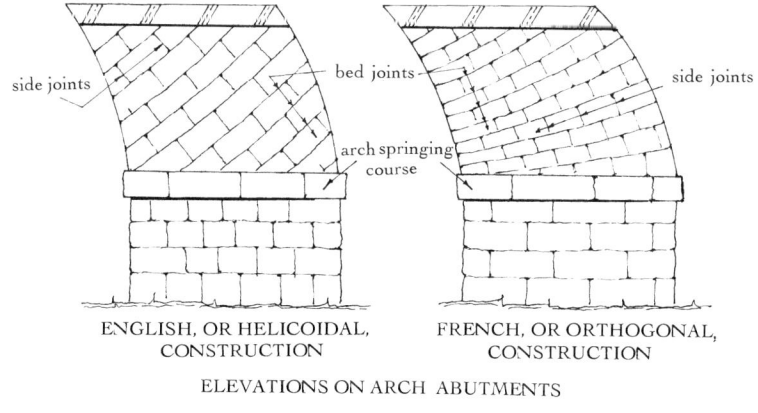

Figure 23 The skew arch bridge.

pendicular to the thrust lines.[1] This became known as the French, or Orthogonal, method but very few such structures were erected in accordance with those principles in the United Kingdom because they were complex and expensive to build (Fig. 23).

William Jessop was the principal engineer for the Grand Canal Company when Chapman's bridges were built, and there can be no doubt that he became conversant with this advance in bridge technology. He first tried out the method for himself during 1797 whilst he was engineer for the Rochdale Canal[2] and it is more than likely that Outram learnt of the technique from his friend and colleague. The Store Street aqueduct crosses the highway beneath at a severe skew angle of 45 degrees and it could not have been built by traditional methods without risk of an early collapse. Outram was no doubt conscious of this danger and he decided to adopt the new method, which proved successful. In July 1798 he hired 200 planks from the Manchester stores of the Peak Forest Canal Company for a period of six weeks[3] and these could only have been used for the temporary falsework for the Store Street aqueduct. In fact the structure was finished by October when Outram was instructed to culvert the Shooter's Brook through the arch.

The aqueduct is of interest because it is believed to be the first major skew aqueduct to be built in Britain on Chapman's principles; it remains the oldest example of its type in use today. Unique though it may be in engineering terms, however, Store Street aqueduct is not the most imposing of bridges and its architecture arouses little aesthetic interest. It is strictly functional in style and its yellow gritstone is now much weathered and discoloured. The structure has a square span of 24 ft 6 in. and a clear height of 15 ft 3 in. Although its skew angle is less than the minimum of 50 degrees recommended by Chapman, it still stands firm, and there are no signs of movement of the arch. Regrettably, the neglect of past years resulted in severe leakage of the jointing and the intrados of the arch has now been rendered, so that the spiral courses and the quality of the masonry can no longer be seen.

As work drew to a close at the end of 1798, Outram was retained to advise on some important tasks which included the adjustment and operation of waste-water weirs on the summit level, the extension of the canal basin at Stockport, and the layout of streets around the canal

---

[1] W.J.McQ. Rankine, *A Manual of Civil Engineering* (1900), pp. 429–32.

[2] R. Russell, *Lost Canals of England and Wales* (1971), p. 201, referring to oblique bridges known as Gorrell's and March Barn on the Rochdale Canal.

[3] PRO, RAIL 856/1, 4 July 1798.

terminus in Manchester.[1] The latter developments, on lands purchased from Lord Ducie,[2] were extensive and of prime importance for the growth of industry and the future of the company, located as they were at an important canal junction.

Outram appeared to take much interest in these developments because he and John Mallalieu, one of the canal proprietors, had bought several parcels of land, stretching almost from the Piccadilly basin to Ancoats Bridge, during August 1796. This land, across which the canal would eventually be built under Outram's contract, was cunningly acquired at a time when the company was in financial difficulties and was in no positon to buy. In fact it was not until 1798 that the company was able to purchase a strip of land for the canal, some 14 yards wide, from Outram and Mallalieu so that the contract could proceed.[3] The minutes neither record the price nor do they comment on the propriety of these dubious dealings but the original purchases almost certainly would have been made with inside knowledge of the company's financial position.

In September of that year Outram, whilst in Manchester, received a letter from his brother Joseph in Alfreton requesting that he attend a business meeting with his father. He wrote back to arrange another date stating that 'There is a meeting of very great consequence here ... if I miss that meeting it may be the greatest loss in my private affairs'.[4] On that occasion it is possible that he was to meet the proprietors of the Ashton Canal Company in Manchester to prepare a report on the state and progress of the works of the canal and to consider the 'probable produce to arise to the said company'.[5] That report was issued to the company 'at large' on 8 November 1798 and made depressing reading.[6] It referred to the shortage of capital and the pressures from creditors, stating that 'the the canal is incomplete, almost without trade, destitute of proper wharfs and warehouses for those inclined to trade upon it and there were no funds to meet these difficulties'. Mention was made of

---

[1] PRO, RAIL 803/1, 5 Sept., 8 Nov. 1798.

[2] Ibid., 25 Oct. 1798.

[3] Ibid., 1 Oct. 1798. It is possible that here was a matter of greater importance. It has long been suspected that Outram and Jessop were involved, about that time, in negotiations and surveys for a railway to link Liverpool with Manchester. A better terminus than Piccadilly would be difficult to find. Below, pp. 318–19.

[4] JRUML, Bagshawe Muniments, 8/4/2881, B. Outram to J. Outram jun., 24 Sept. 1798.

[5] PRO, RAIL 803/1, 1 Oct. 1798.

[6] Manchester Central Library, L.J. Scott Collection, 7/12/1926, Report of the Committee of the Ashton Canal Co., 8 Nov. 1798.

'their present engineer' and of his efforts to calculate trade and profit potential, but the incompleteness of the interlinking network of canals, on which so much depended, gave little hope for the implementation of Outram's forecasts. Today it seems unduly pessimistic; conceivably it was an indirect appeal to the neighbouring canal companies between whose unfinished waterways the Ashton Canal was trapped in its isolation.

Nevertheless, Benjamin Outram was probably disappointed by all this because he was expecting to see some progress in industrial development which would reward his own substantial investment. Shortly afterwards, however, he left for fresh fields and his land acquisitions remained untouched by the time of his death in 1805. His investments were then disposed of by his family and sadly, his hopes for the future were never realised.

The eventual completion of their canal was followed by a period of uncertainty for the impecunious Ashton Company. The final costs of construction have been assessed at about £170,000 when it opened;[1] it was not until 1806 that fortunes began to change for the better and the first dividend of £3 per share was paid. The years that followed are another story but suffice it to say that the earliest troubles of the company were mainly self-inflicted. They were the result of rash assumptions regarding construction costs and trade potential; there was a failure to liaise with neighbouring companies and to co-ordinate completion of their interlinked navigations. Finally, there was a lack of of good engineering design and competent supervision of the works. And yet, in spite of the mistakes and the debts that confused the construction years, the Ashton Canal remains a tribute to those optimistic entrepreneurs who achieved much for the industrial development of the city of Manchester.

---

[1] C. Hadfield and G. Biddle, *The Canals of North West England* (1970), ii, p. 299.

CHAPTER 10

AN INCIDENT AT BARNOLDSWICK

Of the three navigations which crossed the Pennines, the Leeds & Liverpool Canal was by far the longest. It took an indirect and less heavily locked route than did the Huddersfield Narrow and Rochdale canals, and joined the cities of Liverpool and Leeds via Wigan, Burnley, Skipton and Keighley, a main-line distance of 128 miles. The first phase of its construction began in 1770[1] under the direction of John Longbotham of Halifax as principal engineer, and continued for seven years, by which time capital was exhausted. About 70 miles then remained to be completed between Wigan and Gargrave.

The next phase began in 1790 when Robert Whitworth was appointed engineer, with John Harrison as his resident engineer on the Lancashire side and Samuel Fletcher on the other.[2] Both halves of the undertaking were further sub-divided into several sections, each with its own local supervisor who controlled the contractors and direct labour gangs. Communications between this motley assemblage of personnel were always difficult and the problems of management could not have been made easier by the many inexperienced and untrained site staff with whom the engineers had to deal. Whitworth was consultant to other canal companies at the time and could spend only a limited number of days in attendance on any one undertaking. Many decisions on forward planning and engineering were taken with only infrequent reference to him, hence it is not surprising that the construction programme for this major canal progressed in a rather haphazard and inefficient manner. There were many disputes over purchase of land, accommodation works and damages to property. None could have been more protracted nor deplorable than those which took place from 1790 to 1799 between the company and John Bagshawe at Barnoldswick, in the Craven district of the West Riding of Yorkshire.

John Bagshawe (1758–1801), the principal protagonist in these disputes, was a barrister by profession.[3] He had abandoned his chambers

---

[1] J. Priestley, *Historical Account of the Navigable Rivers, Canals and Railways of Great Britain* (1831), pp. 385, 386.

[2] C. Hadfield and G. Biddle, *The Canals of North West England* (1970), i, p. 79.

[3] W.H.G. Bagshawe, *The Bagshawes of Ford: A Biographical Pedigree* (1886), pp. 448–506.

in order to manage four major estates in Yorkshire and Derbyshire which he had inherited from his guardians, William and John Bagshawe, in 1791. Bagshawe felt his responsibilities to the family's inheritance keenly and concerned himself with the day-to-day management of his estates with diligence, aided greatly by his knowledge of the law. He made every effort to improve and extend their acreage wherever possible and to preserve them from the ever-increasing demands of various statutory undertakings. Bagshawe was a kind and good-natured person but, when safeguarding his inheritance, he could be curt and aloof with those whose motives he suspected. With proven marauders, and the Leeds & Liverpool Canal Company were certainly in that category, he was always an unrelenting opponent.

William Bagshawe occasionally lived at Cotes Hall in Barnoldswick until his death in 1785 and had been consulted regularly, by both the canal proprietors and John Longbotham, on the proposed line of the waterway across his lands; relationships then appeared cordial enough.[1] Bagshawe approved of the canal because his estate was one of the nearest, containing limestone, to the Lancashire end of the canal where lime for mortar and tillage was in demand. About that time he had a deep drain made from a limestone outcrop in the fields above Greenberfield Farm with a view to the eventual development of a quarry.

The canal company's plans disclosed that their waterway would pass between the rock outcrop and the highway, hence William Bagshawe insisted on the construction of an 'arched road' under the waterway through which the drain could be extended. There were other advantages to be gained from this in that the underpass would provide access to the fields on his property beyond the canal and to the main highway as well. Longbotham agreed to these measures and wrote to the landowner in September 1770 that: 'After I parted from you recollected that we was under an obligation by promise of making an archway for a road to your lime near Greenberfield. At the next general Committee will lay it before them. I have no doubt but it will be ordered'.[2] The matter rested there until the revival of interest in 1790, a time which approximately coincided with John Bagshawe's inheriting the estates. Bagshawe did not move into Cotes Hall, preferring to continue living in his house at the Oaks, in Norton near Sheffield. He engaged John Lingard, a surveyor and valuer of Astley in Lancashire, to visit Barnoldswick periodically on his behalf in order to collect rents, to keep a watch on the tenant farmers

---

[1] JRUML, Bagshawe Muniments 13/3/451, Leeds & Liverpool Canal Papers.

[2] Ibid., Replication by A. Chambre, York Assizes, July 1798.

and to report local news.

The first formal intimation to Bagshawe of the company's intentions regarding his estates was contained in a memorandum which he received from Robert Whitworth in October 1790. This stated baldly that 'The lengths of the canal through Mr Peter Hartley's Farm will be about 32 chains and through Mr Thornber's about 38 chains. The quantity of land that will be about seven acres'.[1] Soon afterwards, an anxious Hartley at Cotes Flatt Farm sent news of a meeting in Colne on 24 November 1790 where 'The proprietors used the landowners very ill. I have heard they have let the canal to cut through your land but I have sent Mr Whitworth word that he must not put a spaid into the land till it be agreed for. I understand they are not for making a bridge over the canal to Greenberfield, they propose the road to go on the side of the canal'.[2]

The invasion of Barnoldswick village by large numbers of uncouth and aggressive labourers and the establishment of a shanty town nearby[3] must have severely disrupted the rustic calm of that isolated farming community. Then there were the contractors who came to occupy their lands, ruthless and determined to complete their allotted sections in minimum time and at no personal inconvenience. Above them all were the aloof and arrogant proprietors and their officers who, secure in the knowledge of their statutory powers, treated the simple country folk with contempt. Another letter from Peter Hartley tells how his quiet, rural lifestyle had been changed within such a short time by these interlopers and he, and no doubt all his neighbours too, were very worried and troubled by a situation which was beyond their limited experience. He wrote to Bagshawe that

> ... they would have begun a cutting in you[r] estate long since but for my speaking to Mr Whitworth to hould them of. But the[y] certinly will begin a cuting for the[y] are to complete it in such a time. The undertakers and those workmen are fit to stone aneybody for the[y] are the rughest lot of men that ever I did see, for the[y] do not care how much dameg the[y] do to aney bodey ....[4]

Bagshawe now began to take a more studied interest in the behaviour of

---

[1] JRUML, Bagshawe Muniments 8/4/1264, R. Whitworth to John Bagshawe, – Oct. 1790.

[2] Ibid., 8/4/1267, P. Hartley to J. Bagshawe, 26 Nov. 1790.

[3] Private correspondence with Mrs I. Wilkinson, Gilliats, Barnoldswick.

[4] JRUML, Bagshawe Muniments 8/4/1270, P. Hartley to J. Bagshawe, 28 Dec. 1790.

the canal company. His judgement was no doubt coloured by reports from Lingard, as well as by the gossip he was getting from Hartley regarding these unexpected proposals for highway diversions across his land. The scheme, the details of which gradually emerged over the months to follow, was neither discussed, nor agreed, with Bagshawe and Major Farrand, his immediate neighbour. They were to be presented with a *fait accompli* and were obviously expected to like it.

Eventually there came an urgent request from the company to resolve the matter of land purchase for the line of the canal. Bagshawe had deliberately delayed in spite of several reminders and the company were held up as a result. Matthew Oddie, the surveyor for the sector, then wrote, '… hitherto have been fortunate enough to avoid having occasion to call any Commissioners of Jury and we cannot stand still without great injury to the undertakers and workmen as well as inconvenience to the Proprietors and the Public'.[1]

With this it seemed that Bagshawe approached Joseph Outram, who had advised him previously on estate management matters, to negotiate on land valuations. Major Clayton, valuer for the canal company, had carried out an accurate assessment; every field had been itemised, his valuation per acre varying from as little as 10s. for 'highest point Bowker' to 60s. for 'Hall Crofts No. 1'.[2] Twenty-eight years' purchase (i.e. 28 times the annual rent) was offered for the land, which came to £375 for 4.4 acres. As might be expected, Outram argued for £412 10s., based on thirty years' purchase.[3] The company refused to negotiate over the trifling difference of £37 10s. and even threatened to take the matter to a Commission.[4] This was a surprising reaction because they were anxious to begin the earthworks without further delay. In any case amicable agreements had been reached with adjacent landowners and both Bagshawe and Outram were aware of this. The dispute was only resolved several months later, after much acrimonious correspondence, by halving the difference of the two valuations. It also set the scene for an increasing animosity between Bagshawe and the company and in particular with Hardy, solicitor to the company, whose arrogance and sarcastic tongue frequently got the better of him. As a senior officer of the company, he should have done more to avoid the disputes which were to follow, with all the associated wastage of time and needless

---

[1] Ibid., 8/4/1273, M. Oddie to J. Bagshawe, 8 Jan. 1791.

[2] PRO, RAIL 846/4, 24 Dec. 1790.

[3] JRUML, Bagshawe Muniments, 8/4/1273, Valuation by J. Outram, – Jan. 1791.

[4] Ibid., 8/4/1277, J. Bagshawe to P. Hartley, 12 Feb. 1791.

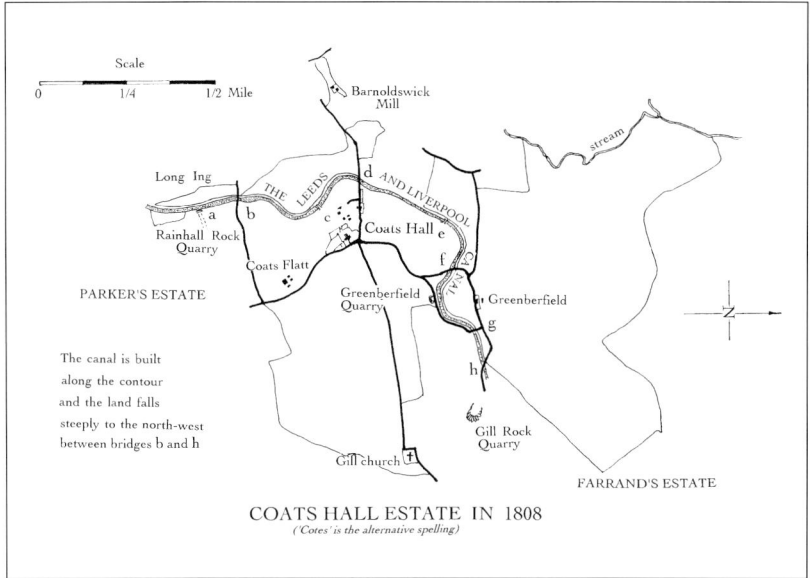

Figure 24 The Coats Hall Estate (as the name was then spelt) in 1808.

expense, but he invariably preferred to do otherwise.

It eventually transpired that the projected accommodation works were quite different from those tentatively agreed with William Bagshawe. The Cotes estate was split into several parts (Fig. 24) by the canal line, and the highways were severed at points (b), (d), (f) and (h).[1] The company intended to build bridges at (b), (d) and (h) and yet refused to build another at (f) in the 'ancient highway', preferring instead to divert the old road some 600 yards along the east side of the canal before crossing again by an overbridge at (h). After a good deal of complaining by Major Farrand, a bridge was built at (g) but Bagshawe was never consulted even though, or perhaps because, his lands were the more seriously affected. A request for accommodation bridges at Eastwood (c) and the Banks (e) was refused, even though an important barn was located on the lower side of the canal at this latter point. The company rejected the earlier agreement for the underpass and drain from the rock outcrop at Greenberfield and in fact the new road diversion crossed over part of the intended quarry site. These plans were passed on to Bagshawe piecemeal, mainly by word of mouth, by his tenants; certainly no written descriptions or drawings of the proposed works were ever presented to

---

[1] Sheffield Archives, OD 1239, Plan of Cotes Hall Estate by Fairbank (1808).

the landowner.

Money for capital projects was in short supply in the early 1790s, as the country was faced with a succession of financial crises. Probably because of this, the policy of the canal company was to eliminate every possible accommodation work. Accordingly the engineering staff set about minimising the number of bridges they were statutorily obliged to erect. A bridge at (f) would be expensive because the steep, sidelong ground would necessitate substantial earthworks. They assumed, therefore, that the public highway could be stopped up at that point and replaced with a diversion. The cost of the latter, which could be made up of limestone excavated from the adjacent canal cuttings, was less than that of a bridge on the original line of the highway. It did not concern them that the diversion would be long and inconvenient for users, nor did it seem to matter that the new road crossed private land. These factors were other people's misfortunes.

When Bagshawe heard rumours of the road diversion he probably began to wonder at the company's motives for their actions. He believed that his land at Greenberfield had the best quality limestone in the district, better even than Major Farrand's successful Gill Rock Quarry, recently opened to the north of his own estate boundaries. Were the company's actions a means of obstructing his efforts to develop a new quarry? In later years Bagshawe openly voiced that opinion, particularly when the news broke that the company had purchased land from Mr Parker at Rain Hall for quarrying purposes. In retrospect, however, that does not seem to have been the company's intention. They were short of capital throughout that phase of the works and were simply out to save as much money as possible. And yet it is not to their credit that these savings must have been used to purchase land from Parker at a cost of 1,000 guineas. These actions, which were in any event probably illegal under their Act, at least secured much-needed revenue for financing what was proving to be a difficult and expensive sector. To that extent their intentions, but not their means, were laudable.

Bagshawe made two visits to Cotes Hall in 1792, on both occasions in the company of Joseph Outram. The September visit was the more significant; Outram assessed and agreed 'damages to land' with Matthew Oddie,[1] but he may have been involved with engineering matters too, because unspecified advice had been sought from his son Benjamin and John Sutcliffe, the Yorkshire engineer, with whom they held a meeting on site. These matters must have been resolved satisfactorily and Joseph

---

[1] JRUML, Bagshawe Muniments, 8/4/2784, J. Outram to J. Bagshawe, – Sept. 1792.

Outram later submitted his account for £21 15s. for the 'whole seventeen days out'.[1]

The first skirmishes over, there began an uneasy peace until early in 1794. Bagshawe did not visit Cotes Hall during this period and the lack of correspondence implies that the company were peacefully pursuing their way; but the disputes were only just beginning. John Bolton, a solicitor from Colne, had been invited to act on Bagshawe's behalf in February of that year.[2] Some time afterwards he rode over to Barnoldswick and what he saw there was enough to bring Bagshawe over from Sheffield in haste. The company, without any warning or formal agreement, had entered the estate at Greenberfield, stopped up the old highway and diverted it across several hundred yards of land alongside the partially finished canal. Bolton was told that the new bridge (at g) would also suffice as an occupation bridge for the Farrand estate. Bagshawe was astonished at this audacity and, as soon as he arrived at Barnoldswick, had an angry exchange on site with Whitworth and Oddie, strongly objecting to the road alignment 'between his Lime Rock and the Canal at Greenberfield'.[3]

The following week Bagshawe, accompanied by Joseph Outram, again met Robert Whitworth on the banks of the canal and repeated his objections. The engineer was accompanied by several proprietors of the company and some site staff, and placatory offers were made to Bagshawe. A bridge could be made at the point where the ancient highway crossed the canal and the road could be turned along the canal on the far side, by which means the new quarry would be left unobstructed and open to the canal. Alternatively they offered to build a bridge which could take a side-cut from the canal under the diverted highway then built, into the new quarry workings.

Bagshawe was amenable to these reasonable offers but at a further meeting next day it appeared that there had been second thoughts and the committee would only 'consider about the road and give Mr Bagshawe an answer'.[4] Feelings ran high following this vague promise; Bagshawe stated that the agents should not have diverted the road and so divided his fields without first inviting his agreement. There was an immediate counter-claim that his leave had been asked for and obtained but, when he asked them to trace the agreement, nothing could be found in writing.

---

[1] Ibid., 8/4/2787, J. Outram to J. Bagshawe, 19 Nov. 1792.
[2] Ibid., 8/4/942, J. Bolton to J. Bagshawe, 17 Feb. 1794.
[3] Ibid., 13/3/451, Replication by A. Chambre; report of a meeting on 28 May 1794.
[4] Ibid., Report of a meeting on 6 June 1794.

The agents then referred from one to another 'until at last it was fixed upon Mr Whitworth who said that Mr Outram had given leave on behalf of Mr Bagshawe'. Joseph Outram utterly denied the allegation, after which the eminent Whitworth weakly said 'he thought he had granted leave'. There was no doubt about it; the committee and senior staff were embarrassed and they were in a mood to compromise. At that juncture Bagshawe should have been magnaminous, gone for a fair agreement on all his accommodation works, and that would have been the end of it. Instead, he enjoyed their discomfiture and he elected to turn the screw. Bolton tried to placate his client but Bagshawe was in no mood to listen and he went on to instruct the lawyer to write to Joseph Priestley, clerk to the canal company, to say that '... no accommodation the company can contrive, can remove the objection to the road passing between the Canal and Greenberfield Lime Rock ... '.[1]

Shortly afterwards Bagshawe again wrote to Bolton:

> ... I believe Mr Outram's son, who is an Engineer rising into fame, will be at no great distance from Colne very shortly. If you think any great advantage can be derived from his going over the line of the Canal with you, I mean in considering of the practicability of building bridges or in any other respect whatever, I will desire him to call upon you. [although] if he goes there perhaps he would wish it should not be made publick ... .[2]

This was a curious statement; perhaps he realised that Benjamin Outram, then 'rising into fame', was associated with several canal companies at the time and could not afford to make enemies at that stage of his career. There were signs that Bagshawe, although unrepentant, was rather unsure as to his next step and in fact it is difficult to deduce just what he did want of the company. About one matter he was sure; he was well versed in the law and always abided by its dictates. He accepted his obligations under the canal Act without question but he also knew that the law was intended for all, hence he could never understand why the company did not as readily observe their obligations to others. Nevertheless, in spite of his contempt for the company's practices, he could still enquire of Bolton on the availability of their shares; rather transparently these were 'for a friend'.

As regards his accommodation works, it seems likely that Bagshawe

---

[1] Ibid., 13/3/451, J. Bolton to J. Priestley, 8 June 1794.

[2] Ibid., 8/4/957, J. Bagshawe to J. Bolton, 25 June 1794.

would have been satisfied with unchanged public highway routes and bridges built at appropriate crossing points. He also wanted occupation bridges at Eastwood (c) and the Banks (e), as well as certain drains and minor works such as cattle-watering places and so on. His rights to an underpass at Greenberfield would have been waived and no doubt he would have agreed to sell a severed triangle of his land at the Long Ing, a field coveted by the company after their purchases at Rain Hall. He was something of an entrepreneur at heart and he knew that the canal would be a good thing for his estate and a splendid outlet for his quarry.

He did not have long to wait in his uncertainty. Bolton was soon to be approached by Oddie in an effort to resolve the matter. The committee had agreed it was reasonable for Bagshawe to have his accommodation bridges but 'wished to know what sum if any, you would be willing to accept to waive your claim to the Company making both, or either of these bridges, and for your consent to the Company completing, and the public enjoying, the road near Greenberfield'.[1] Bolton told them that their propositions were unlikely to be successful and they offered an alternative: a bridge could be built a little nearer to the original crossing of the old highway, hence the road diversion would be mainly along the west bank. Nevertheless, they were adamant that the bridges at (e) and (f) would not be constructed. This ill-considered scheme did not appeal to Bagshawe and so there followed by another period of stalemate.

Some progress was made on other matters relating to the canal. The company were obliged, under the powers of their Act, to receive requisitions for damages and minor accommodation works and Bagshawe submitted a list which included the making and repairing of 'fences' (i.e. dry-stone walls), drains, the sough under the canal to drain the proposed Greenberfield quarry, rubbish removal and repairs to estate roads.[2] The bridges and diversion were again alluded to, as well as the limestone excavated by the company whilst digging the canal and taken for bridge- and wall-building elsewhere. The company agreed to most of the requests except the sough, bridges and diversion. They also refused to reimburse the estate the cost of the limestone excavated, even though their Act stipulated that such excavated materials should remain the property of the owner and left in a convenient place for his usage. They were as petty-minded and intransigent as ever.

On another issue the company was co-operative; Bagshawe was granted the costs of a fence on the edge of their towpath across his land

---

[1] Ibid., 8/4/952, J. Bolton to J. Bagshawe, 20 June 1794.

[2] Ibid., 8/4/953, Requisition delivered to M. Oddie, – May 1794.

but then, in an effort to reduce costs, they offered the return of the land forming the steep canal banks for 'herbage' on the terms that that they would thereby only pay ten years' purchase, or one-third of the agreed price.[1] This posed legal problems regarding bank maintenance but, as can be seen on a site visit today, Bagshawe decided to retain long sections of the banks in this manner and probably did so on the advice of Joseph and Benjamin Outram. The latter drew attention to the financial advantages of retaining the banks for building wharves along them at some future date.

Bolton sounded out the company on the compensation it was willing to pay for not building bridges, suggesting £100 for each structure, but this was rejected. The company refused to bargain, claiming this was twice the proven cost of similar bridges, even though Bolton knew that they had paid others £80 in similar circumstances, and he had learned from an earlier conversation with Whitworth that this was not an unreasonable price. Bolton made interminable visits to meet the proprietors and the engineering staff on this and associated matters but he got nowhere; the company would constantly make proposals which were later withdrawn, all to the ultimate exasperation of the solicitor.[2] Small wonder that eventually the entire matter was to explode into a major but completely avoidable quarrel, from which nobody seemed to benefit. After further arguments about the removal of rubbish from the side of the canal near Cotes Hall, Bagshawe wrote to Bolton to say that 'it is evident that the company trifle with us. I shall put myself to no further trouble or expense'.[3] He then instructed his solicitor to give notice of a Commission, as was his right under the Act, 'to get the business of the bridges settled. I mean the occupation bridges for as to anything that related to them not building publick bridges or having diverted publick roads we will for the present be silent'. Meanwhile a notice was served on Oddie discharging him and the workmen from the diverted road and instructing him to remove rubbish from the estate.[4]

Another eighteen months passed but Bagshawe could have had few complaints about the amount of information forwarded to him in a stream of letters sent by the industrious Bolton. The unfortunate contractors came in for constant criticism over what appeared to the

---

[1] Ibid., 8/4/955, J. Bolton to J. Bagshawe, 21 July 1794, with letter from J. Priestley, 19 July 1794.

[2] Ibid., 8/4/956, J. Bolton to J. Bagshawe, 27 July 1794.

[3] Ibid., 8/4/963, J. Bagshawe to J. Bolton, 16 Sept. 1794.

[4] Ibid., 9/4/969, Copy of notice served on M. Oddie (nd, probably Oct. 1794).

solicitor as rubbish-strewn, untidy workings; half-finished bridges and dry-stone walls; unsoiled, unseeded banks and the like. He showed a complete lack of understanding of the nature and sequence of engineering operations and failed to realise that, to the uninitiated, construction sites always appear an ill-managed shambles until all is complete and finished as the designer intended. The picture of unremitting gloom thus presented by Bolton could only lead his client to further obstinacy: 'if nothing has been done respecting the walls, be pleased to proceed against the Canal Company immediately. This business must now be settled and if I be wrong I must take the consequences'.[1]

As work across the estate progressed to a peak of activity, the number of real or imagined injuries of one sort or another mounted until June 1797, when Bagshawe instructed Joseph Outram to write to the company to tell them that the diversion would be stopped up and destroyed unless agreement had been reached.[2] Evidently the company had just finished the Greenberfield bridge (g) and were openly ignoring the earlier trespass notice served on them. The temporary diversion was then becoming a permanent public highway and the answering letter from Hardy was hardly couched in reassuring terms: 'we are ignorant of the particular cause of the complaint but suppose it may be the road near Greenberfield'.[3]

Subsequently, stern correspondence from Bagshawe and sarcastic responses from Hardy caused further deterioration in their relationship; meanwhile the worthy Bolton tried to press his client to come to some understanding with the company and eventually Bagshawe agreed, but only on the most unusual terms. A meeting was arranged for 11 November 1797 which, hilarious from today's standpoint, was conducted in deadly earnest by the participants. The principal officers of the company, including Priestley, Hardy and Whitworth, congregated at the inn in Barnoldswick whilst Bagshawe established himself in Cotes Hall. Business was conducted through written notes carried by messenger and the sequence of events and summaries were accurately minuted.[4] It does not seem that the parties ever met face to face, even to pass the time of day. Although this gathering was intended to resolve all their differences quickly and conclusively, it soon turned into a complete deadlock. Hardy was on his worst behaviour; good sense was never in evidence and the

---

[1] Ibid., 8/4/1004, J. Bagshawe to J. Bolton, 19 Aug. 1796.
[2] Ibid., 8/4/1017, Copy of letter in J. Outram's hand to canal company, 12 June 1797.
[3] Ibid., 8/4/1018, J. Hardy to J. Bolton, 28 June 1797.
[4] Ibid., 8/4/1333, Minutes of 'Notes and Answers', 11 Nov. 1797.

business was conducted by him on a spiteful, personal level. His colleagues had been too closely involved during the past eight years to view the disagreements objectively and they were similarly unable to act constructively.

'Note 1' from Bagshawe proposed that 'the Company measure up the land required for the Canal and pay the agreed sum less that already received'. The 'Answer' was that Mr Oddie would meet Mr Bagshawe so that they together could measure up, but that there should be 'a reasonable time allowed to take Counsel's opinion upon title and prepare conveyance. Mr B. to furnish abstract of his title'. Bagshawe, not surprisingly, was annoyed by that rejoinder and pointed out that there had been plenty of time in the past eight years for Hardy to check on the title. He also stated that there was no point in measuring land 'until the company are prepared to pay for it'. In any case, if Hardy had not been satisfied over title, why had he previously paid a substantial sum on account? And so it went on; written communications passed to and fro throughout the day and never once was anything but the title to the land discussed. Such phrases from Hardy as 'If Mr Bagshawe is determined to go to Commissioners, they desire he will have the goodness to say so and put an end to this useless correspondence', and from Bagshawe, 'with respect to his own conduct it has been no further hostile towards the Company than to vindicate his own property and to prevent his rights being attacked with impunity', can have done little to improve the atmosphere. At the end of the day the tired and frustrated opponents had not repaired the breach. Bagshawe was clearly winning this battle of wits, no matter its negative outcome, and the despair of the officers can surely be felt in the last note, when they wrote that 'Messrs Priestly and Hardy desire to know whether they are to wait for Mr Bagshawe's answer and beg to observe that they are a long way from home'.

Feelings must have run high at the village inn that day and some unguarded remarks, overheard in the taproom, were passed on to the landlord at the Hall. Those were not recorded but they certainly resulted in the angry letter, barely intelligible in places, sent by Bagshawe to Hardy and Priestley the day following that extraordinary affair. Robert Whitworth came in for an haranguing for some choice comments made from behind his tankard: 'My conduct is under my own immediate protection nor shall anyone attempt to reflect upon it without notice. Indeed with respect to Mr Whitworth one of the servants of the Company, who unwarrantly in publick houses thinks proper to display his malevolence towards myself without rectifying'.[1] Bagshawe went on to note that the agents had always acted in an individual capacity in attacking his property, without coming to an agreement collectively beforehand. Always the lawyer, he had for a long time deplored the improper, bullying methods adopted by the company, their agents and contractors. It was strange that Hardy, himself an able solicitor, had tolerated and, on occasion, even instigated the deplorable behaviour of the staff. The suspicions remain that, far too often, the company's intentions were not entirely honest.

The arguments continued but meanwhile a succession of minor battles was taking place on the diverted highway. The new tenants at Greenberfield Farm, John Waite and his sons, set about obstructing the road with gates and brushwood. The agents retaliated and as quickly as these were erected they were dismantled. The public must have found travel a difficult business just then but no reference was made as to their plight. Presumably they found alternative routes and the opponents were left to fight their battles alone.

However, in April 1798 a writ of inquiry for damages was executed by Bagshawe at York Assizes before Mr Justice Rooke. An abstract of the proceedings stated the pleas in the case of John Waite and his son against company servants George Smith and John Starkie, and reference was made to the history of the conflicts and to the fact that the defendants had pulled down gates on the orders of the company.[2] The question at issue was whether or not the highway was a public highway and the statement referred to the verdict: 'If the Court should be of the opinion that the said road is a public highway then the verdict to stand, if not a verdict to be entered for the Plaintiffs with one shilling damage'. Bagshawe was disappointed by what turned out to be a partial success and he was to write to one of the estate's trustees that:

> but as I know myself to have been very ill used in the business, it should be both ridiculous as well as discreditable to despond as to the future success. The petulancy of the Judge Rook, who by the way is but an old woman, in finding he had an additional Writ of Inquiry to execute was the foundation of the quantum of damages given me which is directly contrary to the evidence. Hardy has put an account in the Leeds and Blackburn papers in a

---

[1] Ibid., 8/4/1333, J. Bagshawe to J. Hardy and J. Priestley, 11 Nov. 1797.

[2] Ibid., 8/4/1351, Abstract of proceedings at York Assizes in King's Bench.

very sarcastic manner.[1]

Hardy had certainly done this and had introduced a paragraph into that factual account which said that the 'action brought about against the Company was a frivolous one'.[2] Not to be outdone, Bagshawe wrote to the Leeds paper asking that 'no one form an opinion as to its merits until a second action had been determined resting nearly upon the same grounds as the former and which will be tried at the next York Assizes'. He then promised to publish a statement of the matters in contention: 'for the benefit of such persons who hereinafter unfortunately have concerns with that body of men from which abatements they will be enabled to estimate the degree of candour they are likely to experience on the part of the Company, or what reliance is to be placed on the assurances that may be given by some of their agents'.[3]

The company persisted in their campaign, in spite of the law, and set about dismantling the road obstructions yet again. John Waite wrote in desperation from Greenberfield that

> ... on Saturday last, being 7th day of April, I sent two of my sons and two other men with them to wall the walls and set the stoops and hing[e] the gates at both ends in the old place where it was before. The[y] came home for something to stapple the gates with. In the meantime the walls, gates and stoops was pulled down and laid by the way side when we went to the place again. We heard that said night that there was sixteen of the Canell Company's men came and pulled it down whilst we unloaded the cart that we had not the opportunity of seeing them. And one of the masters was with them. The Canell people say the[y] will pull it down as often as we make it fast.[4]

These building and dismantling actions were to be repeated twenty times in all and the last occasion must have been a sight to behold. John Waite the younger 'stuck to the gate till it became necesssary to remove him by force'.[5] In fact, a large gang of rowdy labourers from the cuttings dismantled the gate with Waite hanging grimly on to it and threw him,

---

[1] Ibid., 8/4/2096, J. Bagshawe to Micah Hall, Trustee, 6 April 1798.
[2] Ibid., 8/4/1344, J. Bagshawe to *Leeds Intelligencer*, 6 April 1798.
[3] Ibid.
[4] Ibid., 8/4/1345, J. Waite to J. Bagshawe, 9 April 1798.
[5] Ibid., 13/4/451, Replication by A. Chambre; report of assault.

gate and all, to the wayside.

Their glory was shortlived, however, as Bagshawe got to grips with the problems that beset him. He had many sympathisers and one of them, his old agent John Lingard, wrote 'I am sorry that you have these struggles with these people but, you know, they are not spending their own property'.[1] A pertinent remark indeed; unlike the company, Bagshawe was using his own funds and nothing was spared in summoning some of the best men in the business to his aid.

There were two issues to be resolved, namely, the illegal diversion of the road, which was to be referred yet again to the York Assizes, and also the bridges and other accommodation works which had to be considered by a Commission, as laid down by the company's Act.[2] The latter was set in motion by a letter from Bolton to Hardy which advised that he was to proceed immediately to call a meeting of commissioners for enforcing the making of occupation bridges, unless the company agreed to reasonable compensation 'without prejudice to any other matters in dispute'.[3] He asked for 100 guineas in lieu of a bridge 'in the crofts' and another 'in the Banks', besides access along the banks of the canal from Rain Hall Quarry and the severed estate lands, to Greenberfield locks.

The company responded and their comments were passed on to Joseph Outram for consideration. Bagshawe wrote to Bolton that 'Mr Outram is a very sensible man and understands the subject he writes upon much better than I do. He has a son of considerable note as an engineer who will give him any information that may be wanted'.[4] Thus Benjamin Outram, who was then busily engaged on the Manchester narrow canals, was called in to report, albeit with some reluctance, and to support his father and Bagshawe in what he realised was a trivial and grossly exaggerated affair. The two Outrams, both of them prickly characters at the best of times, were not always on the friendliest of terms. That Benjamin was wary of his father's uncertain temper is fairly evident from the correspondence that ensued. They had arranged to meet John Bolton for breakfast in Colne on 1 October, but on 24 September he had to postpone the arrangements because of an important meeting he had with the proprietors of the Ashton Canal. Joseph was indeed irritated

---

[1] Ibid., 8/2/2248, J. Lingard to J. Bagshawe, 9 April 1798.

[2] 10 Geo. III c. cxiv (1770).

[3] JRUML, Bagshawe Muniments, 8/4/1347, J. Bolton to J. Hardy, 10 Aug. 1798; 8/4/1070, J. Hardy to J. Bolton, 13 Aug. 1798.

[4] Ibid., 8/4/1071, J. Bagshawe to J. Bolton, 23 Aug. 1798.

by this and wrote to Bagshawe that 'You see by this letter how you may be acted upon when you are a Father'.[1] The Outrams met shortly afterwards, however, and rode over to Barnoldswick in company with Bolton.

Benjamin's response to Bagshawe's detailed list of instructions is contained in his manuscript report entitled 'Observations on Mr Bagshawe's Estate Upon the Banks of the Leeds and Liverpool Canal in Craven'.[2] This document describes in detail the state of the canal workings across the property and also evaluates each of the established and proposed quarries as regards quality and viability. Outram was evidently conversant with the elements of geology and knew a good deal about quarrying and burning limestone. The different grades of rock and their suitability for various purposes were described and costed in a totally disinterested manner. He seemed, quite rightly, oblivious to Bagshawe's biased instructions and not all his remarks would have pleased his client. Benjamin Outram appears to have maintained a restrained and impartial attitude throughout his involvement with these estate matters. Probably he did not wish to become too deeply involved in a business which might well affect his future as a canal engineer and yet wanted on the other hand to support and help his father and the stubborn but kindly John Bagshawe who by now had become a family friend.

In his forecast of output he noted that the company's new quarry at Rain Hall would yield 300,000 tons at an expense of 6d. to 7d. per ton. But:

> ... the cost by purchase of land and expence of making the side cut will be nearly 2d. per ton, exclusive of the interest of more than 2000£ capital sunk in such purchase and expence. So that at the present selling price of 11d. per ton, the company can't gain more than will repay their capital and interest and expences of agencies.

Outram did not seem satisfied with the quality of stone found nearby in Bagshawe's estate and implied that it had only a limited use as a roadstone. That at Greenberfield, however, was approved of and he gave some practical advice on its exploitation. A good deal of the report refers to the future of the local quarrying industry and its outlets via the canal.

---

[1] Ibid., 8/4/2881, J. Outram to J. Bagshawe, 28 Sept. 1798.

[2] Ibid., 13/3/451, Legal and other papers: Mr Outram's report, 19 Oct. 1798.

A most interesting document, it nevertheless did not refer to the important matter of the occupation bridges on which Bagshawe needed advice. At the latter's subsequent request, Benjamin Outram wrote to his father regarding costs and although, he avoided controversy, he confirmed that the occupation bridges

> ... done in the rough manner that several are which we passed, would cost from 60£ to 70£ each bridge, exclusive of the earth banking and rough stone to raise and form the roads to approach them, which I think in the one situation would cost 30£ and in the other 50£, or upwards, on the supposition they are for occupation only.[1]

Bagshawe's estimates of 100 guineas were seemingly not too far from the correct value after all. However, Benjamin concluded by saying: 'I wish you may settle matters with the Committee amicably ... I hope by conversing with Mr Hardy and some of their moderate men you will settle matters on a fair and liberal footing'. But it was not to be; his very words had confirmed Bagshawe's views and disproved the company's estimates. The negotiations with Hardy were abandoned and the matter would go to a Commission after all.

On the issue of the diverted highway, matters were already proceeding and the case was heard at the York Assizes of autumn 1798. There was no underestimating the opposition on this occasion, and the elaborate case prepared by counsel, A. Chambre, was indicative of these efforts. In defending the construction of the diverted road, the company contended that this was a more convenient and commodious route and sought to justify it under the General Highway Act 'as having been diverted for twelve months and been acquiesced in'.[2] Chambre pulled few punches, pointing out that the canal Act permitted the company to divert roads,[3] so that there was no justification for introducing clauses from an entirely different enactment. Moreover, their Act clearly stated that 'the Company are directed not to make the Canal across any public highway until they have made at their own expence a proper bridge over or a convenient road under the Canal, to be kept in repair by them'. Indeed, 'the present Accountant General of the Court of Chancery, who had, by experience, learnt how shamefully they used the public', had

---

[1] Ibid., 8/4/2885, B. Outram to J. Outram, 4 Nov. 1798.
[2] 13 Geo. III c. LXXVIII (1773).
[3] 10 Geo. III c. cxiv (1770).

applied for the inclusion of further regulations in the company's two subsequent Acts.[1]

In a telling statement, Chambre turned the court's attention to the company's purchase of Rain Hall. He submitted that their Act enabled the company only to apply their funds to the cutting of a canal

> just as with as much propriety and reason, as it is conceived might the Trustees of a public Turnpike Road apply money borrowed for the purpose of making it, in setting up a set of teams and waggons to travel upon it, in opposition to other carriers, as this Company thus apply their money borrowed for the express purpose of completing the Canal, in schemes and trading speculations upon it to the ruin of other competitors. But they do at a great loss.

The case drew steadily to a conclusion and a verdict in favour of Bagshawe, who wrote to a trustee of his estates that 'yesterday the Court of the King's Bench determined unanimously in my favour and reprobated in very strong terms the conduct of my opponents which they said had been arbitrary, vexatious and oppressive. Lord Kenyon said further that it was a Job'.[2] His lordship also observed that 'the inconvenience of having public roads made over private property was obvious to everyone. It would be giving tenants an opportunity by collusion of injuring their landlords'.[3] Such was the attitude of the landowning classes.

The Commission met in November 1798. Their decision was not entirely satisfactory for the estate's well-being, but at least it confirmed that Bagshawe should have his two occupation bridges after all, as well as various minor works.[4] And yet, in spite of their defeat, the company still wrangled over such matters and the unfortunate Bolton had to bear the brunt of the struggles. That he was discouraged is shown by a letter to Bagshawe in which he stated that he was

> unhappy about your concerns. Mr Hardy has at last condescended to favour me with a reply, or rather the receipt of it. I feel myself very inadequate to the exertion and more than ordinary care and

---

[1] 23 Geo. III c. xlvii (1783); 30 Geo. III c. lxv (1790).

[2] JRUML, Bagshawe Muniments, 8/4/2100, J. Bagshawe to Micah Hall, Trustee, 2 Feb. 1799.

[3] *Leeds Intelligencer*, 7 April 1799.

[4] JRUML, Bagshawe Muniments, 8/4/1084, Proceedings of the Commission, 9 Nov. 1798.

alertness which are necessary in all proceedings with them. In fact a person who contends with them should have nothing else to do.[1]

There was a further Commission held early in 1799 when Thornber, the former tenant at Greenberfield Farm, was awarded damages. Bolton wrote again to say that

> the Company's people behaved I think, worse than ever, especially Hardy who I now think is one of the greatest brutes in existence. I must confess I had a very different opinion of him ... They were completely defeated. The Commissioners awarded Thornber more for every specific damage which they determined he had asked.[2]

Congratulations came from all directions; Bagshawe was obviously well-liked and his tenants and professional advisers all sent letters of evident goodwill, including from John Waite, his loyal tenant at Greenberfield Farm who wrote:

> I am very glad what a complete victory you have gained over the Canal Company. We have stopped the road according to Mr Bolton's directions and no persons have attempted to break it open ... I do not hear that the Agents of the Co. say anything in consequence of their defeat the Company have taken an Estimate for a Bridge in the old Road. My son John keeps the gate locked near Coats Ford leading to Gill Church ... The lime rocks at Greenberfield turns out very well and is equal in Quality with Major Farrands or the Companys or better and hath delivered three boat loads and as many more gotten but for the snow and watter that the workmen could not work.[3]

In spite of their reluctance to comply with their legal obligations, the company had no option but to build the occupation bridges and other works at Cotes Hall, although it seems likely that they were able to avoid bridging the canal for the old highway after all. Local hearsay suggests that a swivel bridge was built and survived until the twentieth century. Whilst it is true that an order was made by Quarter Sessions at

---

[1] JRUML, Bagshawe Muniments, 8/4/1116, J. Bolton to J. Bagshawe, 17 Oct. 1799.

[2] Ibid., 8/4/1097, J. Bolton to J. Bagshawe, 8 April 1799.

[3] Ibid., 8/4/1357, J. Waite to J. Bagshawe, 18 March 1799.

Pontefract in April 1799, for a 'Wooden or Swivel Bridge' with ramparts 'of an ascent of four inches and a half a yard in length and no more',[1] the evidence to be seen on the ground today suggests otherwise.

John Bagshawe had staunchly defended his inheritance, and had conscientiously developed and extended his estates. These actions did not secure his personal fortune nor his health, which had been anything but robust for some years past, and he died of consumption in August 1801.[2] His brother, the Revd William Bagshawe, was the executor of his will, but the conflicts which had prevailed on the estates at Cotes Hall for so long were of little importance to him, and it seems likely that renewed machinations by the company at this critical time resulted in his agreeing to the retention of the highway diversion. Documentary evidence has not survived but the diversion remains, linked with the structure which also was to serve as Major Farrand's occupation bridge. The line of the old highway can still be seen leading to the banks of the canal but a bridge, if ever built, has long since gone. Thus it seems that the canal company was able to defeat their old adversary, even though they had to await his death before succeeding.

In retrospect, the conflicts at Cotes Hall may seem trivial, pointless affairs which reflected little credit on most of the people involved, but such a view overlooks the limited business experience, the resources and attitudes of those days. To Bagshawe, his lands represented the family's future and to him there was a clear duty to defend the estate, always using the correct legal procedures, against all-comers. There can be little sympathy, on the other hand, for the canal proprietors and their officers. The forerunners of a new industrial race, they were thrust into situations of which they had little experience or understanding. They were men in a hurry, secure in the supposedly unimpeachable authority of their Acts of Parliament. Of human relationships, of compromise, they knew or cared little and the clauses and powers of the statutes were twisted and abused, mainly because they were dogged with problems of management and an acute shortage of capital. They nevertheless completed their great enterprise successfully and so much is to their credit. Their methods would not be replicated nowadays but of course the practices of today were learned only after many years of experiences not dissimilar to those disputes which involved John Bagshawe with the Leeds & Liverpool Canal Company.

---

[1] WYASW, West Riding Quarter Sessions Order Book, 1 April 1799.
[2] Bagshawe, *Bagshawes of Ford*, pp. 448–506.

CHAPTER 11

BENJAMIN OUTRAM & CO.
THE IRONWORKS AT BUTTERLEY

In his report on proposals for the Cromford Canal, William Jessop wrote that 'I know of no better Situation for establishing considerable Iron Works; as there appears to be great plenty of Iron-stone; which with the Coal and Lime-stone for fluxing, may be brought together by the Canal, to the Banks of the Derwent; where powerful Falls of Water may be procured, and Machinery conveniently constructed'.[1]

It was common knowledge, as a result of many years of local prospecting and mining, that the region was rich in mineral deposits and Benjamin Outram, when presenting his evidence in Parliament in support of the Cromford Canal Bill, knew all about it.[2] He claimed that the canal would open a communication with 'a great Number of Mines of Coal' at its eastern end, quite apart from numerous veins of iron ore in a district of some 20,000 acres. He also stated, without going into much detail, that there was a great number of lead mines which would likewise benefit from a canal and stressed that 'there is an inexhaustable Quantity of Limestone at the West End of the Line; a great variety of Marble and some of the best Building Stone I ever saw'.

Neither engineer was then thinking much beyond the task of building the waterway but minds perhaps changed during the driving of the Butterley tunnel. For example, Joseph Outram's letter to Philip Gell in January 1790 gives rise to the suspicion that his son was beginning to take a more circumspect view of the potential mineral wealth of the district with the remark that 'the undertakers wear a favourable opinion of the measures in the Tunnel and are in good spirits ... an excellent Prospect of Coal ...They have discovered Coal at the West End of the Tunnel by Boreing under their deep cutting'.[3] More detailed evidence of the minerals discovered during the excavations of the tunnel and pits was provided by a longitudinal tunnel section which was plotted, probably by the geologist William Smith, and retained by William

---

[1] DRO, D258/41/32q, Report of Mr Jessop, engineer, on a design for a canal from Langley Bridge to Cromford, 31 Dec. 1788.
[2] HLRO, Evidence on Cromford Canal Bill, 1 July 1789.
[3] DRO, D258/50/15p, J. Outram to P. Gell, 10 Jan. 1790.

Jessop.[1] Several coal seams, from 12 to 30 in. in thickness, had been located. The dip of these measures varied between 1 in 8 and a maximum of 1 in 1 and there were several outcrops and frequent faults. These factors did not promise easy or economic mining but the discoveries nevertheless generated much interest.

It was a fortunate coincidence that the Butterley Hall estate, the 200 acres of which crossed the line of the tunnel, was put up for sale in 1790 by Edward Thomas Warren Horne.[2] Shallow coal-pits had been worked on the estate for many years and it would have taken little imagination to assess the future prospects of deep mining for iron ore and coal, quite apart from the advantage of the property being so close to the line of the Cromford Canal. No doubt Benjamin Outram, then only 26 years of age, recognised this splendid opportunity of a lifetime. In spite of his lack of capital, his already proven ability as an engineer and entrepreneur must have readily persuaded the influential and prosperous Francis Beresford, clerk to the canal company, to purchase the freehold of the estate. He then leased a moiety to the impecunious Outram until such time that the young engineer's funds were sufficient to purchase 50 per cent of the holdings; thus Benjamin Outram & Co. was established by June 1790.[3] Benjamin Outram acted as managing partner from the inception of the company. He was, apparently, in sole charge of engineering at the ironworks soon to be built on the estate, as well elsewhere on the company's holdings, although agents were eventually appointed to oversee the operation of the outlying coal mines and quarries.

Outram probably moved into Butterley Hall soon afterwards, although he still continued with his engineering work on the unfinished Cromford Canal. He was thus more easily able to supervise coal and ironstone mining, which immediately began on the estate. The earliest financial accounts of the company date from 1790 and these record some of the work in progress in June that year; for example 'to the Miners for sorting Ironstone at the Pit in Wolley Meadow ... 17s. 6d.' and the following November, Outram himself was paid over £15 for 'drawing 1446 tons of Coal out of the Pits with his Horses at 2½d. per ton'.[4] The latter was likely to be in fulfilment of a contract with Sheasby and Dadford, then at work on the Cromford Canal, who had agreed to pay 1s. per ton for

---

[1] J. Farey, *A General View of the Agriculture and Minerals of Derbyshire*, iii (1817), pp. 343–5.

[2] S. Glover, *History and Gazetteer of the County of Derbyshire*, ii, pt 1 (1833), p. 200.

[3] DRO, D503/4, Articles of Partnership of B. Outram & Co.

[4] DRO, D503/35/1–2, Cash Book; D503/99/14, Notebook.

1,018 tons of coal, presumably to fuel steam pumping engines during the driving of the Butterley tunnel.

All production was then entirely for local consumption and so it remained until the canal was finished and reliable transport to other regions, which was so essential for the company's development, could be provided. Nevertheless, some important preparatory work continued on the estate for several months; exploratory boring for coal and iron ore was put in hand, as well as the sinking of new pits. Beresford, on behalf of the company, then secured a future for the ironworks by the acquisition of land for limestone quarrying at Crich[1] and also for wharves on the canal close by, only three miles from Butterley. The first furnace was under construction by November 1791 and permission was also granted by the Cromford Canal proprietors to drive 60 yards of the Butterley tunnel immediately below the site of the new works 'of such width as they shall think fit'.[2] Two or more shafts were sunk at much the same time as part of an ingenious method of gaining access down to the canal (see pp. 218–19). This was followed in January 1794 by the purchase of a small water-engine,[3] which had been used by the canal company on the tunnel at the eastern end; almost certainly this was used for operating the hoists in the shafts.

Iron smelting was not new in Derbyshire and Outram would have been familiar with the basic raw materials and techniques before he embarked on this enterprise, which was clearly intended from its inception to manufacture iron. Most of Britain's indigenous iron ores are carbonate of iron which, when mixed with clay and sand in the natural state, are known as clay ironstones and, when coloured black by carbonaceous matter, are commonly called black-band ironstones. These ores, frequently found among the coal measures of Derbyshire, were described by Farey as 'black Binds and Shales, which contain numerous Layers of Nodules of different sizes'.[4] The earliest workings in the county were shallow opencast 'rakes' and bell-workings, which provided ore for numerous bloomeries and later charcoal blast furnaces. The number of charcoal furnaces steadily declined, to be replaced by more modern furnaces fired by coke. The earliest of these were at Morley Park near Belper, Chesterfield and Wingerworth, all of which were in

---

[1] DRO, D503/3a, Lease, 20 April 1792.
[2] PRO, RAIL 819/1, 8 Nov. 1791.
[3] Ibid., 23 Jan. 1794.
[4] Farey, *General View*, i (1811), p. 393.

production by 1780.[1]

The first blast furnace at Butterley was of stone, 40 ft high.[2] When such a furnace was in blast it was charged with ore, mixed with limestone to act as a flux and coke to provide the necessary heat. Two holes near the base admitted blasts of cold air from cylinder bellows worked by steam engines. Lower down again was another hole through which the slag, formed from the incombustible impurities in the ore, could be drawn off. Below this was another orifice, or a spout, through which molten iron was run off at intervals and discharged into sand moulds as pig iron. The latter contains between 3 and 8 per cent carbon and other elements, and can be modified in structure by re-melting in a cupola before being run into moulds as cast iron. Farey quoted returns for the various ironworks in 1806. By then, the annual output at Butterley was the highest in Derbyhire with 1,766 tons. This compared favourably with Ebenezer Smith's Griffin Works at Chesterfield which produced 1,700 tons; both companies then had two furnaces in blast.

There was a significant reorganisation of the management of Benjamin Outram and Co. in the year before the first furnace was put in blast. It must have been clear that, in spite of the vigour and enterprise of Outram and the limited capital available to Beresford, these factors alone would not take the company to a prime position in what was a highly competitive industry. Accordingly, the partnership was increased by the inclusion of John Wright (1758–1840), a wealthy Nottingham banker, and also the engineer William Jessop. This was formalised by a deed dated 10 December 1792,[3] although it seems that an informal partnership of the four men had been in existence from the previous year. There was nothing unusual about this arrangement; partnerships in such industries were then commonplace, whereas joint-stock companies were rare.[4] John Wright, who had married Beresford's eldest daughter in the recent past and was then a partner in his father-in-law's practice, was to bring substantial capital to Benjamin Outram & Co. This, coupled with his ownership of land nearby which contained considerable mineral deposits, helped to provide a long-term stability for the company.

The partnership, aims and objectives were defined in a detailed indenture; the four men were in business as co-partners and traders for

---

[1] Ibid., p. 397.

[2] P. Riden, *The Butterley Company 1790–1830* (Derbyshire Record Society, 1990), p. 53.

[3] DRO, D503/4, Articles of Partnership.

[4] S. Pollard, *The Genesis of Modern Management* (1965), pp. 24, 25.

a period of sixty years for the purpose of 'getting and smelting Ironstone, Casting and manufacturing Iron, getting and burning Limestone and getting Coal and Slack' and they were not to hold interests in any similar business in Derbyshire or Nottinghamshire. The original company name was to remain unchanged and business still centred on the Butterley Hall estate, where royalties were payable on coal and ironstone extracted (at rates of 8d. per ton for coal and 1s. per ton for ironstone) to Beresford and Outram, both of whom retained ownership of the property. Furnaces, forges, limekilns and buildings could be erected on the site but none was to encroach nearer than 200 yards to the Hall; mining was permitted everywhere except under, or within ten yards, of any building. The deed also recited the details of the land leased for quarrying by Beresford at Crich and the later purchase of adjacent property by the partnership for the use and benefit of the company.[1]

The indenture stated that the capital stock was to be £6,000 'to be brought in equally when wanted' and if found insufficient the partners would 'bring in more in like proportions'. By August 1794 the company was valued at £10,000, still in equal shares, and when the partners began a private ledger in April 1796,[2] the stock was raised to £16,000. Two years later it was increased to £32,000. All these sums included annual dividends, set arbitrarily at 5 per cent of the partners' investments initially and increased to 10 per cent after 1796, an indication presumably of the improving profitability of the company. This curious and simplistic assessment, which was obviously unrelated to the company's market performance, was a widespread practice at the time.[3] Profits were considered to be the surplus after dividends had been credited to each partner's capital account; losses occurred if net earnings were less than the assumed dividend. Thus at the end of the year the partners would each be credited with the dividend and the surplus, or loss, was divided between them. For many years these sums were not paid out but were retained within the company.

The investments of the partners were listed and rationalised in their private ledger in April 1798. Beresford had contributed £8,000 and John Wright £14,000, £6,000 of which was allocated to Jessop and Outram, each of whom was able to subscribe only £5,000. Consequently it was agreed that when Outram and Jessop had each paid £3,000 to Wright, they would be considered as having an equal share in the company. Until

---

[1] DRO, D503/3a, 9a, 16.
[2] DRO, D503/28/1, Private Ledger, 1790–1856.
[3] Pollard, *Genesis*, pp. 272–6.

then, Wright would be entitled to 8 per cent per annum return on the monies invested on their behalf. The agreement added the proviso that so long as any part of the £3,000 remained unpaid, the persons concerned would not be entitled to receive any profits for themselves. Moreover, the latter would continue to be paid to Wright until all debts, with the accumulative interest at 8 per cent, were fully paid up. As a result of Outram's successful management (and apparently at his instigation)[1] the capital was increased annually, to £52,000 in 1800 and to £80,000 in 1802, which was the last time that cash payments were invested by the partners. Outram was unable to meet any of these demands from his own purse and continued to borrow heavily from Wright for his contributions. At the time of his death in 1805 he had not settled the debts due to Wright (except for the annual interest and his share of the profits) or to the company, from which he had also borrowed significant amounts after 1800.[2] Jessop, on the other hand, had cleared all liabilities by 1805, at which time he became a full partner in the firm.[3]

Outram was confirmed in his appointment as managing partner from the date of the partnership deed, although written terms of his appointment have not survived. His salary during those early years is unknown but after 1800 it seems to have been £500 p.a.,[4] plus expenses incurred when on company business. It was a reasonable remuneration for the times, even considering his prodigious responsibilities and contributions that he made to all aspects of the company's business.

The Cromford Canal was opened in October 1794 and, just one month later, Benjamin Outram & Co. were charged £36 for canal tolls;[5] the company was now in business. By the following May, their mining activities were in full swing but were damaging the Butterley tunnel. It was reported to the canal committee that a failure had occurred in the lining during February and yet another in April.[6] The workings of the colliery (presumably from Butterley Carr) of 'Messrs Beresford and Outram' (the canal company frequently referred to Benjamin Outram & Co. as such) were 'too near the said tunnel'. Under the Cromford Canal Act, the mining company was liable for the cost of repairing the tunnel

---

[1] DRO, D503, Cause heard before the Master of the Rolls, 16 May 1811.
[2] DRO, D503/42/1, Furnace Ledger, 1800–22.
[3] DRO, D503/28/1.
[4] DRO, D503/42/1.
[5] PRO, RAIL 819/1, 18 Nov. 1794.
[6] Ibid., 28 May 1795.

due to subsidence and also for losses incurred through stoppages of trade. Accordingly, Outram & Co. paid £100 for repairs on that occasion but the canal company also had obligations under their Act, in that they were required to purchase any pillars of coal left under the tunnel during mining operations. Hence the coal remaining under the tunnel was conveyed to the canal company for £90 in compensation. This was a substantial pillar, some 22 yards wide and stretching 155 yards east from the company's furnace.[1] In spite of these safeguards, the tunnel and other parts of the canal suffered greatly from subsidence in later years and this was ultimately the cause of closure early in the twentieth century.[2]

Once the company was established and connected to the growing canal network, orders were soon forthcoming from canal companies for which both Outram and Jessop were principal engineers. Orders were also received from coal-owners who wished to link their collieries with nearby canals.[3] During 1795 large quantities of cast-iron gang rails, waggon wheels and wrought-iron axles, as well as fittings for several swivel bridges, were sent to Braunston on the Grand Junction Canal. From the following year, coal-owner Joseph Wilkes was provided with a complete railway for his Brinsley Colliery, as were the Morewoods of Alfreton for their Hermitage Colliery, both lines leading to the Cromford Canal. Pig iron was sent to London on receipt of orders from the engineer John Rennie and in later years further orders for railways and miscellaneous cast-iron fittings came from the Huddersfield Narrow, Peak Forest, and Trent & Mersey canals, the London Dock Company and even the Caledonian Canal. Although railways based on Outram's preferred designs were a speciality of the firm during his lifetime, production was by no means restricted to these. As the years went by and experience grew in the foundry, a wide variety of quality goods was turned out. From 1800 there are numerous references in the ledgers to the sale of pipes and branch fittings for pumps and water mains; reservoir cocks; beams, stanchions and window frames for buildings; weighing machines; and even more basic commodities such as stoves, fire grates and garden gates. Purchasers for these goods included turnpike trusts, collieries and waterworks companies such as coal-owners Barber & Walker and the London Bridge Waterworks Company. Limestone from Crich and pig iron from Butterley were also often sold to other ironworks, notably to

---

[1] Ibid., J. Radford and T. Walker to canal company, 21 June 1796.
[2] C. Hadfield, *Canals of the East Midlands (including part of London)* (1966), p. 193.
[3] DRO, D503/42/1.

Cort & Co. of Leicester. Lime for agricultural purposes and coal, although much used by the company, were likewise sold in considerable quantities.

It does not seem, however, that the company costed and priced manufactured products as accurately as they might have. Close attention was given to price lists of rival concerns, especially for their standard products manufactured in quantity, and prices charged for goods were often comparable. Thus Joseph Outram junior wrote to the coal-owner Thomas Walker that 'I am sorry to hear a complaint from you of the prices of your Pipes as they are charged precisely the same prices as at Chesterfield Griffin Foundry & lower considerably than you have them from Coal Brook dale & I think equally good articles … '.[1]

The company was not averse to manufacturing more complex equipment (which was probably costed rather more carefully), even in the specialised fields of steam engine and pump manufacture. Writing to Messrs Gorton & Thompson, of Ollerton in Nottinghamshire in early 1799 Joseph Outram jun. stated that 'Our Engineer [presumably his brother Benjamin] being absent has prevented me answering your favour … He informs me that a 4 Horse Power rotative Engine complete including House & all Materials would be 280£. An 8 Horse Do Do Do 520£'. Then followed his extraordinary definition of the unit of horsepower: 'You will understand what is here meant by 4 Horse & 8 Horse power is what 4 or 8 Horses would do night and day constantly therefore may properly be called an 8 Horse and 16 Horse power, as they will do the work of that number. We hope to receive your order … '.[2] It was many years before the unit of power came to be universally defined in numerical terms; until then engineers described horsepower in various ways, but none so vague and inaccurate as this definition.

It was not surprising that the company was also manufacturing cannon by the turn of the century for both the government and merchant services. In response to an enquiry from Messrs Dickinson, Maillett & Shore, merchants of Birmingham, for example, Joseph Outram jun. mentioned that the company was then casting 4, 6 and 9 lb cannon for the Liverpool market and went on to outline the manufacturing process:

> Much depends upon the manner in which they are executed and something on their dimensions, those used by Government are bored out of the solid & turned on the outside, those for Merchant

---

[1] DRO, D503/12/1, Letterbook, J. Outram jun. to T. Walker, 8 Dec. 1798.

[2] Ibid., J. Outram jun. to Gorton & Thompson, 21 Feb. 1799.

service are seldom finished in that Stile generally not turned and often cast with Cores to reduce the expence of boring. We wish to know what are the sorts you want, whether you expect us to prove them and whether you would provide models, or we must make them to drawings ...

He went on to say that the company had, until then, been entirely involved in the home trade 'but having lately extended our Works are prepared to make all Articles used in the Sugar Islands or any other Articles for the export Trade. Our Metal is of very superior Quality and strength ... If we have the pleasure to serve you we will do it in a manner that will merit the continuance of your favour'.[1]

As regards payments for goods received, a discount of 10 per cent was allowed for sales to other iron manufacturers as well as to wholesale dealers but generally three months credit only was allowed to all others.[2] A 5 per cent discount was given for immediate cash payments but this varied, depending on the customer. Payment by bills of exchange was common and often traders settled accounts by bills they had received from others. This system was usually acceptable but it was not unknown for the persons from whom the payments were to be drawn to prove unreliable. When writing to the Mansfield Foundry in January 1796 regarding such a bill, Joseph Outram junior remarked that 'We will do our utmost endeavours to get its Value from the person upon whom it is drawn but as he is far from being a responsible person we dare not venture at present to give Mr Abbott credit for the amount in his Accnt'.[3]

The early years of the Napoleonic Wars proved difficult for Benjamin Outram & Co. which, in common with other businesses, was periodically troubled by cash-flow problems. The shortage of cash in circulation resulted in frequent, urgent appeals for payments, particularly during 1797 when the Bank of England suspended cash payments; this led to a desperate crisis for industry generally. Thus Benjamin Outram had to admit how critical the situation had become when he wrote to the Morewoods of Alfreton stating that 'We have depended on a remittance from you on account of the Gang Road but have not yet received it, we beg to remind you that tomorrow is our pay day and hope you will

---

[1] Ibid., J. Outram jun. to Dickinson, Maillett & Shore, nd, probably early 1799.

[2] J. Lindsay, 'The Butterley coal and iron works, 1792–1816', *Derbyshire Archaeological Journal*, lxxxv (1965), p. 37.

[3] DRO, D503/12/1, J. Outram jun. to WA. Davenport, 22 Jan. 1796.

favour us by the bearer or early in the morning with 200 or £300 or we shall be greatly disappointed'. That appeal did not succeed immediately because he wrote next day, rather more sternly and anxiously, that 'the terms on which we offered to furnish you with rails and on which you gave us orders were for ready money' and, rather unreasonably considering that a mere twenty-four hours had passed since his first request, 'if you mean to persist in your refusal of payment [I] must take steps that will to me be very disagreeable to procure Justice for Myself and Partners'.[1]

Transport too, could create difficulties during the Canal Mania years before sections of the national network were complete. For example, goods sent from Butterley to London and the south first made an uncertain passage through the shallows of the Trent down to Gainsborough for transhipment to coastal vessels. Before the Derby Canal was completely finished by mid-1796 it was likewise difficult to move goods to the north and west. Hence, when corresponding with a client in Burton, Joseph Outram junior could only voice his frustration by stating that

> I am sorry and ashamed the execution of your order should have been so long delayed. The difficult and tedious passage for our Narrow Boats up the Trent has almost solely occasioned it and I have waited for alternate opportunities of sending it through ... till wearied by continual disappointments I am grown as impatient for its being sent off as you can possibly be to receive it — You may depend on its being sent off here tomorrow and of course (barring further unforseen difficulties) it will reach Shobnal by Friday or Saturday. We shall shortly have a communication with the Grand Trunk Canal by Derby, when our Narrow Boats can with ease go from hence to Burton and your future orders you may rest assured shall be executed on the shortest notice ... [2]

In 1800 Benjamin Outram claimed that a labour force of 500 men was employed by the company in all their concerns, a number which was to rise to 800 within five years.[3] Some of these men were unskilled labourers of local origin but others, particularly the skilled craftsmen, were recruited from afar. This impressive number of employees further

---

[1] Ibid., B. Outram to Mrs Morewood, 3–4 Feb. 1797.
[2] Ibid., J. Outram jun. to Messrs Thornewells, 25 April 1796.
[3] DRO, D503/82/2, Debts of Workers.

emphasises the steady expansion of the company during those early years; the total annual wage bill was then approximately £26,000. Most of these men worked a twelve-hour day, six days a week, and wages or 'reckonings' were usually paid monthly. At times during the Napoleonic Wars, however, the scarcity and rising cost of provisions necessitated the company advancing £40 or £50 per week in small amounts, as subsistence for their workmen. The situation was particularly critical at the turn of the century, a period of acute distress when the company found it impossible to find the £250 to £300 per week in small change required for this purpose. Sometimes the men were given written orders to local shopkeepers for goods to specified amounts, for which the company would be accountable. To an extent this system worked satisfactorily at first but matters got out of hand and local shops contracted many bad debts. As a result, the company lost a number of men who had absconded from the district, unable to meet arrears of payments.

In common with other similar companies, skilled workers were often hard to come by, particularly when manufactories such as Butterley were established in country districts. The company was in some difficulties during 1796 because even Jessop joined in the attempts to engage good moulders when he wrote to Boulton & Watt at the Soho works in Birmingham. He was unsuccessful because James Watt junior replied, most courteously, that 'I am sorry to inform you that we ourselves are so much in want of workmen of that kind, that we have been obliged to repeat our advertisements in different papers. Should any supernumary hands offer themselves, of which from the want of success of our former advts we have but little expectation, we shall with great pleasure address them to you'.[1]

Such key workers, once employed, were generally bound to their companies by contract. Skilled men might be bound for periods from one to five years but apprentices, not surprisingly, were tied for up to eight years. Terms of employment for a capable journeyman might include free housing and fuel and even 'land to keep a cow on easy rent if they wish it',[2] but conditions also included heavy fines (half of which were paid into the works sick fund) for such offences as neglect of duties or drunkeness at work. A common practice was to engage contractors to undertake a particular job for an agreed rate or lump sum, the contractor to be accountable for the management and wages of his workforce. This practice relieved the company of any responsibility for the diligence or

---

[1] BCL, Boulton & Watt Collection, J. Watt jun. to W. Jessop, 2 Jan. 1796.
[2] Lindsay, 'Butterley coal and iron works', p. 32.

discipline of workmen, as well as much administration. Thus, for example, one Timothy Green was engaged in October 1800 'to find and provide sufficient number of assistants along with himself, to coke all the coals for the purpose of supplying Butterley Furnaces that shall or may at any time be wanted for the wages or sum of Nine Guineas for every Four Weeks. Each Furnace shall be at work for the term of five years from the Date thereof'.[1] Such practices continue to this day, particularly in the construction industry, although contractors so engaged are mostly those with highly specialised skills, or those which use products protected by patent.

Some firms were unscrupulous in their attempts to poach good operatives from rival organisations and the Butterley letterbook contains several references to such matters. Outram was always quick to admonish others about such malpractice, notably when he wrote to Raby's at Dale Abbey Furnace complaining that one of their workmen had been persuading Butterley moulders to join Raby's and that his 'Temptations' had created much dissatisfaction among their workmen 'which may prove very injurious to us'.[2] Often enough, Outram too was accused of luring good men away from their employers to work at Butterley. He wrote to John Curr, the colliery agent of the Duke of Norfolk in Sheffield, on one such occasion after he had been accused by Curr's assistant of 'an attempt to seduce from Employment some of your Workmen'. He went on to claim that it was 'a practice which I hold in detestation', explaining that Samuel Wragg, one of his own agents, had visited Sheffield on business and, being on the look-out for a modelmaker and moulders, had been introduced to one Holden who claimed to be 'at liberty'. A wage of one guinea per week was agreed but it was not until Outram had received Curr's letter of complaint that he realised that Holden had been untruthful. Moreover, as a result of Wragg's enquiries yet another of Curr's employees had approached Joseph Outram junior to seek employment, although he was honest enough to admit that he was already engaged by Curr. The latter was assured that the man had been told that

> we should have nothing to do with him as long as he was under any Engagement to you, nor till he had a Certificate that he was fairly at Liberty. It is not any apprehensions of Law but my Ideas of Justice that have ever determined me never to employ a Workman that another has a claim to. Workmen are already too

---

[1] Ibid., p. 31.
[2] DRO, D503/12/1, B. Outram to Messrs Raby, Dale Abbey Furnace, 27 March 1798.

independent of their Employers.[1]

Yet another letter of complaint was received in 1801 from Longdon Chambers of Sheffield, which stated that one of their moulders had been engaged at Butterley and that the man had been coerced into leaving them. This was hotly disputed by Outram who denied that there had been any 'undue influence by intoxication'. He wrote that the man had told the Butterley agent that he was out of employment; in any case, the company's agents were always instructed not to approach those already engaged by other firms.[2] No doubt many employers were, to a degree, guilty of such underhand behaviour, particularly at peak production periods, and it is more than likely that the Outrams at Butterley were just as deeply involved as any of the others, in spite of loud protestations of innocence.

By 1796 Outram was simultaneously principal engineer for several canals, a practice quite separate from his management functions at the ironworks. His responsibilities must have been very demanding and, probably as a result, his health began to suffer. In February of that year he wrote to a client of 'my long illness' but in spite of this, he continued to deal with a 'multiplicity of engagements from home',[3] quite apart from his exacting task of running the company. He probably had no option because he had managed the business until then almost single-handed (his partners did not seem able, or willing, to take up the reins of the company at any time, even during his illness), confirmed by the fact that he was eventually forced to turn to his younger brother Joseph to 'manage our affairs at Butterley'.[4] Joseph obliged and moved into Butterley Hall. He was conversant with estate management practices but it is fairly clear that initially he had to struggle with the engineering problems he encountered in this new appointment. Hence Benjamin, often enough from his sick-bed, had to advise him and also issue instructions for the day-to-day operations at the works. At the same time, Joseph continued to assist his father in their very busy surveying practice; he too must have been greatly stretched in having to cope with his two quite disparate jobs.

At first, Joseph Outram junior's strengths were in administration and he soon showed competence in this respect. The nature of his responsi-

---

[1] Ibid., B. Outram to J. Curr, 20 Oct. 1796.
[2] Lindsay, 'Butterley coal and iron works', p. 29.
[3] DRO, D503/12/1, B. Outram to S. Lloyd jun., 24 Feb. 1796.
[4] Ibid.

bilities can be gleaned from a letter to one of his father's clients, shortly after his appointment, in which he wrote that

> Our Partners have engaged to meet here on Monday next week but one for the Purpose of taking Stock and examining the Accounts of the last Lady Day and I having entered the Books in an unsettled form require not every day but every minute I have to spare from my occasional vocation of looking over the different workmen, travelling etc.[1]

Up to that time it does not seem that Benjamin Outram had appointed any senior staff who could manage the works during his frequent absences and deal with engineering issues as well. This was not unusual in the iron industry; traditionally there was a lack of trust between owners and subordinates and a delegation of duties to salaried managers was commonly believed to be a practice which invited trouble from theft and fraud.[2] It seems clear that Outram subscribed to this view and was thus reluctant to take on staff for responsible appointments, preferring to do everything himself. The practice must have led to a weakness in the management at Butterley and would have been most evident during his illness because, the following year, George Goodwin was appointed as assistant to Joseph Outram junior, probably in the capacity of accounts clerk. Goodwin was only in his early twenties and Joseph was to keep a close and suspicious eye on Goodwin's activities for some years,[3] but his proved to be a successful appointment and he continued to serve the company well in senior management positions until his death in 1848.[4]

Joseph Outram junior was to continue apologising to clients of the surveying practice throughout 1796 for his inattention to their business 'in consequence of my Bro's indisposition'. The nature of that indisposition is uncertain; the fact that Benjamin was sent off to Buxton where Joseph hoped that 'he will meet with every requisite Assistance' for what was believed to be rheumatism, suggests that the affliction was not unduly serious.[5] However, the period of Benjamin's confinement to a sick-bed (at intervals over fifteen months) and the tone of his father's

---

[1] JRUML, Bagshawe Muniments, 8/4/2864, J. Outram jun. to J. Bagshawe, 8 June 1796.

[2] Pollard, *Genesis*, p. 23.

[3] DRO, D503/12/2, J. Outram jun. to G. Goodwin, – Aug. 1800.

[4] Riden, *Butterley Company*, p. 72.

[5] JRUML, Bagshawe Muniments, 8/4/2860, J. Outram jun. to J. Bagshawe, 31 April 1796.

letter to a client that 'Benjamin continues very poorly and unless he will be be very carefull — *may be no more*',[1] lends credence to the view that the illness was more serious than rheumatism, although some of the symptoms might have resembled that. Whatever Benjamin's malady might have been, the suspicion remains that this could have led to his demise just nine years later in 1805.

Benjamin Outram seems to have recovered by early 1797, or at least sufficiently so as to take over from Joseph junior some of the more demanding aspects of the business at the works, because he was to write to an acquaintance in January of that year that 'ill health has prevented me from calling upon him lately but I hope soon to be able to go from home as I gain strength very fast. It will give me great pleasure to hear that Mrs and Miss Arden are well. I have experienced much ill health since I left them at Buxton … '.[2]

Benjamin Outram resumed his civil engineering career about that time and not long afterwards he met Dr James Anderson (1739–1808), Doctor of Laws and Fellow of the Royal Society, a distinguished Scottish economist. The acquaintance probably had much to do with their interests in improving public transport and more particularly the promotion of a national system of railways on which Anderson was to publish an article in his magazine *Recreations in Agriculture, Natural History, Arts and Miscellaneous Literature* in November 1800. Anderson, a widower, had settled with several of his eight surviving sons in Isleworth, near London, three years earlier. He was joined there in 1798 by his only daughter Margaret (1778–1863), a high-spirited and intelligent young lady who, until then, had lived all her life in Scotland. The family came from a long line of Scottish gentry but the Andersons were by no means wealthy, and the children relied much upon the precarious earnings of their author father.[3]

Benjamin Outram married Margaret Anderson on 4 June 1800 at St Magnus the Martyr Church,[4] located by the old London Bridge, and the happy couple returned afterwards to Derbyshire, to reside as gentlefolk at Butterley Hall. It seemed a promising marriage and Margaret in later years was to write that

---

[1] Ibid., 8/4/2854, J. Outram to J. Bagshawe, 20 April 1796.

[2] DRO, D503/12/1, B. Outram to Mr Arden, 23 Jan. 1797.

[3] M.F. Outram, *Margaret Outram (1778–1863). Mother of the Bayard of India* (1932), p. 67.

[4] *Monthly Magazine*, 1 July 1800.

> As Mr Outram was full of hope and energy, and the picture of health and with promise of long life, our prospects were great and, in place of subduing expectations of grandeur and prosperity, his sanguine disposition and knowledge of his own successful projects made him flatter me into the belief that fortune would place us on the highest pinnacle.

At that time Margaret believed that her husband was then earning two or three thousand pounds a year and indeed the future did seem most promising. Regrettably, her situation was to change quite dramatically within five years.[1]

Outram abandoned most of his interests in canal engineering during 1801, partly no doubt as a result of his new marital responsibilities and the birth of his first child, Francis, born in 1801, who was followed by Anna in 1802, James in 1803, Margaret in 1804 and and Eliza in 1805. Much of his future work was to lie in pioneering the design and construction of railways and the work that this enterprise brought to the Butterley works, where most of his considerable talents were subsequently applied. Joseph Outram junior was still retained as manager, however, having proved to be a capable administrator, and becoming well versed too in the engineering activities of the company. At about this time the sons of William Jessop, namely Josias (1781–1826) and William (1783–1852), were introduced into the company with part-time, minor appointments. The amiable Joseph Outram junior seemed to get on well enough with them both and indeed, the Jessop and Outram families continued to be on the best of terms during this period. Joseph, nevertheless, was ambitious to advance his own career and may well have left Butterley about 1803 for employment at the Clyde Ironworks in Glasgow. During November 1804 he became set on acquiring a partnership there,[2] in the purchase of which he was assisted by his father. Unfortunately the move to Scotland did not prove very successful and he left Clyde four years later.[3] Joseph may then have returned briefly to assist the Jessops in the management at Butterley but eventually he was to go back to Glasgow to establish a new career for himself in commerce.[4]

---

[1] Outram, *Margaret Outram*, p. 93.

[2] JRUML, Bagshawe Muniments, 8/5/99, J. Outram to Revd W. Bagshawe, 27 Nov. 1804.

[3] GRO, D2159/19, A. Faulds to D. Mushet, 18 April 1808.

[4] National Archives of Scotland, SC 36/48/22, pp. 489–91, inventory of Joseph Outram's estate.

During the first few years of the partnership, most mining activity had centred on the Butterley Hall sites but a splendid opportunity to expand came in 1796 with the acquisition of a lease of the mineral rights under a substantial part of the Codnor Park estate, some two miles downstream on the Cromford Canal from Butterley.[1] More important still was the lease of five acres of land on the same estate which was adjacent to the canal, just below the junction with the Pinxton branch. Although, as lessees, the partners covenanted to erect industrial buildings on the latter acreage within ten years (and they had every intention of building an additional works there eventually), initially they wisely concentrated on exploiting the coal and ironstone deposits. Their immediate plans for developing the sites were prepared by the end of the year and since these would affect operations on the canal, proposals were submitted to the annual general assembly of the Cromford Canal Company in May 1797.[2]

There were fourteen locks from the junction with the Pinxton branch down to the Erewash Canal at Langley Mill and the principal coal seams in Codnor Park lay between locks seven and eight, on what was known as the Long Pound. Outram devised a neat plan which he submitted to the canal company in the hope that they would promote and pay for it. He proposed to gang coal from pits on the steeply rising ground lying to the west downhill to a wharf on the pound. This position would be within the twelfth mile from Cromford and within the third mile from Langley Mill. He also intended to cut a branch canal, one mile in length, at a cost of £800, from the centre of these pits, to join the Cromford Canal summit level close to the Pinxton junction, such that the distance to Cromford from the pits would also be within the twelfth mile. Hence, downgate traffic would take coal from the east of the branch through seven locks to Langley Mill and beyond; meanwhile upgate traffic would serve consumers along the summit pound with coal from the western side of the branch. The branch canal would thus provide the same tolls for the canal company as if coal had moved upgate from the wharf, through the locks into the summit pound, but this system would save wear and tear on the locks and conserve water as well.

Another feature of the design was that a steam engine with a pump would be erected opposite to the Long Pound for draining the pits. The pump was designed to discharge into a culvert under the canal but if the 50-yard lift were increased by another 3 yards (the company paying for

---

[1] Riden, *Butterley Company*, p. 58.
[2] PRO, RAIL 819/1, 31 May 1797; DRO, D503/12/1, nd (early 1797).

the extra costs), the water would eject into the canal pound instead and thus aid the downgate traffic operations. Joseph Junior wrote to the company about the time of the submission for permission to cut a channel from the pound above the fifth lock to feed a 'Water Whimsey' (the hoist and water-balance system) for lifting coals from the Codnor Park workings. He promised that the channel invert level would be high enough to take water only when the overflow weirs were discharging and, in any case, all the water would be returned by the steam engine and pump. What was more, it would be discharged higher up the flight of locks from the level at which it was drawn.[1]

All these plans were sound enough but the company were rather reluctant to support the scheme. Nevertheless, Thomas Walker was appointed to report on the advantages of the project for the canal company and also to check Benjamin Outram's estimates for the branch. Walker claimed that the estimates were far too low, and this led to certain committee members accusing Outram of deceiving the company as to the real costs, which would only be discovered after work had been finished—and when Outram had presented his bill. Joseph Outram junior politely replied on Benjamin's behalf, explaining that a substantial quantity of rock would be excavated along the branch but this was not included in the estimate because it was earmarked for use elsewhere by Outram & Co.[2] He stated that there were other savings as well for the canal company but Walker had apparently failed to recognise them.

Disagreements continued until Benjamin Outram took over the correspondence with a vengeance.[3] Although he regarded Walker as a friend, he would not let him get away with false accusations. 'Much as I value Mr Walker's friendship and much as I am obliged to him for his professions of readiness to serve me—I must accuse him of want of candour in the Business that has brought me to a charge *I despise*, that of an attempt to deceive the Cromford Canal Co.'. Outram reiterated the statements made by his brother and went on to say that 'I never made claim on Mr Walker's friendship in this or any other business between myself or others, nor will I ever ask, wish or expect more from any Man than that *strict justice* I should feel myself bound to administer to another. The Friendship of honest minds will never expect more'. It is difficult to understand why Walker had not accepted the explanations of Joseph Outram junior but it seems from this and similar correspondence

---

[1] DRO, D503/12/1, J. Outram jun. to H. Cutts, 24 March 1798.
[2] Ibid., J. Outram jun. to T. Walker, 21 Sept. 1798.
[3] Ibid., B. Outram to T. Walker, 8 Oct. 1798.

that the slightest attempt to impugn the honour of Benjamin Outram always met with a tongue lashing. This probably discouraged all but the bravest either to argue with him or to dispute his methods. It is likely that most of the employees of Benjamin Outram & Co., and probably his partners as well, would think twice before disagreeing with him on any matter relating to the management of the company, however insignificant. It was obviously much easier to go along with him.

As regards the projects in Codnor Park, some discussions continued for a while between the two parties, but nothing ever came of it and the plan for a canal branch was abandoned in favour of a complex of railways. These were built and maintained by Benjamin Outram & Co. and served much the same purpose as would a canal branch, but at no cost to the canal company.

The company relied almost entirely on the Cromford Canal and the national network of waterways for both the supply of raw materials and the distribution of finished goods, quite apart from the transport of substantial quantities of lime and coal. These vital communications, coupled with the extensive use of the company's own railways as feeders, proved most satisfactory and would do so for many years ahead. But such means were not enough by themselves. An efficient road system was also essential for local communications, as the company clearly found as it expanded the scale and location of its operations. In the absence of effective local and national highway authorities which were suitably financed and staffed to undertake such tasks, the company, not unusually for the time, became involved, albeit reluctantly, with the repair of local highways and the construction of turnpikes.

The task of highway maintenance was then the responsibility of the parishes and neglect was common. This seemed to be the case locally when Benjamin Outram wrote, late in 1796, to Pentrich highway surveyors, 'Wishing much to have the Roads about us put in such a state as that this neighbourhood might have tolerable Communications with the rest of the World in Winter'. Outram referred to 'the very bad state of the Lanes from hence to Pentridge and Swannick' and that he had offered cinders from the company's furnace free of charge (in spite of the fact that he had nothing to do with the expense of repairing roads), provided that the 'the inhabitants would lead them upon the road before winter'. Nothing had been done to take up his offer and he told the surveyors that 'if they would not exert themselves to improve the lanes I would take the means pointed out by Law to Compel their repairs'.[1]

---

[1] Ibid., B. Outram to surveyor of highways, Pentrich, nd (probably late 1796).

Rapid industrialisation meant that the region could no longer be described as rural and in these changing circumstances it was reasonable to expect the parish officers to be less casual and neglectful of their duties. Outram bluntly offered his advice:

> Would it not be better for every farmer to employ his horses for a week at the dead time of Business to make the roads good to his fields and Markets, rather than Let them lie idle in the Stable or Yard & be Obliged to travel up to the neck in Dirt whenever a wet season presents itself? The Roads I am Anxious about are not in the way of being cut up by Coal Trade or Foreign Business but are Chiefly used by the Neighbours only & if once made good would require but little Labour to keep them so. Set your own Teams to work, give a proper example ... I am Determined to be no longer laid up in the Dirt.[1]

The reaction to that letter is unknown but such complaints were certainly common; some time later Joseph Outram junior was to send a strongly worded letter threatening to bring the law to bear on the surveyors in Heage regarding the road linking Bull Bridge and Belper as it extended through their district, stating that 'we beg to remind you that this is the second and last application'.[2]

The need for improved communications resulted in the Alfreton to Derby Turnpike Act of 1802 which authorised the reconstruction and improvement of the existing road through Swanwick, Butterley, Ripley and Denby, which was 'greatly out of repair, and some parts circuitous, narrow and inconvenient for travellers and cannot be effectively amended, widened, improved and kept in good repair by ordinary course of law'. Naturally enough, the trustees included Benjamin Outram, along with his father and brother, and also John Wright, John Beresford and William Jessop, who no doubt were principal promoters of the scheme. There were numerous other trustees listed, which was perhaps indicative of the local concerns and frustrations that had been aroused by the deplorable conditions of the road system. Not surprisingly, the contract for the work was awarded to Benjamin Outram & Co. in the sum of £7,288.[3] Although no details remain of the work done, the company supplied 961 cart-loads of slag, probably for the northern end of the job.

---

[1] Ibid.

[2] Ibid., J. Outram jun. to surveyor of highways, Heage, 14 April 1801.

[3] DRO, D503/41/1, p. 152.

The contract was completed between October 1804 and December the following year and Edward Banks, already well known to Outram for his work on the Huddersfield Narrow and Ashton canals, was engaged for the task and acted either as subcontractor or agent for the company.

The shallow seams at Butterley were steadily being worked out by the turn of the nineteenth century and about this time John Wright sold valuable parcels of his land to the company and also leased 'all stone, ironstone and minerals' under both his Butterley Park and Knowts Hall estates, which were adjacent to Butterley Hall.[1] For as little as 30s. an acre compensation and eventual restitution of the land for agriculture, the company was permitted to 'dig, get, manufacture and carry away same, to dig clay, make bricks, sink pits, make drains, erect engines, kilns, furnaces, hovels, sheds, stables, make roads' and 'to do all other works and things necessary for the obtaining, using and disposing of said materials'. Wright also leased 361 acres of land to the company in Somercotes and conveyed a small acreage by the Cromford Canal for lime-kilns and a wharf. Both localities were known to be rich in ironstone and this certainly proved to be the case in the ensuing years.

This was yet another occasion in the early days of the company that Wright proved a munificent benefactor in the exploitation of his estates, apart from his generous loans to Outram and Jessop. As the company's banker he provided support with overdrafts and there can be little doubt that his unstinting investments helped the company to develop into the major enterprise it had become by the time of Outram's death. Altruistic though his earliest motives may have been, Wright was to benefit greatly from his involvement with the company. Francis Beresford died in 1801 and his share passed to his son John, who sold out to Wright five years later for a mere £10,000. The partnership of 1792 was terminated in 1815, retrospectively from March 1813, shortly after the death of William Jessop.[2] His son William Jessop junior succeeded to the partnership and continued the management of the Butterley Company, the title adopted after the death of Benjamin Outram. John Wright had never taken an active part in the management of the firm but remained the senior partner until 1830, when he retired and transferred his assets to his son Francis.[3]

---

[1] DRO, D503/M1/5/1, Description of roads, lands and minerals sold by J. Wright to B. Outram & Co.

[2] DRO, D503, Judgment of the Master of the Rolls, 8 Dec. 1813.

[3] DRO, D503/28/1.

CHAPTER 12

ON RAILWAYS AND TRAMROADS

Primitive forms of railway, in which vehicles with flat-rimmed wheels ran in pre-formed ruts, were not unknown in ancient civilisations but modern European experience of such self-steering systems only stems from about the sixteenth century. Agricola's well-known book, *De Re Metallica*, published in 1556, illustrates with annotated diagrams typical cases which must have been commonplace in Central Europe at that period. These were of pit tubs with flat-rimmed wheels, which ran on wooden rails of rectangular section, controlled by vertical iron pins which projected down from the tubs to fit into a groove central to the rails. It is likely however, that the flanged wheel was introduced into Central Europe before the end of that century.

In Britain, the first rudimentary railways of the pre-locomotive era were laid for the transport of coal between Strelley and Wollaton, near Nottingham, and from Broseley in Shropshire to the River Severn, during 1603 and 1605 respectively.[1] Thereafter began a widespread but uncoordinated development of this mode of transport,[2] notably in the north-eastern coalfields, where railways were commonly known as waggonways, or Newcastle roads;[3] in Shropshire, where they were first used underground; and also in South Wales. Generally those built at ground level were short-haul tracks laid from pit-head to nearby river-borne transport.

Rectangular wooden rails were the norm, as were self-steering waggons with flanged wheels, but some of the earliest waggons were fitted with narrow wooden rollers backed by wheels on the same iron shafts to act as flanges, and other railways used wheels entirely cut from timber blocks. Alternatively, flat-rimmed wheels, guided by continuous timber scantlings nailed to longitudinal timbers, were also used.

The advent of the canal era saw futher rapid extension of such railways, many in the Midlands and South Wales, as feeders from coal mines and quarries. There was very little uniformity of design or gauge

---

[1] R.S. Smith, 'Huntingdon Beaumont: adventurer in coalmines', *Renaissance and Modern Studies*, i (1957–60), p. 121; idem, 'Britain's First Rails: a reconsideration', ibid., iv (1960), pp. 123–31.

[2] A.R. Griffin, *Coal Mining* (1971), pp. 11–25.

[3] J. Curr, *The Coal Viewer and Engine Builder's Practical Companion* (1797), p. 8.

on these systems and careful attention was not always given to the important questions of alignment and gradient. Wooden rails of rectangular section, nailed to tranverse timber sleepers, were still common by the late eighteenth century but cast-iron flanged wheels were more widely employed after 1753.[1] In fact it seems likely that these were manufactured even earlier because the Prior Park Railway in Bath almost certainly had waggons with flanged iron wheels, judging by contemporary drawings dating from 1734.[2]

Marked improvements to the wooden railway were made by Richard Reynolds, a partner in the Coalbrookdale Ironworks, when 'between five and six tons of rails were cast on the 13th November 1767, as an experiment ... '.[3] These rails were flat strips of cast-iron, each 5 ft long, 4 in. broad and 1¼ in. thick, which were nailed to existing oak rails to form a running surface. Curiously, the practice was said by a contemporary observer (J.C. Hornblower) to be only a means of storing iron in times of slack trade,[4] but, whatever the reason, more than twenty miles of such tracks were built in and around the Dale works by 1785.[5] This simple means of reinforcing timber rails and reducing surface wear spread to other industrial regions. As previously noted, Outram had considered such a system whilst designing the Derby Canal railways but eventually opted for the plateway.[6] In this he was probably influenced by John Smeaton's experience, dating from before the completion of the Eddystone lighthouse in 1759,[7] as well as by the successful adoption of the plateway in Derbyshire, during 1788, by Joseph Butler to link ironstone pits with his Wingerworth furnace.[8]

From the evidence available it is impossible to attribute, with certainty, the original design of angle-iron railways, or plateways, to any one inventor. In common with other useful industrial devices it was probably developed gradually over many years by artisans seeking to

---

[1] N. Wood, *A Practical Treatise on Rail-Roads and Interior Communication in General* (1825), p. 12 (Report by R. Stevenson of Edinburgh); G. Jars, *Voyages Métallurgiques* (1765), i, p. 200.

[2] J.T. Desagulier, *Course in Practical Philosophy* (1734), pl. 21, 22.

[3] Wood, *Practical Treatise*, p. 12.

[4] C.F. Dendy Marshall, *A History of Railways down to the year 1930* (1938), p. 143.

[5] Ibid., p. 55.

[6] Above, p. 61.

[7] J. Smeaton, *A Narrative of the Building and a Description of the Construction of the Eddystone Lighthouse* (1791), pp. 196, 197.

[8] J. Farey, *A General View of the Agriculture and Minerals of Derbyshire* iii (1817), p. 288.

ease and speed up the carriage of bulk loads. There can be no doubt, nevertheless, that some admirable refinements were made by John Curr, viewer for the Duke of Norfolk's collieries in Sheffield, in so far as he enhanced their practical value for underground workings. Curr described these railways in his illustrated book *The Coal Viewer* (1797), which was published just ten years after the first of his underground railways was installed in 1787. He claimed that the mode was 'superior to anything hitherto practised, as the results of seven years experiment informs me'. Curr did not claim to be the inventor of the angle-iron rail and was probably referring only to the track layouts, his designs for 'wheel corves' (or pit tubs) and his ingenious underground haulage systems.

The standard plates used for Curr's track were 6 ft long and 3 in. wide 'on the trod', with countersunk nail-holes drilled one inch from each end. The flange was 2 in. high above the flat along the entire length. Rails were supported on lugs which fitted into timber sleepers and the gauge was either 22½ or 24 in., depending on the size of corve in use. Some additional play of the wheels was recommended for negotiating tight bends. The cast-iron was ½ in. thick and a standard unit weighed from 47 to 50 lb, although shorter and thicker plates were recommended for heavier loads. Curr proudly claimed in later years that his methods were widely adopted nationally and there seems no reason to doubt him.

Outram knew of Curr's plateways because he wrote to him in October 1796 requesting a copy of his treatise as soon as it was published.[1] He also mentioned the system some years later (but in no sense disparagingly) when he wrote of the practice

> in mines, and other works underground, where very small carriages only can be employed, very light rails are used, forming what are called Tramroads, on a system devised by Mr Curr; and these kinds of light Railways have been much used above ground in Shropshire and other counties where Coals and other Mineral are obtained.[2]

Thus Outram implied that his own far-reaching ideas for regional networks of railways, built according to the principles of his own

---

[1] DRO, D503/12/1, B. Outram to J. Curr, 20 Oct. 1796.

[2] J. Anderson, *Recreations in Agriculture, Natural History, Arts and Miscellaneous Literature*, iv (1803), p. 476.

'improved plan', were not unduly influenced by Curr's ideas.[1] However, although Outram's system was satisfactorily established and refined by the date of publication of *The Coal Viewer*, the practices in the Sheffield mines would have been common knowledge throughout the region for many years beforehand; it would have been surprising had Outram been unaware of them.

It is difficult to avoid mention of the myth which links the words 'tram' and 'tramroad' with the name of Benjamin Outram. In spite of its rejection by several authorities over the years, belief still persists in the story, which stems from the 1857 edition of Samuel Smiles's biography, *The Life of George Stephenson, Railway Engineer*. Here it was stated that 'tramway' was derived from the second syllable of Outram's name. Although this fiction was withdrawn from later editions, the story was given fresh credence by Clement E. Stretton even as late as 1905 when, in presenting an historical paper in Little Eaton on the Derby Canal railway, he blatantly described the line as the 'Little Eaton Outramway'.[2] Stretton invented some extraordinary railway history during his lifetime, much of it a mixture of accurate facts, where there was ample contemporary evidence to support his views, with some imaginative falsehoods where there was not. Hence his description of the Little Eaton railway was correct because it was fully operational at the time of his visit and lecture, but the historical facts he quoted as to its origins were largely inaccurate. Unfortunately, too many of his fables still survive. Outram never described his works as 'tramroads' or 'tramways'; invariably he referred to them as 'railways' and very occasionally as 'railroads'. He never claimed 'tram' was derived from his surname.

In fact the word 'tram' derives from the Old Scandinavian *traam*, meaning a baulk or beam of timber. 'Tram' was commonly used in Britain to describe the longitudinal members forming the frame of a waggon or barrow. The word was used in the coal mines of the North East of England as early as 1708, probably to describe a sledge on which corves, or baskets of coal, were drawn from the mine workings.[3] It does not seem to be until 1734 that wheeled trams were introduced[4] but the word 'tramway' first occurred in 1775 and 'tramroad' in 1796.[5]

---

[1] Ibid., pp. 473–7; M.F. Outram, *Margaret Outram (1778–1863). Mother of the Bayard of India* (1932), Appendix.

[2] C.E. Stretton, *The Little Eaton Outram-way. A paper read at Little Eaton on 29 July 1905* (1905).

[3] J.C., *The Compleat Collier* (1845), p. 39.

[4] M.J.T. Lewis, *Early Wooden Railways* (1970), p. 312.

[5] Ibid., p. 314.

'Dramroads' (a term which seems to originate in South Wales) first appeared in the *Journals of the House of Commons* in 1796 and as 'tramroads' on several occasions thereafter.

This supposed connection of tram or tramroad with Outram's name was apparently unknown among civil engineers of a later generation. Thus Thomas Tredgold, when discussing the various types of railway in his book of 1835, describes the plateway as 'having the guiding flanges upon the rails instead of upon the wheels of the carriage; it gives advantage of employing carriages that can be used where there are not rails laid down. Railways of this type are called tram-roads, from their being first used for running trams upon ... '.[1]

And if these accounts are insufficient to dispel the myth, let Outram have the last word on the subject. In his report to the proprietors of the Brecknock & Abergavenny Canal in July 1799 Outram wrote: 'On the Railways which I have constructed (and others copied therefrom) the Expences of Conveyance and repairs are far less than on any other kind of Railways; they are made nearly on the Principle of those called Tram or Dram Roads in this Country; but the Rails are stouter, have higher Flanches, and are cast in shorter lengths ... '.[2]

Much of Outram's thinking on the design of railways and their possible use for longer distance transport can be gleaned from a paper 'On Cast Iron Railways' written by Dr James Anderson, his father-in-law, in *Recreations in Agriculture, Natural History, Arts and Miscellaneous Literature*, published in 1803 but written two years before. Anderson began this interesting article, substantial sections of which must have been contributed by Outram, by stating that 'Every one will admit, that the number of horses kept at present in this country is in itself an evil of a very serious nature, which ought to be removed if that can be done without producing a greater evil in its stead'. He claimed that the current extensive reliance on the horse (he presumed there were at least 20,000 heavy draught horses employed in London alone)[3] had been mitigated in part by the introduction of canals but the economic extension of the latter was often limited by the physical characteristics of the country through which they passed, quite apart from the lack of water to supply them. Anderson concluded that, although railways at that time were essentially local affairs and 'have not yet been introduced into

---

[1] T. Tredgold, *A Practical Treatise on Rail-Roads and Carriages etc.* (1835), p. 29.

[2] NLW, Maybery 14, B. Outram, Observations on the Brecknock & Abergavenny Canal and Railways, 1 July 1799.

[3] *Recreations in Agriculture*, iv (1803), pp. 198, 208.

general practice', they were not affected by terrain to the same degree as canals. He believed that a network of railways, properly managed, could supersede canals in the future, particularly in regions where canals could not be conveniently constructed. This could only help reduce the soaring costs of basic provisions which prevailed at that period and result in the greater internal prosperity of the country.

Obstacles to the greater use of railways were the widespread reliance on timber rails and the crude nature of their construction. Anderson was dismissive of the improvements that had been made to such wooden railways over the years, including the reinforcement of running surfaces with iron bars. This type of railway was suspect, in terms of long-term reliability and expense of maintenance, for the universal network which he had in mind. A realistic alternative was the adoption of the improved system devised by 'Mr Outram, engineer' who had taken the entire business in hand and had 'contrived at the same time so far to diminish the expence and improve the strength of the road as to bring them to a degree of perfection that no one who has not seen these can easily conceive [what] could have been done'. Outram's design had been 'carried much beyond the limits of what was for many years conceived to be possible ... It may be in future employed to a wider extent still, to which no limits can at present be assigned'.

Seven railways already built by Outram on his improved plan were then listed in order of their construction. These were: the Derby Canal railway (5 miles); the Crich railway, linking limestone quarries with the Cromford Canal (1½ miles); Barber & Walker's railway from the Beggarlee Colliery to the Cromford Canal (1½ miles); the Peak Forest railway, from the limestone quarries near Buxton to Whaley Bridge (6 miles); the Marple railway (1½ miles); the Blisworth railway on the Grand Junction Canal (3½ miles of double track); the Ashby de la Zouch railway (4 miles of double and 8 miles of single track). At the time of writing, all but the last were operational and that was expected to be finished by summer 1801; in fact it was only completed in September 1802.

Dr Anderson went on to applaud the works of Joseph Wilkes of Measham in Leicestershire, 'a spirited and judicious agriculturist whose name is well-known to all those who attend to rural affairs'. A railway had been built for him to link a coal mine with local markets in a manner similar to Outram's improved plan, although 'from local circumstances [it was] laid upon wooden sleepers and is not as perfect as those done on stone'. Wilkes apparently 'found it so fully to answer his expectations after it was finished that he communicated to the Society of Arts an account of some trials he had made of it'. Joseph Wilkes was a wealthy coal-owner and shareholder in several canal companies. He knew both Jessop and Outram and in fact all the cast-iron components for his railway had been supplied by Butterley. As noted later, Wilkes was to recommend Outram and his improved railways for use on several waterways, notably the Ashby, Worcester & Birmingham and Grand Junction canals.

Anderson's article included a section obviously written by Benjamin Outram entitled 'Minutes on the Construction of Railways', in which the engineer outlined his basic design principles. Gradients were of prime importance and whereas, obviously enough, he advocated a level track where trade was about equal in both directions, a downward gradient of 1 in 150 was recommended where trade was one-directional; this was generally the case where mines or quarries were linked with navigations. Provided that gradients were not greater than 1 in 100, it had been found that horses applied much the same effort on a loaded, downward journey, as they did on the return when there was little or no upgate loading. Such gentle descents were not always practicable, however, and if gradients were of the order of 1 in 50, six or eight waggons, each loaded with 30 to 40 cwt, would place an intolerable burden on a horse in preventing an uncontrollable run. On such steep inclines, the train of waggons had to be braked by 'sledging' some of the wheels. This process was achieved by chaining iron slippers on one or more waggons. These were placed under the wheels on approaching a steep incline to prevent their rotation, thus forming a sledge. Gradients along parts of a railway could also be eased by utilising steep inclined planes at intervals where loaded, descending, waggons could be used to raise gangs of empty, or partly filled, waggons quickly and efficiently, although additional static machinery and haulage ropes were necessary on such structures for the purpose of controlling acceleration.

It was often necessary to cross valleys by bridges and embankments and to cut through ridges of land in order to ensure the best possible gradients and to avoid sharp bends and circuitous routes. In rugged country, even short tunnels might have to be driven. The breadth of bed recommended for single lines was four yards, and six yards for double lines, exclusive of fences, side drains and and side slopes. Outram advised that the formation surface of the earthworks should be covered with a six-inch thickness of compacted broken stone, or gravel, upon which the sleepers were laid. He strongly recommended the use of stone sleepers 'in all places where [they] can be obtained in blocks of sufficient size'. It does not seem that he ever resorted to the use of timber for sleepers, except for temporary railways on construction sites.

The specifications for both sleepers and rails were unchanged from

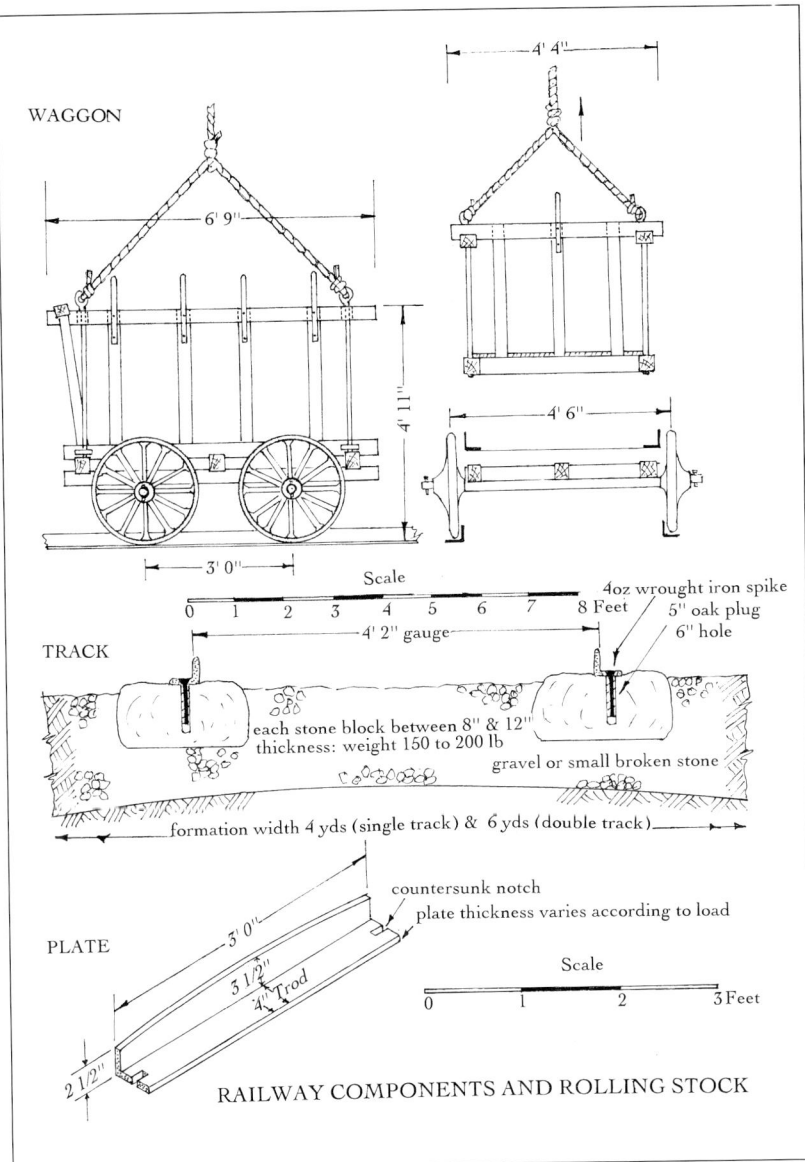

Figure 25 Railway components and rolling stock.

those that Outram had laid down for the Derby Canal railways some years previously, except that he now advocated a gauge of 4 ft 2 in. (a convenient 50 in., which he also described as the 'width of road') between the flanges of the rails (although it is not clear whether Outram meant that he measured gauge between the inside or outside of the flanges), such that the wheels of the carriages would run 'in tracks about 4 feet 6 inches asunder'. Each cast-iron rail was 3 ft in length and varied in weight according to the loading; for heavy work, constant use and durability, Outram recommended rails of 40 lb (these would be approximately $^5/_8$ in. in uniform thickness) but in some cases, even rails of 56 lb would be necessary (approximately $^7/_8$ in.). It was essential to super-elevate the outer rails on horizontal curves and also to reduce the gauge, all in proportion to the radius of curvature. Special rails were needed at locations where lines branched off, or intersected with other lines, and also where railways crossed over highways. After the blocks and rails were positioned, the 'horsepath' between the rails was filled with gravel. Some of this ballast would also be placed outside the blocks so as to prevent movement and misalignment, but it was also important to ensure that the finished level was below that of the tracks so as to keep these free from dirt and other obstructions (Fig. 25).

This plateway system, with only minor modifications to the design of the rails, was also adopted by Jessop on major railways, notably for the Surrey Iron Railway and the associated Croydon, Merstham & Godstone Railway; also for the Kilmarnock & Troon Railway in Scotland,[1] in spite of the fact that Jessop, in conjunction with the engineer Christopher Staveley, designed and supervised the construction of the Loughborough and Nanpantan railway with edge-rails and waggons with flanged wheels.[2] This railway, opened in 1794, was part of the Forest line of the Leicester Navigation and, although the technology proved successful, it seems to have been the only occasion when Jessop used the system which, of course, eventually displaced the the much cruder plateway. Farey visited the site during 1808; unusually for him, he described the rails and wheels in the vaguest of terms, stating that the rails 'had bars flat on top and the wheels are cast with flanches inside for keeping the trams upon them'. Structurally the edge-rail is far superior for this purpose than the angle-iron plate would ever be. At the time, however, Farey clearly preferred the plateway because he stated that ' ... the flanches of iron railways are almost universally cast on the bars, and the wheels are plain, by which they are fitted for being occasionally drawn off the rails on to common roads'.[3] This view was shared by others, including Jessop, who implied as much when reporting on the proposed

---

[1] C. Hadfield and A.W. Skempton, *William Jessop. Engineer* (1979), pp. 175–82; C.E. Lee, 'Early railways in Surrey', *Trans. Newcomen Soc.*, xxi (1940–1), pp. 71–3.

[2] Hadfield and Skempton, *William Jessop*, p. 171; R. Abbot, 'The railways of the Leicester Navigation Company', *Trans. Leics. Arch. and Hist. Soc.*, xxxi (1955), pp. 51–61.

[3] Farey, *General View*, iii, p. 288.

Derby Canal railways.[1] He suggested that waggons could be of a type which readily transferred to the highway at the railhead for the local distribution of goods. Perhaps this was why he did not develop the edge-rail system at a later stage of his career.

Outram saw things differently, judging by the ideas discussed in Anderson's paper. First, he considered that railways should be developed beyond their current uses for the local transport of coal and quarry products. By way of an example it was explained how railways might link the London docks with 'commodious' parts of the city. It would be easy to transport goods to various fixed points by railway, in much the same manner as with canal operations, from where parcels of goods could be distributed to nearby destinations. This process, however, which involved much separating and handling of goods, would prove costly. It could not be justified for the relatively short hauls envisaged, hence an alternative procedure would have to be adopted. This necessitated all the waggons of the railway being of the same design, such that 'each [would be] capable of containing one ton of sugar or other goods of similar gravity' and 'the body of each of these waggons be put upon a frame that rests upon the two axles of four wheels that are calculated to move only upon the railway'. Each wagon would be loaded with goods intended for the same destination where the body of the wagon could be lifted by crane from its frame and transferred to another, similar frame fitted with road wheels and shafts for horses. Thus the carter could take the load to the street and house to which he was directed and, after unloading, the body could be returned to the railway. On heavily trafficked routes the tracks might, with advantage, be doubled up to prevent undue congestion and to speed up the process. At no time did Outram propose transferring waggons directly from railway to highway, as did other engineers of his day.

Using the same basic strategy, it seemed practicable enough to construct railways from 'London to Bath, and every other part of the country, applying the same construction principles of Outram's improved plan'. Such railways would be used exclusively for the transport of weighty loads, leaving the highway solely for coaches and light carriages. Anderson, in conjunction with 'the inventor', costed a single railway at £1,000 per mile but, because single lines were often an inconvenience, only double tracks should be considered for such inter-regional railways. These could be expected to cost about £2,000 per mile, except

---

[1] DLSL, DCCR, W. Jessop to chairman of subscribers of the intended Derby Canal, 3 Nov. 1792.

in the vicinity of London where prices were higher than elsewhere and where the railways 'should be of the strongest sort', because these would be heavily trafficked routes costing about £3,000 per mile. Taking maintenance, interest and other charges into account, it was reckoned that such double railways were cheaper to maintain by £100 per mile per annum over the turnpikes. Moreover, it was estimated that goods transported by the highway would cost a contractor, transporting 600 tons of goods each day, 3s. 4d. per ton against 4d. per ton by railway, comparing costs on a similar basis.

It was claimed that, as a result of these calculations

> few measures that can be proposed will hold forth such an important national movement as this would be, and if applied to almost every part of the country it would reduce the number of heavy road horses to one-eighth of what they are at present ... and augment the number of cattle or other consumable provisions in a proportionate degree.

The price of the carriage of goods of all kinds would be reduced to an 'amazing extent', would induce a cheapness of provisions, encourage manufacturers and agriculture, and 'would produce a general prosperity which, by augmenting the consumption of taxable commodities, would augment the public revenue'.

A significant factor in these arguments was that the railways must be held in public ownership, retained on the same footing as public highways, under the direction of commissioners appointed for the purpose of their proper superintendence. Hence they would be 'kept open to all who shall choose to employ them as a King's Highway under such regulations as it shall be found necessary to subject them to by law'. This was the only means by which the control of railways could be free from 'a set of knowing, artful and too often unprincipled persons' because, like the canals before them, the railways would be a tempting bait to the 'avaricious money-holder, who will contemplate with a greedy joy the opportunity that is presented to him of augmenting his all too abundant stores'. Anderson invited 'the attentive consideration of every individual who shall read this. Let him not take the above facts upon my authority, but scrutinise them with the most rigid attention so as to be sure he is and cannot be deceived'.

This far-sighted paper, containing much practical reasoning and common sense, was a portent of what was to happen within a few decades, with the advent of the steam locomotive and the railway era. As Outram was to find out during the years he spent unsuccessfully promoting his ideas to unduly cautious canal proprietors and industrialists, he was too far ahead of his time. John Francis recognised this in 1851 when writing on the ideas expressed in Anderson's paper.[1] He stated that 'the notions were sagacious and instructive but the time had not arrived for the thorough and earnest consideration of the railways' system'. Furthermore, the avarice of man during the coming railway era was also correctly anticipated by the authors but their warnings in that respect, not surprisingly, came to naught. It is remarkable, nevertheless, how the problems and the possibilities to which Anderson and Outram referred are paralleled by those matters which preoccupy and perplex the present generation of transportation engineers.

---

[1] J. Francis, *History of the English Railway (1820–1845)* (1851), i, p. 171.

CHAPTER 13

THE CROMFORD CANAL RAILWAYS

Several railways were listed in Dr Anderson's article, all of which can be confidently attributed to Outram. All have been referred to elsewhere except the Crich and the Beggarlee Colliery Railways, both of which connected with the Cromford Canal.

The Crich Quarry Railway (The Butterley Company Gangroad)

The railway at Crich was operational by early 1793 when the western end of the Cromford Canal was nearing completion. Farey briefly described the line as he saw it in 1808 when the limestone quarries (owned by the Butterley Company) which it served were in full production.[1] The railway ran from the Bull Bridge wharf on the canal, 1½ miles to limestone deposits situated south-east of Crich and north of Fritchley village. It entered the quarry by a tunnel which was driven through 100 yards of shale strata before penetrating the limestone. By the time of Farey's visit the tunnel had been extended into an open pit some 150 yards by 80 yards wide; it was 24 yards deep in several places, such had been the demand for the stone. The tunnel into the quarry was worked out by 1900.

From the quarry floor the railway rose with a steady but moderate gradient for the first 700 yards, followed by a long straight run down to the wharf, with gradients of between 1 in 80 and 1 in 50.[2] Undulating ground necessitated the building of a stone-walled embankment through Fritchley, below which was a shallow cutting crossed by a masonry arch bridge. In Farey's day the waggons used on the line were of 30 to 35 cwt capacity and carried stone in blocks weighing up to 4 cwt each. Five loaded waggons could be drawn down by one horse which could apparently return, drawing the empty train without difficulty. The waggon wheels were cast with round holes in them instead of spokes so that wooden truncheons could be thrust in to jam the wheels and control

---

[1] J. Farey, *A General View of the Agriculture and Minerals of Derbyshire*, iii (1817), p. 336.

[2] 'Dowie' (A.R. Cowlinshaw and J.H. Price), *The Crich Mineral Railways* (1971), pp. 5, 6.

speeds down the steep hurries which led from the higher parts of the quarry. Once at the wharf, tipplers located on a high bank above the canal overturned the waggons, shooting the contents directly into boats, or on to the wharf. To withstand the violent shock of loading in this manner the boats had to be made from wrought-iron plates, lined with thick deal planks. In later years a self-acting inclined plane replaced the earlier method.[1] The original gauge of the railway was 3 ft 6 in., which Outram eventually considered was suitable only for mineral lines. It was increased to 3 ft 9 in. in the mid nineteenth century when plates were replaced by edge rails. The line and quarry were closed in the early 1930s.

The Beggarlee Colliery Railway

In May 1797 Thomas Walker of Bilborough approached the proprietors of the canal company to discuss a possible tramroad to link Beggarlee colliery, newly leased by him and his partner Barber, with the canal. He was considering building the line to join the Cromford Canal at Langley Mill, just above the junction with the Erewash and Nottingham canals. It would serve his purpose equally as well, however, were the line to link up with both the latter canals and so avoid the Cromford Canal altogether. Naturally he would like to have a reduction in tolls were he to proceed according to his former plan, and he revealed that the Erewash proprietors were also willing to pay a modest tonnage to the company for his coal entering their waterway from the Cromford Canal. This crafty plan was, not surprisingly, successful. The line ran 1¾ miles from the canal up to the colliery, with a branch leading three-eighths of a mile from there to a pumping station;[2] it was built with iron products manufactured at Butterley.

It was late in 1798 when Barber & Walker set about building their railway. The job was not contracted out to Outram but it seems likely that the principles of the improved plan were followed fairly closely during construction. Advice was requested on rail design and so Joseph Outram junior sent a messenger to Walker with a model of a rail of 36 lb (to carry a load of 2 tons) for his approval. Walker then suggested a few alterations which obviously complicated the standard design. Besides a wider sole and larger notches, miscellaneous beads and tabs were

---

[1] Farey, *General View*, iii, p. 339.
[2] Ibid., p. 337.

required on the castings, of which Joseph did not approve. He replied (no doubt on the instructions of his brother) that 'the Tabs will be found very inconvenient in turns and if the Rails are well laid I never knew one unsteady on its bearings on the end and if you have them broader, unless they are exceedingly well bedded, it will have more a tendency to break the Rail than otherwise'. He went on to point out that the changes added to the weight of the rails without increasing their strength and this would render a rail of 36 lb (of Walker's design) only equivalent in strength to one of 32 lb (of Outram's design). He equated this weight decrease to a load reduction of 4 cwt. Accordingly, if Walker insisted on his changes then Outram & Co. would either have to make rails of 40 lb, or else guarantee the 36 lb rails for loads of 4 cwt less than the 2 tons Walker planned. This would be necessary because the company guaranteed their products for three years and they had no intention of guaranteeing a rail of 36 lb to do the duty of one of 40 lb.[1] It seems, however, that there was some compromise because the final weight adopted was 38 lb! All these arrangements were successful because, according to Farey, when finished it was 'a neat and perfect railway'.

Farey mentioned both the Crich and Beggarlee lines among the twelve railways he had observed branching from the main line of the canal, besides five others on the Pinxton branch. Not all of these were constructed during Benjamin Outram's lifetime, nor is it certain which were built using Butterley iron rails and rolling stock, and in accordance with his improved plan. The exceptions are those railways for which accounts appear in the furnace ledgers or are mentioned in the company's letterbook.[2] In particular, these included railways for Mrs Helen Morewood, the Revd D'Ewes Coke and Joseph Wilkes. None of these railways was built under the personal direction of Benjamin Outram, although he generally took an interest in the setting out and alignment of them to ensure that they were laid to his specifications. If he found any imperfections then, ever jealous of his reputation, he was quick to criticise and sometimes was very cutting in his condemnation.

Morewood's Hermitage Colliery Railway

Squire George Morewood of Alfreton Hall had long been in dispute with Joseph Outram senior over mineral rights of their adjacent properties.

---

[1] DRO, D503/12/1, J. Outram jun. to Thomas Walker, 30 Nov. 1798.
[2] Ibid.

After his death in 1792, the hall and lands were bequeathed to his wife Helen but the controversy continued, which eventually resulted in her appearing as defendant in a court case during 1793, brought by both Joseph and Benjamin Outram.[1] She was accused of extracting coal from under their property in Alfreton. The Outrams were successful, both in the first trial and also on appeal, but animosity between the two families lingered for years afterwards, as their business correspondence often showed.

In spite of these wounding disagreements, Helen's second husband, the Revd Henry Case (his surname was changed to Case-Morewood on marriage)[2] called at the Butterley Ironworks during the summer of 1796 to discuss the purchase of rails for a railway to connect his wife's Hermitage Colliery with the Cromford Canal. His wife had applied to the canal company six months earlier, asking them to build a railway from a colliery near Swanwick to the western end of the tunnel, but the company declined, stating that it was not their policy to construct railways.[3] Land for wharfage was nevertheless available and she was assured that space for this purpose would be allocated at each end of the tunnel if she so wished. That was as far as the company were prepared to go. The Morewoods decided to accept the offer of facilities at Greenhill Lane wharf, just east of the Butterley tunnel (a site shared with lime kilns erected by Outram & Co.) for a railway leading to their Hermitage pit, situated a short distance north of the canal. Their line was built by their own labour force and a contract was made with Outram & Co., but only for the supply of railway track and other components. In later years a branch from this line was to run north to pits at Greenhill and Alfreton Common.

Work was already in progress on site preparation in October that year when Joseph Outram junior went to visit, only to find that some unsatisfactory earthworks had already been finished by the Morewoods' workers. On the instructions of his brother, Joseph checked the existing route. Although the matter had nothing to do with him, he strongly disapproved of the setting out and decided to select and level a much better line. Joseph junior then wrote to Mrs Moorewood in his usual courteous and conciliatory manner and recommended abandoning the existing section in favour of his preferred alignment.[4] He was sorry that

---

[1] DRO, D184M/WB1–58; *Derby Mercury*, 27 March 1793.
[2] B. Nicholson, *Joseph Wright of Derby. Painter of Light* (1968), i, p. 323.
[3] PRO, RAIL 819/1, 30 Nov. 1795.
[4] DRO, D503/12/1, J. Outram jun. to Mrs Morewood, 3 Oct. 1796.

'so much time and expence has been applied to so bad a purpose in forming the ground'. His levelling of the line had shown that a suitable falling gradient of 1¼ inches in a yard was possible and this meant that all her coal could be brought down an iron gangroad at a 'very easy expence'. His checks of the work already completed showed that in some places, the gradients were anything between 3 in. to the yard and on the level. In others, it was 'absolutely ascending the wrong way. There the Horse would be overloaded & in the other extream would be overpowered by the push of the Waggons'. Furthermore, he considered that the present cuttings would turn into ditches in wet weather and the only way to remedy all this incompetent work would be to form an entirely new road. He finally stated that it would be 'inexcusable to lay a good Road on so ill formed a basis' but he would do all he could to comply with his brother's instructions to 'make you a compleat Road', and would gladly meet her representative to discuss the matter. He met her husband on site a few days later when he explained his proposals.

In spite of Joseph's explanations, Morewood's supervisor completely ignored the new alignment and carried on as before. He may also have removed Joseph's setting-out pegs. Hence, Benjamin Outram wrote to Morewood several days later, bluntly complaining that the man had done so, 'without making any application to learn my Bros intention or apology for his interference'. He also noted that Mrs Morewood had sent a labourer along to say that 'she is persuaded to adopt the existing line at present Cut', but to continue would make the present line 'a bungled piece of work neither profitable to Mrs Morewood or creditable to us and if Mrs Morewood is determined to have it done in that imperfect manner I hope she will excuse us for declining to have anything to do with it and accept our thanks for her kindness in offering us the order for the Castings'. He stated that many rails and waggon wheels had been cast but these could be disposed of elsewhere and he would rather 'advise her to use common Carts than be at great expense to make an imperfect Road. I find her Agents strive to excuse errors by persisting rather than correcting them'. And also 'that we have business of our own to prevent our intrusion into other people's and the ready mode would have been to deliver the Rails and let the Agents lay them down as they please but I cannot help considering our Credit as outballancing the Profit in a business so near home'.[1] Outram then asked for Mrs Morewood's 'final determination' which must have arrived almost immediately because he was writing again two days later.

---

[1] Ibid., B. Outram to Revd Morewood, 22 Oct. 1796.

Mrs Morewood had asked for his reasons for refusing to provide rails and Outram responded with more restraint than previously.[1] His objections had been 'far from a desire not to accommodate you' and were made strictly on engineering grounds. He explained that, whilst the first section from the colliery was satisfactory, the rest crossed steeply sloping ground and it would be costly to excavate there to the slack gradients necessary for an efficient railway. The line should be moved over some 40 yards on to much easier ground where it would be far cheaper to dig out a better route. Outram did not approve of the layout at the wharf but because her agent wished it to remain as it was then 'it may do very well'. Of the parts 'that were so irregularly formed' he was adamant; there were 'insurmountable objections'. As regards the rails, there were already 300 or 400 rails ready cast 'and the remainder may be made as fast as the Workmen can lay them down'. If Mrs Morewood agreed, he could send men over when needed to start laying rails from the pit in the direction of the canal. Mrs Morewood's own labourers could work ahead of these platelayers, preparing the ground, spreading and consolidating a bed of stones. Outram said that several sets of waggon wheels were already cast in the works and these could be bored and fitted with axles so that one or two waggons could be used immediately to convey and distribute coal slack which formed the path for the horses. It was a most amicable letter, offering some sound advice. Outram also reminded her that under the powers of the canal Act she was entitled to lay a railway, subject to agreement with landowners over damages; he even offered to negotiate with the owner of the land and wharf on her behalf if Mrs Morewood so wished.

Work on the railway was finished by the end of the year, probably much as the Outram brothers had intended. Boats for transport of her coals were provided by Outram & Co. but it was not long before Mrs Morewood was complaining again to Outram about their too infrequent visits to the wharf, as well as his credit terms. He in turn was 'much astonished at her note' and once more they were at odds with one another.[2] Outram, not to be outdone, countered with a series of complaints and criticisms, not all of them germane to the business under discussion. He wrote that there were always insufficient coal stocks on the wharf to provide economic loads for the boats, because only four waggons were at work on her line; even these would not last long because of the rough manner in which they were used. And 'if a proper

---

[1] Ibid., B. Outram to Mrs Morewood, 24 Oct. 1796.
[2] Ibid., B. Outram to Mrs Morewood, 11 Jan. 1797.

number of Waggons had been made as directed 3 months ago more coals might have been ganged down than all the boats could have taken'. He was critical of her prices, saying that coal from the Ripley and Pentrich pits was 7d. cheaper than hers when sold at Cromford. Moreover, he was ordering all his boatmen not to load any more at the Hermitage wharf but to go where their custom was welcome and at whatever credit terms he chose to make. As if that was not enough, he claimed that 'there is not another Colliery in the Kingdom but would be glad to pay their Foundry Bills in Coals', alluding no doubt to the Morewoods' negotiated method of part-payment for their railway. As for the rest of the cash owed to Outram & Co., 'it is hoped that Mrs Morewood will direct it to be discharged by the end of the Month as the Money will then be wanted & an extra charge of Interest will not compensate the disappointment of delay'. As a parting shot, which recalled his earlier criticism of the construction of the approach to the wharf, his postscript noted that 'The manner in which Mrs Morewood's Wharf is now made & approached by the Railway will be inconvenient to stocking coals unless another Branch of Railway is laid on the top of the Bank as B. Outram wished it to have been done at first. If something is not done & a 2 months frost or a slackness of trade should happen the inconvenience will be severely felt'.

This must have invited a vigorous and immediate riposte from Alfreton Hall because Benjamin Outram wrote again next day in more placatory mood 'to remove some false impressions my last Letter seems to have occasioned'.[1] He never meant to say that her coals were inferior to others but he pointed out that they sold at her wharf at 7s. 3d. per ton, whereas other pits nearer to Cromford sold at 7s. Hence her coal was at a disadvantage at the western end of the canal but she had the advantage for all downgate traffic to Leicestershire, Derby and Nottingham. Mrs Morewood was assured that 'the people who have tried them near here give a good account of them' and also that 'so far from wishing your colliery should not answer or that you should have any uneasiness about it I have done all I could do to promote its welfare for the sake of the Country'. Outram went on to explain some of her queries on the railway. He stated that the cost of the rails was not computed from the length of road but on the number supplied. As regards breakages of rails in service he made some interesting points. If any rails broke through defects in casting (and this could easily be proved on examination) then they would be replaced free of charge. The company would guarantee rails for three years for an additional price of 6d. per yard; this did not

---

[1] Ibid., B. Outram to Mrs Morewood, 12 Jan. 1797.

mean 6d. per yard per annum but '6d. per yard paid at first for all three years'. However, 'the short steep part where they lock the wheels does not come under this insurance'. And finally, unusually appeasing, he assured Mrs Morewood that 'I wish your Colliery complete success'.

There were some further eruptions before the business of the Hemitage railway was concluded, none more so than when Joseph junior wrote an anxious letter in early February 1797 to request payment for the supply of rails and other goods.[1] Rails for 1,190 yards of track had been supplied, plus waggon wheels and many ancillary fittings, and the order was complete; over £400 was now due to the company. 'We beg to remind you that tomorrow is our pay day and hope you will favour us by the bearer or early in the morning with 200 or 300£ or we shall be greatly disappointed'. Mrs Morewood returned a note, presumably to say that payment would not be forthcoming until the railway was actually finished and fully operational. Benjamin Outram then took over the correspondence in his familiar terse, aggressive style, his former pleasantries forgotten.[2] 'You was at first and repeatedly informed that the terms on which we offered to furnish you with Rails & on which you gave us orders, were for ready Money … every Contract we have yet executed for Companys or individuals will prove that they are so'. Moreover, all the work that he or his brother had done as individuals was in an effort to help and save her from the consequences of bad workmanship and the fact that the road was still in an unfinished state was not the company's fault; these matters were nothing to do with her contract with the company because this was strictly for the supply of the railway's components. Outram was quite correct in this respect and Mrs Morewood was at fault. Regrettably she did not express any appreciation of the efforts that the Outram brothers had made to ensure that her railway was constructed properly.

Unfortunately for the company the account was still unsettled by early June 1797 and this resulted in another anxious letter from Joseph junior.[3] He wrote to Morewoods' agent because the family were away from home but were expected back that week. 'We were in full expectation of receiving the balance of Mrs Morewood's Account against our last pay day & were put to great inconvenience by the disappointment … we have many heavy payments to make the latter part of this Week we hope we shall not be again disappointed & beg you will

---

[1] Ibid., J. Outram jun. to Mrs Morewood, 3 Feb. 1797.

[2] Ibid., B. Outram to Mrs Morewood, 4 Feb. 1797.

[3] Ibid., J. Outram jun. to W. Parker, 14 June 1797.

endeavour to hand us the money as soon as possible'. Mrs Morewood swept aside that timid request and responded with a demand that the company make up two separate accounts; one for her and the other for her husband. Joseph was baffled by this further evasion and when replying to the agent, wrote that he scarcely knew what was hers and what was her husband's and, in any case, that was not how business was done at Butterley.[1] This of course was just a delaying tactic of the Morewoods but no further reference is made to this business in the letterbook, and so it is likely that the account was settled in due course, presumably in Mrs Morewood's own good time. Not all customers were as recalcitrant as the Morewoods, even during those times of acute financial crises, and yet in the years following Outram's death in 1805 an assessment of the company's finances showed bad debts in excess of £7,000.[2] It is hardly surprising that Benjamin Outram could be so determined, and often so impolite, when dealing with such customers.

The Butterley Carr Railway and access to Butterley tunnel

One of the railways mentioned by Farey was that leading from the Butterley furnace three-eighths of a mile to Butterley Carr Colliery, situated on the Butterley Hall estate.[3] Nothing is known of the design of this line but it must have compared closely with that at Crich which was built about the same time. The furnace, which was located over the line of the Butterley tunnel about 170 ft below, was close to two large-diameter shafts which had been sunk down to a recess adjoining the canal level. Farey stated that tram boxes of coal, ironstone, limestone, fluor and other materials for use in the works were drawn up one of the shafts from boats below, and pig iron and cast goods were simultaneously lowered down the other shaft. Originally a water bucket, supplied from a nearby reservoir, descended a third shaft as a counterpoise for raising or lowering goods in the main shafts. In this respect the system was similar to that adopted by Outram when driving the Standedge tunnel on the Huddersfield Narrow Canal, although the method was commonly applied in Derbyshire mines of that period. In later years a steam engine was substituted and guide chains were then installed in the

---

[1] Ibid., J. Outram jun. to W. Parker, 29 June 1797.

[2] DRO, D503/103/B1/3A, Report of J.S. Harvey to the Master of the Rolls, 8 Dec. 1813.

[3] Farey, *General View*, i, p. 102; iii, p. 337.

shafts to steady the frames holding the boxes. When the latter were lifted to ground level a staging, on which a set of tram wheels were backed into position by a horse, was slid under the suspended frame. A box was then placed on the wheels and the contents were moved off to an unloading bay, after which the tram was drawn to the second shaft and the box lowered once more to the canal level. Farey also mentioned that a short tunnel was driven from the base of the shafts into the workings of the Butterley Carr Colliery; the mouthing of this survives to the present day.

The Brinsley Colliery Railway

Mention has been previously made of Joseph Wilkes and his enthusiasm for railways.[1] A progressive industrialist, much interested in all modes of transport, he owned prosperous collieries at Brinsley (Nottinghamshire) and Measham (Leicestershire). Farey referred to the former pit which was served by a railway beginning at the Brinsley Wharf on the Cromford Canal. This ran north-east about a mile to New Brinsley village and then to the colliery, some three-eighths of a mile beyond. All the railway components were provided by Outram & Co. and a full range of prices was submitted to Wilkes during mid-1796 for rails, waggon wheels and axles.[2] Wheels were priced at 14s. per hundredweight; rails varied from 18 lb per yard for waggons of 1½ tons loading, up to 40 lb per yard for waggons of 2½ to 3½ tons loading. The heaviest rails were offered at 10s. each and all were guaranteed for three years at a premium of £5 per cent of the contract price 'provided they are properly laid on a bed of stone and fairly used with carriages not exceeding the above weights'. Benjamin Outram wrote to Wilkes from Manchester in May 1797, presuming that 'our People have delivered your whole order of Rails at Brinsley before this' and noting that when he had visited the site two weeks previously, coal was already being ganged down from the pit to the wharf.[3] He nevertheless regretted that 'time would not allow the laying of a bed of broken stone under the sleepers', which he rightly felt was essential for the stability of the road. Presumably he was then indirectly hinting to Wilkes that this would have ramifications for any future claims for broken rails.

---

[1] Above, pp. 203–4.

[2] DRO, D503/12/1, Price list, nd (probably 1796).

[3] Ibid., B. Outram to J. Wilkes, 9 May 1797.

This railway was obviously designed in the main in accordance with the improved plan, although some sections were set on wooden sleepers which Outram and Anderson did not think were entirely appropriate for railways. Outram mentioned in passing that he hoped that a 6 in. foundation bed of small stones would be placed when the line at Measham was laid because construction of this railway was apparently due to begin and was to be finished by the end of the year. Having diverted attention to this proposed railway he then, good salesman that he was, took the opportunity to canvass for a contract for the supply of rails 'as you intimated to me when I last saw you, the earlier you do it the greater the favour as we can do them better & to more advantage having more time to cast them'. He invited payment for the Brinsley contract of '350 to 400£ at two months Credit or, if agreeable please to remit us Bills you may have on hand for that amount. Our usual Credit for rails is Cash in 2 Months from delivery of each considerable parcel, or Bills due in that time'.

Other Railways

Other railways, referred to briefly by Outram in correspondence, were built at the western end of the Butterley tunnel to serve collieries at Hartshay and Pentrich, no doubt with materials provided by Butterley ironworks. They were probably completed by the time the canal from Cromford to the tunnel was finished in late 1794. Several other colliery railways were in course of construction in the early years of the nineteenth century. One of these was for the Revd D'Ewes Coke on the Pinxton branch of the Cromford Canal for which Edward Banks was the contractor.[1] Little is known of the original construction of this line but the rails (3,762 in number) were purchased from Butterley, in addition to miscellaneous fittings. The length of main line was 1,665 yards and, unusually for that district, timber sleepers were laid on part of the road.

It is likely that Outram developed his improved plan from early experiences gained on the Cromford Canal railways and particularly on the Crich line. Then followed the the Derby Canal railways, the success of which may have more widely publicised his ability as the leading railway engineer of his generation.

---

[1] DRO, D503/42/1, Furnace Ledger 1800–22, p. 101.

CHAPTER 14

THE ASHBY DE LA ZOUCH CANAL AND RAILWAYS

Proposals for an Ashby Canal had been aired on several occasions prior to the Canal Mania years.[1] The purpose of the canal was to exploit the coalfields at Ashby Wolds in Leicestershire but coupled with this was the need to improve transport for the export of lime from quarries north of Ashby de la Zouch. Conflicting interests ensured that these plans came to naught until 1794.

Early that year a scheme was prepared jointly by William Jessop and Robert Whitworth for a waterway which began at Marston Junction on the Coventry Canal, some three miles from Nuneaton, to follow a 26½-mile meandering, but level, easterly route by Hinckley and Market Bosworth to the Ashby Wolds, where there was a rise of 139 ft over four miles to the summit level.[2] The water supply to the latter was provided by steam-driven pumps from a reservoir of 50 acres, 51 ft below. The waterway then ran for a further 4½ miles through Ashby de la Zouch and a 700-yard long tunnel at Old Parks, before locking down 84 ft to a junction, a quarter of a mile beyond. A branch ran from this junction to the right for five miles to Coleorton, with a further extension of 2½ miles to the limeworks at Cloud Hill. The left-hand branch ran for 1¼ miles to yet another junction, from which one branch led nearly one mile to Staunton Harold and the other 3½ miles to Ticknall. The Act[3] was obtained in May of that year and authorised a waterway of nearly 50 miles in length, with capital of £150,000; the engineers' estimate was £138,238.

Jessop then withdrew, his part in the task completed; Whitworth and his son took over as principal engineer and resident engineer respectively.[4] The contractors concentrated on the lower canal section as far as the locks at the Wolds, when works started during the autumn of that year. Construction also began on major earthworks and the tunnel at Old Parks on the upper level, so that these would be finished in advance of

---

[1] C. Hadfield, *Canals of the East Midlands (including part of London)* (1966), p. 147.
[2] Ibid., p. 148.
[3] 34 Geo. III c. xciii (1794).
[4] PRO, RAIL 803/2, 29 July 1794.

the approaching waterway.¹ The lower section had been costed at the surprisingly low figure of £27,317 but after only two years' activity it was seen to be seriously under-priced; the actual expenditure would be closer to £100,000. Many shareholders were significantly in arrears with their calls; it was a time of national financial crisis and by 1797 the proprietors were sufficiently alarmed to stop work on all but the lower canal. Their attention then turned to the possibility of constructing railway branch lines up to the quarries instead of canals, because the latter would be 'too expensive to be executed with any hope of final success'.² Robert Whitworth was asked to report and prepare estimates. Fortunately for the company, their Act of 1794 had been drafted with clauses empowering the company to transport goods by 'Rollers, Inclined Planes, or in any other Manner than by Water'. Not unreasonably, this was later interpreted as permitting the substitution of branch canals by railways.

It is surprising that railways were not chosen for all the branches above Ashby Wolds at a much earlier date when the proprietors and their engineers were preparing their petition and bill. Even from a cursory study of the deposited plan, it seems obvious that the winding courses of the upper branch canals, as well as the expensive and vulnerable means of providing the water supply to the summit by steam engine, were uneconomic engineering solutions. It was a classic case of terrain which was unsuitable for inland navigation but appropriate enough for railways. The failure to select this alternative technology also questions William Jessop's complete commitment to railways at that time. However, this was not the first occasion that railways had been proposed for the limestone workings above Ashby. Jessop had, in earlier years, prepared a rather limited scheme for the Earl of Stamford for a railway, or stone-road, in addition to a short canal, which would lead from the limeworks at Breedon Hill down to the River Trent. He had reported his findings in September 1787.³

In his report, Jessop drew attention to the increasing demand for lime for agricultural purposes, noting that 'No limeworks that I remember to have seen appear to afford so much room for an increase of sale with so little difficulty in the way of it, as those in the neighbourhood of Breedon'. Land carriage costs (as much as 12s. per ton) and the

---

¹ Ibid., 15 June 1795.

² Ibid., 15 Jan. 1797.

³ PRO, RAIL 803/15, W. Jessop, On a Proposition for Making an Easy Communication between the Earl of Stamford's Lime Works to the River Trent, 10 Sept. 1787.

inaccesibility of the quarries were obstacles to sales, in spite of the benefits the products could bring to farming. The charges for the 2½-mile run down to the Trent would not cost more than 8d. per ton and from there the lime could be sent to Warwick, Stafford, Derby and even into Lincolnshire, at cost savings of 8s. to 10s. per ton. 'Many will use four times the quantity they do at present and others within the same district will get it, who never before got it at all'. He anticipated that the current production of 20,000 tons per annum from Stamford's Breedon quarries could increase fourfold and, upwards of 60,000 tons of this would go off by water. If that were so 'I should not think it matter for astonishment'.

Jessop's proposals were for 1,800 yards of double railway (he did not state whether this was a plateway or edge-rail) from Breedon Hill to connect with a 4 ft deep canal section, 1½ miles long, leading in a straight line to the Trent. There were to be three locks, each capable of holding two Trent barges, and the water supply would be provided from nearby brooks. The canal would terminate at Weston Cliff and he further recommended one or two locks in the opposite bank of the river to allow access into the Trent & Mersey Canal, which was some 14 ft above normal river level. The latter proposal was in fact incorporated into the later plans of Benjamin Outram when building the Derby Canal, although these locks were rarely used in later years and eventually became derelict.[1] Jessop also suggested, that a 'stone-road' might be built instead of a railway at only half the cost. Formed with small stones and used by carriages with cylindrical wheels, it could be maintained at nominal expense. The estimate for his plan was £6,216 and yet for all its advantages for increasing trade, Lord Stamford took no further action on the matter. Perhaps the fact that the route passed through Lord Huntingdon's deer park, as well as Lord Melbourne's meadows, on its way down to the Trent influenced that decision.

Nine years later, when Robert Whitworth was faced with the committee's request to design railways as substitutes for the upper branches of the Ashby Canal, his solution was wholly inappropriate in that it failed to address the purpose of the partially completed waterway as well.[2] Whitworth decided on two completely separate systems, neither of which was connected with the Ashby Canal: one was for a railway to run from Ticknall to the Trent at Ingleby; the other, almost identical with Jessop's earlier plan, was for a combined canal and railway to link

---

[1] Above, p. 63.

[2] PRO, RAIL 803/15, R. Whitworth, Report to the Committee of the Ashby de la Zouch Canal, 6 Feb. 1797.

Breedon Hill with Weston Cliff. He recommended that the railways, in view of the likelihood of very heavy traffic, should be of iron (or wood covered with iron) for the downward journey, although wooden rails should prove satisfactory for the return trip. Presumably he was thinking in terms of carriages with flanged wheels. Whitworth's report of February 1797 included estimates for the two systems, amounting in total to £12,218. Rather unwisely, he stated at the end of the report that 'The above are the best estimates of the several works mentioned in the preceeding Report that I can at present make; not being much aquainted (sic) with making Railways'.

The proprietors were probably not much impressed by the clumsy, unimaginative plans presented by their elderly engineer, but other issues diverted them at the time and so the question of railways was left in abeyance until June 1798. By that time, both Whitworths had been dismissed from the company, perhaps on cost grounds, to be replaced temporarily by Thomas Newbold, a local man.[1] The latter, in company with Thomas Jewsbury, a committee member, was instructed to view the proposed canal branch lines and to take levels so as to ascertain the best courses for railways to Ticknall, Breedon and Cloud Hill.[2] Common sense prevailed on this occasion, in that they were told that the lines must link up with the main canal at Ashby Wolds. The two men were also instructed to deliver detailed estimates by the next meeting; a tall order indeed. That onerous task must have been too much for Newbold and Jewsbury because neither reports nor estimates were presented as requested. Instead, at committee meetings on 21 August and 6 September discussions centred on the need to seek the advice of Benjamin Outram. Initially there seemed to be some uncertainty about what Outram's brief should be, but eventually he was asked to survey and level the best routes for railways from the works at Ticknall, Staunton, Cloud Hill and Breedon, to communicate with the main canal on Ashby Wolds. Money was short and he was asked to deliver 'separate and distinct estimates' for each of the branch lines.

Outram lost no time in visiting the locality and his preliminary written report was presented to the committee on 25 September. He was politely critical of the canal system originally proposed, noting for the Cloud Hill branch, for example, that '[I] find it so very unfavourable and circuitous and its termination at Cloud Hill so much above the quarries of limestone that I can decidedly say that the railway will be greatly

---

[1] PRO, RAIL 803/2, 14 March 1798.
[2] Ibid., 5 June 1798.

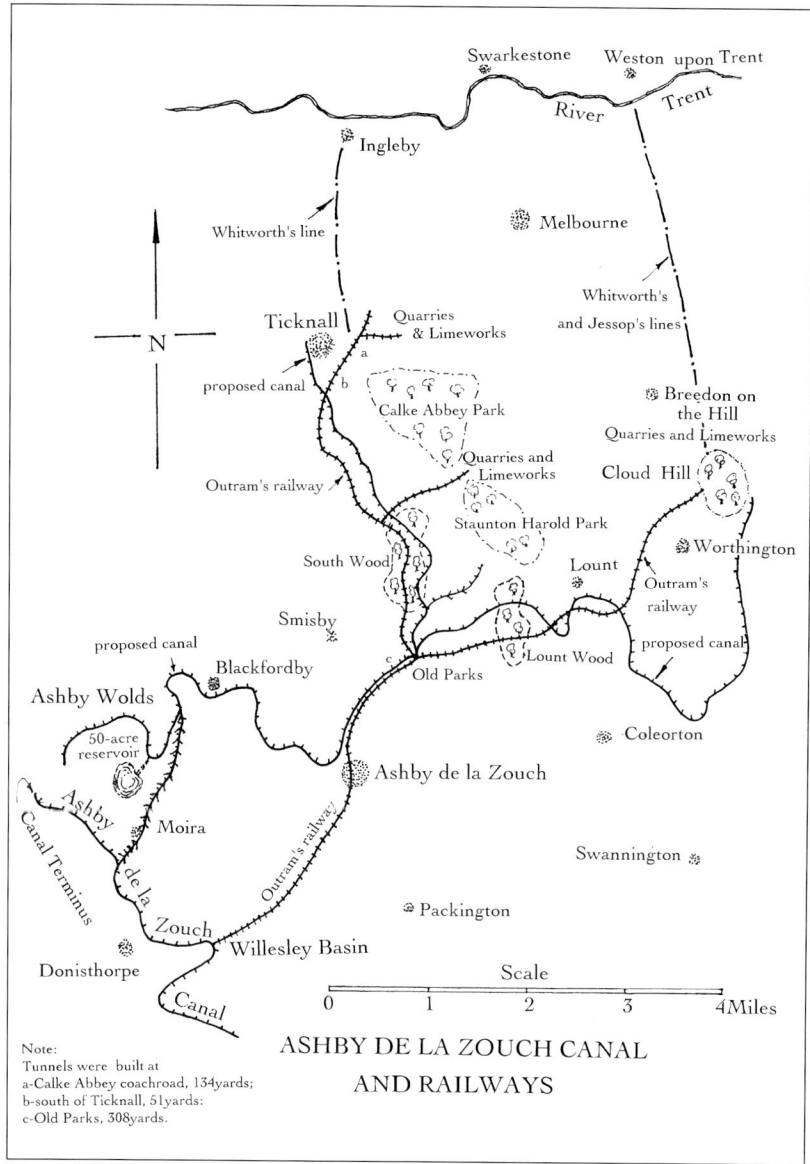

Figure 26 Ashby de la Zouch Canal and railways.

preferable to all parties interested, for it would be but half the length, would cost but one fourth of the expense and, if completed, would do the business cheaper by half'.

Outram outlined the practicalities of building two separate railways which some persons unknown had apparently discussed with him during

Plate 18 Cutting and overbridge near South Wood on the Ashby railways.

his site visits. One line was from Cloud Hill, passing by Coleorton and Packington to connect with a cut to the main canal, and another was from Ticknall and Staunton to join with the canal close to the Wolds. This made for railways totalling 18 miles in length but 'for the most part out of the [canal] line authorised by Parliament'. Outram clearly preferred a different plan, stating that; 'If therefore the company wish to make communications by railways between the limeworks at Ticknall, Staunton, Cloud Hill and Breedon and the canal now completed, the cheapest and most equal mode appears to me by uniting the whole in nearly the same way the proposed canal was meant to have united them' (Fig. 26).

From Ticknall his line would run south through a short tunnel under Sir Henry Harpur's coach road which led into Calke Abbey, ascending gently to the top of Ashby Old Park, just north of the tunnel formerly projected. A short branch from Staunton would join this line near to South Wood. Because a 'most favourable line presents itself from Breedon between the Smoile and Lount collieries to join the above in Ashby Old Parks', then 'From these branches ... a railway might be extended on a gentle and gradual descent westwardly and southwardly to the west of Ashby town, crossing the Burton road ... and extending from thence westwardly to the Wolds south of Blackfordby and thence descending by the line proposed for the locks to the present termination of the canal'. The gradients on the lower parts of the route being rather steep, he felt that 'a more favourable line might be formed if, after

Plate 19 Central section of tunnel under Calke Abbey carriageway.

crossing the Burton road, a southwestardly direction were pursued down the Willesley valley west of the fish ponds and pleasure grounds, to join a short branch from the canal below Willesley'.

Outram claimed that the railways 'would not collectively exceed fifteen miles in length' (final measurements totalled just over twelve and a half miles) and 'would be completed in one-third of the time and for less than one-third of the money the canal would have required'. A major advantage was that by building railways instead of canals, 'the limeworks would be brought closer by 6 miles to the markets, the sales would therefore be extended 6 miles farther down the canal and the trade proportionally increased'. Outram also proposed to the committee that the main canal should be lengthened as far as possible northwards on the same level, and a railway built beyond which would serve the collieries on the northern extremity of the Wolds. That latter recommendation was investigated[1] and eventually accepted by the committee. The canal was thus extended to colliery wharves at Moira, and so the Ashby Canal was built free from lockage over its entire length of 30 miles.

Outram deferred preparation of an accurate survey and detailed design of the lines until the proprietors decided on which plan they liked best, but he assured them that his preferred scheme would cost no more than £30,000 and the alternative, with separated lines linking with the canal,

---

[1] Ibid., 1 April 1799.

would cost between £8,000 to £10,000 more. He also argued against the choice of canals for the upper branches as originally planned, referring to their vulnerability in times of drought and flood, and estimated that the cost of their completion would be about £100,000. Outram was now a confident and determined advocate of the 'improved plan' and he must have been delighted at this splendid opportunity to construct his first major railway. Perhaps it was as well that he had no inkling of the many disputes he would have with the company, on both financial and contractual matters, in the years that lay ahead.

In October 1798 the general assembly of the company accepted Outram's plan, subject to the approbation of the affected landowners. These included Lord Stamford, Lord Moira, Lord Ferrers, Sir Henry Harpur, General Hastings and others.[1] No serious objections were forthcoming at the time and accordingly the engineer was instructed, in December 1798, to proceed with detailed surveys. Just prior to starting field work in mid-January, Outram conducted the committee along the route and also explained that it might be better to lay double lines along the entire length, except through the tunnel at Old Parks.[2] The members accepted all his recommendations and seemed anxious that he should get on with the job with all haste.

By the end of January Outram had further modified his ideas. He wrote to the company clerk that, 'Knowing the necessity under present circumstances of reducing the expenditure of the Ashby company within as narrow limits as possible, I have been considering how single railways between Ashby Parks and the different limeworks might be regulated'.[3] He thought there was no need for passing places but difficulties were bound to arise on the branches where the lines served different interests unless bye-laws were framed so as to control the times which the various gangs could enter the tunnel. South of the latter there was less of a problem because double railways would run down to Willesley basin. By the adoption of single rails as described, he reckoned to reduce expenditure by £5,000. These savings were adopted but passing places were preferred by the proprietors, rather than his recommended method of traffic control, and in fact several such loops were subsequently laid during the progress of the works.

Shortly afterwards, Outram's plans, estimates and specifications were

---

[1] Ibid., 1 Oct. 1798.
[2] Ibid., 14 Jan. 1799.
[3] Ibid., 29 Jan. 1799.

submitted to the company.[1] The estimates consisted of the usual lists of priced items of work to be executed on the main line and the various branches. They are summarised thus:

| Willesley line | 7,100 yards of double railway | £14,315 10s. |

Branches to:

Cloudhill	7,260 yards of single railway	£5,878 0s.
Ticknall	7,200 yards of single railway	£7,432 0s.
Margets Close	420 yards of single railway	£350 0s.
Sir H. Burdett's	622 yards of single railway	£740 0s.

| *Total* | | £28,715 0s. |

At first sight, Outram's contract documentation seems impressively detailed. However, the quantities of the extensive earthworks differ from those obtained during independent engineering inspections three years later, which suggests that Outram's calculations were not as accurate as they might have been. This can be blamed in part on the fact that the engineer had to do the fieldwork, surveys, cross-sections and computations of quantities in a great hurry (less than one month). Nevertheless, it has to be said that everywhere these figures, no doubt deliberately, erred generously on the top side; the estimate was thus boosted by about 10 per cent.

The consequences of this would depend on the preferences of Outram's clients as to the method of payment for the work. On the one hand they could agree to accept the estimate of £28,715 10s. (or some lesser, negotiated figure) as a fixed-price quotation, paid out in instalments as work progressed. Alternatively, they could insist on an accurate re-measurement of each item of completed work, which could then be costed according to the price per item quoted by the engineer in his estimate. The latter method was preferable for a project of this size and complexity, and it is likely that Outram expected to be paid on this basis. On the other hand, if the proprietors preferred a fixed-price contract and were to accept his estimate as a fixed quotation for this purpose, then Outram would be well placed to make a good profit.

Besides the earthworks listed in Outram's submission, other major items included in the scheme were a 500-yard long, 11 ft wide tunnel at Old Parks, priced at £3,300; a double railway along the main Willesley

---

[1] Ibid., 1 April 1799.

line which, including rails, plugs, nails, stone blocks, gravel and labour, was rated at 22s. per yard; and the several branch lines, which were single railways, priced at 11s. 6d. per yard. The accompanying, and very brief, specifications were for railways built according to Outram's improved plan, with sleeper blocks of not less than 150 lb weight and cast-iron rails, one yard in length, weighing on average 38 lb, with none less than 35 lb. For level crossings on public highways, the rails were to be of double thickness, but with low flanges, and proportionately less on private carriage roads. All rails were guaranteed for three years from 25 March 1800, this presumably being the date when Outram expected to begin work, laying rails on the complete earthworks.

In several respects the specification documents were vague, particularly when excavations and drainage were described. For example, earthworks would be formed 'to the proper inclinations' and, 'a trench to be opened on the upper side of each road to carry off the the surface water and land floods and such culverts, soughs and other drains to be made as shall be necessary to drain the whole completely'. The only safeguard for the company was that Outram stated that the earthworks would be properly levelled and soiled to the satisfaction of committee representatives Greaves and Wilkes. The latter was the coal-owner from Measham whom Dr Anderson held in such high regard and who had probably recommended Outram and his system to his fellow proprietors. There was nothing unusual about these arrangements—Outram was simply conforming to the standard practices of the day—but the obscure descriptions and lack of dimensioned construction details were to lead to some serious disputes, much to his disadvantage, before the works were completed.

Outram had hoped for an early reply and acceptance of his tender for a 'design and build' contract, because at the end of February he sent a special messenger with a letter for the company clerk, to enquire of progress.[1] He explained, quite reasonably, that he was unwilling to seek other contracts for the Butterley foundry because the Ashby railways would tax the capacity of the works to its limits of output, once an order for the casting of rails began. In the meantime he had a large number of men who had little to do; he was polite but clearly anxious to expedite matters.

Two weeks later he wrote again, assuring the committee that he had repeatedly examined the ground and had modified the line and estimates

---

[1] Ibid., 26 Feb. 1799.

so as to overcome sundry objections.[1]

> I hope I shall not be thought to ask too much when I request to be informed this day whether the committee are satisfied with those estimates and whether they wish to enter into contract, either positive or conditional, with me for the execution of the works, the magnitude of which are such as would almost wholly occupy our foundry for 15 months. Being under the expectation of, for some months back, entering into engagements with you I have avoided entering into any large contracts for cast iron. To do so would be highly imprudent in a concern where nearly 300 men are looking for their daily wages.

Outram wrote that 'foresight and fixed markets for the articles manufactured are necessary'. He also stated that he had the opportunity to accept large orders from the West Indian plantations and America, quite apart from sundry orders from other canal companies and industries, and reiterated that he dare not accept any of them in the present circumstances. The canal company was assailed with these, and other arguments and hints, in order to get a decision. At that stage, Outram was unwise to pursue the matter so urgently; time was needed for the proprietors to absorb and to discuss the proposals and quite properly, they saw fit to ignore his promptings.

The general assembly met on 1 April 1799 and unanimously approved the scheme and its specifications, directing the committee to enter into contracts for the 'speedy construction and completion of same'. No such action followed, however; the committee seemed eager to proceed but there was an unwillingness to enter into a contract. The members felt they had a good reason; they were acutely short of funds and were reluctant to commit themselves to the responsibilities and legal obligations of a formal contract.

Further letters from Outram, advising the proprietors that he had a good company of unemployed miners ready to take on the Ashby Old Parks tunnel and requesting instructions, were once more disregarded by the proprietors.[2] On 19 April, however, they must have surprised him by ordering rails for five miles of railway (about 100 tons of double line).[3] The first delivery of 70 tons was made by mid-July and Outram could

---

[1] Ibid., 15 March 1799.
[2] Ibid., 3 and 8 April 1799.
[3] Ibid., 19 April 1799.

reasonably have expected his contract to be signed to coincide with this date, but it was 6 August before the committee again considered the matter. Even then, it was only vaguely resolved that Outram 'be directed to proceed in the construction of the said railways according to the instructions he shall receive from the committee', and that was not good enough.

Outram received a copy of the resolution and with it was a report of a special meeting of the committee held on 22 August to discuss a letter from Lord Stamford who, at this late stage, wrote that he could not agree that a railway could be made without the authority of a fresh Act. Moreover, he had no intention of consenting to the proposed changes unless such necessary parliamentary procedures were put in hand.[1] Stamford even went so far as to say that if this were not done then he would insist on the building of a canal to his works at Cloud Hill, as originally planned.

It was fortunate that the chairman at the time was the Earl of Moira and he was given the task of writing to his neighbour. His letter, written with all the aplomb and confidence of the aristocrat, noted that the gentlemen of the committee

> are happy in the consciousness of having upon every occasion paid the most respectful attention to your Lordship's wishes but they are limited in that disposition by what they owe towards those for whom they are acting and cannot think that they should be justifiable were they to load their constituents with the expenses of an application to Parliament which appears to the committee to be totally unnecessary.[2]

The committee had interpreted the Act entirely differently, believing that they were entitled to employ either canal or railway to meet their objectives.[3] In any case, funds were inadequate for the adoption of the original plan and trials had shown that there was insufficient water to supply both the Cloud Hill and Ticknall canal branches. Moreover, the company was obliged under the Act to supply lime at a cheap rate and transport by a railway could undercut that by canal by at least 1s. per ton. There was no intention of reverting to a canal and, in a telling

---

[1] PRO, RAIL 803/3, 17 Aug. 1799.

[2] Ibid., 22 Aug. 1799.

[3] Ibid., 1 Oct. 1798. Certain clauses in the Act were considered appropriate to justify a railway instead of a canal. This issue had been considered by the general assembly.

conclusion, Moira agreed that Stamford could certainly prevent the railway entering any part of his land, but the consequences of that would be that it would stop at the boundary. The company would thus be deemed to have complied with the Act in that they had brought the railway 'near to Cloud Hill'. That seemed to quell the opposition, although the company saw fit to engage a lawyer shortly afterwards to check on the legalities of this stance, and to guard against any further objections.

Outram had been puzzled by the resolution of 6 August and so he wrote to Piddocke, the acting clerk, in early September to state that he was in want of directions as to how to proceed.[1] Neither a completion date had been specified nor a method of payment agreed, and he had not even been told where and when to begin work. He guessed that these decisions rested upon the funds available and he therefore took it on himself to comment on the company's shaky finances, saying his was 'sorry to find [it] had so little in hand. Surely some efficient steps will be taken to get in the arrears or it will be attended with the worst of consequences. I cannot be bound to the terms I offered if I am not furnished with money to forward the works with regularity and dispatch'. Outram claimed that, since submitting his tender, cast-iron prices had increased by 5 per cent, labour costs had risen and, up to that time, he had only been paid £1,000 for the goods and services he had provided. It was hardly surprising that he was now pressing for a formal arrangement before starting work on the railways and, in his frustration, he felt able to 'urge the committee to take the necessary steps to put their finances in a proper state'.

Shortly afterwards the committee resolved, at long last, that Outram was to begin work on both the Ticknall and Cloud Hill branches and also in the deep cuttings at each end of the tunnel, besides various sections between Ashby and Willesley.[2] It seems likely that these instructions were only given verbally to Outram on site by Greaves and Wilkes who were nominated to act on behalf of the committee. Even though these may have been informal arrangements, they nevertheless constituted a legally binding contract. It is difficult to believe that the proprietors were unaware of that fact and yet there was still no mention of the need to settle the all-important terms of that contract. It was as surprising as it was unwise for Outram to agree to follow this uncertain path but he may well have felt obliged to do so because of the commitments he had

---

[1] Ibid., 4 Sept. 1799.
[2] Ibid., 13 Sept. 1799.

already made so far. He was under pressure to bring work into the foundry and here was an opportunity to keep the workforce busy for several months. His better judgment should certainly have prevailed, as it always will after the event, but then he believed that he was dealing with honourable and reliable men. He could have few fears when Lord Moira and other aristocrats, besides several distinguished county gentlemen, were members of the company; surely everything would shortly be put to rights? He was soon to be disabused of any such notion.

Although the proprietors had now made some sort of progress and had, intentionally or otherwise, successfully evaded their responsibilities in so doing, those arrangements could not serve the company's interests any more than they would Outram's. Neither the contract period nor the method and rates of payment had been confirmed and, as previously noted, many details relating to the construction of their railways and the materials used, were only vaguely specified. But quite apart from that, the proprietors should have known better than to enter into a contract on any terms at all, because they were constantly short of funds and must have known that regular payments to their contractor could not be guaranteed. This was mainly due to the company's inability to persuade its shareholders to pay their regular calls and, although other canal companies were similarly afflicted during the Canal Mania years, it still emphasised their inexperience in promoting major enterprises of this kind. The company lacked a strong, unified management team and, as a result, members of the committee were unwilling to take firm decisions, always preferring to put matters off for another day. Although those gentlemen of the committee would no doubt be horrified to be accused of dishonesty by their evasions, their corporate behaviour might well be considered sharp practice today. Regrettably, the uncertainties described were never resolved satisfactorily throughout the entire period of the contract. Inevitably there were disputes which reflected little credit on either party, although sympathies must surely be with Outram rather than the proprietors of the Ashby Canal Company.

In the meantime, Outram got on with the job as directed. He regularly visited the project but he did not devote all his time to the contract. His agent on site was the 26-year-old John Hodgkinson, a family friend, who was probably assisted by Walter Kenear. Both men were regularly employed by Outram as engineers for his railway work and were well able to define lines and levels accurately. The workforce was mainly drawn from the local farming community, strengthened perhaps by experienced itinerant labourers from other Midland canals. The reputation of any civil engineering company is largely dependent on the quality of the management and staff on site and Outram would be fully aware of this fact and of the necessity of forming the men into efficient gangs. Both Hodgkinson and Kenear were capable of organising and supervising labour but on the Ashby contract the task was made more difficult by their having to manage gangs over several miles of railway. This was due to the insistence by the proprietors that most sections of the work were under construction simultaneously. As a result, it was not surprising that the final inspection of the lines revealed a number of deficiencies in construction which were directly caused by a lack of supervision.

In December 1799 Outram was having second thoughts regarding the gauge of the railway and recommended to the committee an increase from 3 ft 6 in. to 4 ft 2 in.[1] His remarks confirmed his growing conviction that railways were to become a principal mode of transport in the near future. He wrote that the smaller gauge was only suitable for limestone and mineral lines but:

> it is exceedingly probable that railways will soon become general for the transport of merchandise through the commercial parts of the Kingdom, and it appears that many hogsheads and packages require carriages 8 inches wider than those used on the railways at Derby and Crich, and that carriages fitted for railways 4 feet 2 inches in width between flanges would be wide enough for all sorts of loading. It therefore seems desirable that all extensive railways should be of the same width and that width should be sufficient to suit all the purposes of trade.

This major design change would increase the width of bridges and tunnels by 18 inches and, including for the costs of modified earthworks, his revised estimate would be £29,500, an increase of £784 10s.

In that letter, Outram yet again reminded the committee that they should now 'consider and thoroughly settle the terms of his contract and give instructions for the draft to be prepared and the rates of payment to be specified, the company on their part engaging to furnish certain sums monthly and the contractor engaging for the time of completion in proportion to such payments'. Quite rightly, he pointed out that it was in the company's interests to furnish the capital because a protracted execution of the works would be injurious to their interests and, of course, it would also increase his overheads. Outram hinted that he expected extra payments, not only for the trouble and expenses of his

---

[1] Ibid., 3 Dec. 1799.

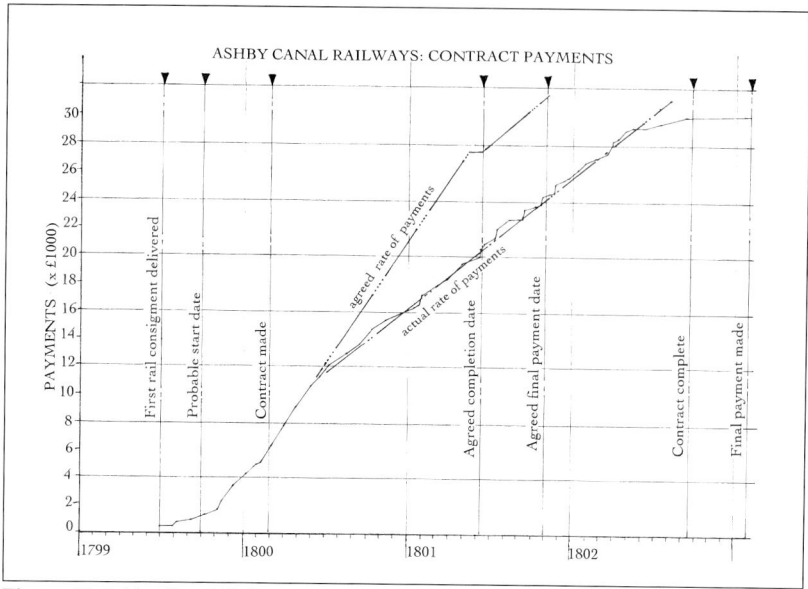

Figure 27 Ashby Canal Railways: contract payments.

many journeys to site, but also for the services provided when negotiating with objectors to the scheme, and for overlooking some of the construction work in progress on the main canal and the associated reservoirs. Obviously some of these tasks were completely unrelated to his railway contract but all, nonetheless, were ignored by the committee.

Further solicitations caused the committee to pass a resolution accepting, at last, Outram's estimate and also his specifications.[1] They agreed to pay him £1,400 on 28 February 1800 and £21,300 by 15 other instalments, subject to him finishing the works by 1 May 1801. Thus it seemed that they had agreed to a fixed-price contract of £29,500, and, although it was never clearly stated as such, nowhere in the minutes was it suggested that the company had done other than accept Outram's terms in their entirety. Outram was present at the meeting; he verbally undertook to construct the railways according to the resolution and that was how the terms of the contract were finally settled. Nothing was ever confirmed in correspondence.

A monthly rate of payment of £1,400 had already been maintained

---

[1] Ibid., 17 Feb. 1800.

since the previous November,¹ and so Outram had reason to believe that finances were now secure and that his works could be programmed on that basis. Unfortunately for him, the company was already falling behind steadily on their payments by June 1800. They were never able to provide more than £1,400 per month on more than two occasions during the next two years; generally, payments were between £600 and £800 per month.

It seems likely that the construction programme was also on target until June but afterwards it continued only at the pace dictated by the monthly payments (Fig. 27). Outram continued to show dissatisfaction over the lack of progress and in October he urged the clerk 'to explain to the gentlemen the necessity of furnishing the means of finishing the works'.² He also reckoned that the job was half-finished then and that he could easily meet a midsummer deadline, 'if money were to be advanced in proportion as works shall be done'. Judging by his performance up to that time, as indicated on the graph, there seems little doubt that this could have been achieved. The committee promised to make larger payments in order to accelerate progress but, true to form as ever, this did not happen.

Whilst it seemed that some payments had been made for work on site, a separate account for rails purchased from the Butterley ironworks had not been settled in December and a very alarmed Outram was prompted to write to them that 'their last engagement has been less adhered to than the former and on the last pay day no money was paid to the foundry where they had to borrow to pay the work people ... I must stop the whole works unless the committee will make their payments'. About that time he also let slip to the committee that this business had placed him 'in a most unpleasant situation with my partners at the foundry'.³ The desirability of a written contract was never more obvious than at that time. Outram wrote a similar letter in February 1801, followed by two more during April, but the proprietors did not bother to answer any of them. He stated that the company's erratic payments were causing him acute difficulties for two reasons. firstly, 'The increased expense of agency by so great a prolongation of the time necessary, is very considerable', meaning that the extension of the contract period caused by the slowing down of his operations was adding greatly to his

---

¹ PRO, RAIL 803/15, An Account of Monies paid by the Treasurer to B. Outram from 25 June 1799 to 14 Jan. 1803.

² PRO, RAIL 803/3, 3 Oct. 1800.

³ Ibid., 22 Dec. 1800.

overheads. Secondly, it was 'impossible to know how to provide for this or to undertake other engagements'. Understandably, his future plans were jeopardised by their failure to settle accounts promptly.

Outram's reasonable remarks fell on deaf ears because the proprietors were finding some funds every month and so they seemed to expect work to continue without disruption. Either they failed to understand that their fluctuating and irregular rates of payment were causing planning and cash-flow problems for their contractor, or else they were not in the least concerned. On the other hand, Outram's several sub-contractors for quarry products and transport services, as well as his employees on site, always expected prompt payment; the difficulties that Outram was having with his clients were of no concern to them. The Ashby Canal Company offered him little solace except that it 'had good reason to believe that much more money will soon be subscribed'[1] and it would remit as the state of the finances would permit. Almost as if to rub salt into the wound, the company then went on to order John Hodgkinson to construct four passing places between the tunnel and Cloud Hill, each not exceeding 45 yards in length.

Work continued steadily, but slowly, until December 1801 when the committee, suddenly eager to see the completion of their enterprise, interviewed Outram who consented to finish all the railways by 1 May 1802 provided that he was paid £3,600 by instalments. These were to be £600 on 1 January 1802, £1,000 on 29 January, £1,000 on 26 February, and £1,000 on 23 April. Outram agreed that, if he failed to meet the deadline, then he would transport limestone from the quarries to the canal basin at Willesley, and also carry slack coal to the various limeworks, at a price of 18d. per ton, until the time when the railways were fully operational.[2]

Came 19 May, the railways were still unfinished and the committee declared at a special meeting that their contractor had defaulted, and resolved that the company had 'sustained and are still hourly sustaining a very great and serious loss and Mr Outram be immediately required to convey the lime and slack on the terms and conditions specified'.[3] Ostensibly this seemed a reasonable enough stance to take but the proprietors also failed to mention that Outram had never been paid in accordance with the agreement! Only £300 had been paid over by 1 January, with two further payments totalling £700 later that month. £400

---

[1] Ibid., 21 April 1801.
[2] Ibid., 18 Dec. 1801.
[3] Ibid., 19 May 1802.

Plate 20 Low embankment with original stone blocks, Ticknall line.

was paid out on 20 February, £1,400 by four instalments in March, and £600 by two instalments in April. The total received by Outram on the settled date was £3,400, just £200 less than they had promised.[1]

The committee further resolved that Outram be summoned to a special meeting to discuss the matter, although there is no record to show that this happened. More than likely it did not and relationships, not surprisingly, deteriorated rapidly from that time onwards. On 15 June 1802, the committee was informed by their agents (presumably Greaves and Wilkes) that the railways were still in a 'very imperfect state' (Hodgkinson had more than 1,000 rails to fix, even a month later). Accordingly the treasurer was told not to disburse more funds until further notice. Curiously, however, Outram was not informed of this decision in a letter that the clerk was ordered to write to him. Probably more by oversight rather than by design, the clerk requested only an account and valuation of the works executed, extra over contract.

It was difficult to keep secrets and Outram soon found out, through informants, of the committee's actions. Now very annoyed by that discovery he wrote to Smith, the clerk, on 5 July noting 'the disgraceful order which from another channel I learn is entered on their books'.[2] The engineer then rightly pointed out that an account of all his additional works could not be made up until the railways were finished, but Smith could tell the committee that these would exceed £2,000. This did not include for additional costs which ran into several thousand pounds resulting from the company's 'non-performance of the stipulations'. He also expected a return for the increased price of labour and materials which had occurred over the duration of the contract, plus the burgeoning costs of his overheads. His letter concluded with a broadside: all along he had assumed that he 'was dealing alone with men of honour and gentlemen but late proceedings induce me to expect I may have to seek justice from the laws of my country'. This impetuous response with its exaggerated claims was not the way in which to deal with the committee which, in spite of its wayward disbursement of funds, had paid out £29,463 by the beginning of June.[3] Outram was clearly being unreasonable; it is quite likely that he was now deliberately holding up work on site.

John Hodgkinson then joined in the fray by writing to the committee late in August requesting a delay to the final inspection of the lines by Greaves and Wilkes. He claimed (without fully knowing the facts) that

---

[1] PRO, RAIL 803/15, Account of Monies.

[2] PRO, RAIL 803/3, 5 July 1802.

[3] PRO, RAIL 803/15, Account of Monies.

this was due once more 'for want of regular payments'.[1] Outram was then requested by the company's clerk to inform Wilkes and Greaves of the date when the system would be finished and in a fit state to be inspected. This happened late in November when Outram arrogantly claimed that 'I have viewed the whole of the roads and am happy to say that all the material parts stand better than I ever saw new railways. They will do me great credit with the public whatever the sentiments of the committee'.[2] The latter were to think otherwise.

Wilkes and Greaves reported back to the committee following site visits on 15 and 16 November 1802. Their collective views differed markedly from those of Outram. They had found some serious deficiencies in the formation widths along the single-track lines; these were generally 3 yards wide rather than 4 yards as stipulated in the specification, and that led to many of the stone sleepers having insufficient support on the outside edges. They declared that the embankments along the Willesley valley were not in a fit state for inspection and elsewhere embankments were too narrow. The approaches to public highways and accommodation roads crossing the railways were not satisfactorily finished in general; the rails at these places were not always of the specified heavier types, with lower flanges.

Even at that late date, ballasting was still in progress in many places along the Ticknall branch and the inspectors were critical about the thickness of this layer and also of the quality and size of stone blocks used along the southern end of this line. Their report implied that there had been some strong words with Outram whilst they were examining the latter. In mentioning their concerns about unsatisfactory blocks, they reported that the engineer had retorted that 'we should have examined the blocks at the time they were set down'. The inspectors rightly disputed this because the contractor had specified the size and quality of the blocks and in no circumstances could he blame the client for any shortcomings in this respect. Hence they were quite correct in stating that 'we believe that Mr Outram was bound by his contract to make a road of good and sufficient materials wherever he might find them under the power of the Act and having so completed it, it then becomes our office to inspect it'. No doubt Outram was already fully aware of this, but his uncharacteristic intransigence and improper remarks owed much to his exasperation at the proprietors' past failures to agree terms which were still, in several respects, in doubt. Now they were telling him about his

---

[1] PRO, RAIL 803/3, 24 Aug. 1802.
[2] Ibid., 10 Nov. 1802.

Plate 21 Railway bridge over Ticknall main street.

obligations under the contract. It was perhaps not surprising that his temper got the better of him.

The two committee members admitted that their report was sketchy and limited in its scope but nevertheless it was honest and objective enough. It was obvious that much work was still needed to be done before the railway system could be accepted by the company and so, in late January 1803, the committee decided to appoint John Warner, the engineer of the Coventry Canal, and one George Parramore, to inspect and value the works, and to submit a report.

The terms of reference of these inspecting engineers were not clearly defined and when they got down to the task of evaluating the works they had clearly assumed it was an admeasurement contract. Outram would no doubt be concerned to learn of that because he had written to the clerk the previous October that he 'wished to know what the committee mean, whether to act as if a contract had been made at the time I so much desired it to be drawn, or if no contract had ever been meant but the works left to valuation'.[1] This meant that he did not know whether the company intended this to be a fixed-price contract or an admeasurement contract. The clerk did not seem to know much about it either because an answer was never given to the question.

---

[1] Ibid., 3 Oct. 1802.

Throughout the contract Outram had been paid in instalments (erratic though those payments had always been) as though the price was fixed at £29,500, and he had organised progress of the works to suit these payments. The company had never disputed with Outram about this practice and had left him to it, only occasionally carrying out checks to ensure that he was not being excessively overpaid at any time. Surviving documentation supports this view; finished work was never re-measured and costed in accordance with the contractor's list of priced items, as would happen in an admeasurement contract. It is conceivable that the valuation by Warner and Parramore was simply intended to be a check on whether the company was getting value for its money and to ensure that, on concluding the contract, Outram had not omitted to complete any items listed on his original schedule of works. But if this was the intention of the committee, then nobody seemed to have told Outram and the inspectors about it.

The inspecting engineers' formal report, presented to the committee on 1 March, brought no good news for Outram.[1] It was critical of most aspects of the railways' construction and scathing about workmanship and the quality of much of the natural and manufactured materials used. The engineers began with some remarks on earthworks, noting that the formation widths specified for single and double railways were 4 and 6 yards respectively but on average these were found to be only 3.83 and 5.70 yards respectively. The side slopes of banks and cuttings were less than 1 in 1½ and they considered 1 in 2 was more appropriate (earthworks, nevertheless, are generally stable with side slopes of 1 in 1½). A major criticism seemed to have been the shaping of the formation levels for the railways, which were so irregular in places that the finished levels were made up with differing depths of ballast. Banks had settled badly in some sections and, as a result, the rails were frequently out of gauge. Drainage was poor, soughs were too few and standing water covered the blocks in many places, particularly on the Ticknall line where some of the deep cuttings were impassable because of the inadequacy, or absence, of drainage channels. Many of the imported stone blocks were said to be defective, and yet nearby quarries could easily have provided better ones. Finally, the plug holes in the blocks were often cut incorrectly, or carelessly, and plugs were found to be loose, the spikes too small and, often enough, the rails were insecure as a result. The inspecting engineers recommended to the committee that all trade be stopped for a month until these defects were made good along the main and branch

---

[1] Ibid., 1 March 1803.

lines—which their measurements aggregated to 12.9 miles approximately. Regrettably, the inspecting engineers did not quantify the remedial tasks to be undertaken, hence it is difficult to judge the scale of the defects over the entire length of the railway.

Warner and Parramore also submitted two valuations to the committee.[1] Both were based on an admeasurement of the entire works and, not surprisingly, their calculations exposed Outram's inflated assessments of the quantities. The first valuation was stated to be a 'Measurement at Mr Outram's prices in his Estimate' and this amounted to £25,618 17s. The second 'Measure and Value' also listed the same items but at 'prices fixed by them'. This was probably an attempt to include for the inflationary trends in labour and material costs since Outram's submission of four and a half years earlier. They inadvertently omitted two major items of earthworks from this second statement but even so, the final sum amounted to £26,782 9s. which included contingencies of 11 per cent. Extra works, which were necessary variations to the original plans, were also measured. They took into account the monies that Outram owed the company for the use of their bricks in tunnelling and culverting, besides various minor omissions and damages to lands. These sums hardly affected the issue because they were evenly balanced at only £717 19s. and £679 11s. respectively. In concluding their critical report, Warner and Parramore stated that 'every allowance [is] made for, [and] proper extra charges in price, so as to afford a fair profit to the contractor; nor do we consider him entitled to any claim beyond his estimate for work done further than we have estimated herewith'. Outram would not have liked that.

Quite apart from these monetary considerations, with which Outram could reasonably disagree, the criticisms in Warner and Paramour's report could not be ignored or easily explained away. It emphasised the unsatisfactory quality of the workmanship everywhere and, as often is the case, this could only be attributed to a deplorable lack of management supervision. It has already been noted that the site must have been difficult to supervise closely but there were other contributory factors too. At the time, Outram was very busily engaged in promoting his railways in several regions of the country. His site agents, Hodgkinson and Kenear, were also involved with those works and all of them must have been absent from the Ashby sites for periods when the chargehands and labourers would be left much to their own devices. In these circumstances it is hardly surprising that the quality of workmanship suffered

---

[1] PRO, RAIL 803/15, Account of Monies.

and did not match the high standards which Outram had contracted to achieve. Three weeks after receiving the report, the committee instructed Outram to complete the works prior to another inspection, otherwise they would take the matter of reparations into their own hands 'without prejudice to any demand the company may have upon him for those deficiencies'.[1]

As regards the financial issues raised by Warner and Parramore, the company's minutes made no reference to the marked difference between Outram's estimates and each of the two valuations of their inspecting engineers. At the time, Outram's threats of litigation over payments of fees, unrelated to his railway work, were very real and it was probably for this reason that the company saw fit to arrange a further check and valuation of their railway system. Accordingly, they did not await the completion of repairs before they approached Christopher Staveley, who was then engineer to the Leicestershire Navigation, on 4 April. He was requested to repeat the inspection and valuation of the works and was pressed to submit a report quickly. With some urgent prompting, Staveley's report arrived for the committee's consideration in early July, at a time when much correspondence was passing between the disputants.[2] The proprietors must have been relieved to note that Staveley's report on the valuation and condition of the works was similar to that of his predecessors. His valuation was £27,511, based on his admeasurement (his calculations of earthworks differed from both Outram's and those of Warner and Parramore) but applying Outram's original rates. Staveley had included and costed 20 additional items claimed by Outram and he considered that the contractor was entitled to payment for 23 others, implying that these were matters for further negotiation with the company. He also noted that 'In the execution of works of this nature and magnitude there are always many occurrences which create incalculable expences. It is therefore reasonable to add 5£ per centum ... for contingencies of different descriptions'.[3]

Whilst these inspections were in progress, Outram's dispute over the costs of his rendering extra services was being taken very seriously by the proprietors. Outram had listed over fifty tasks, separate from the railway work, which he and his engineering assistants had performed for the company, the bills for which had been steadfastly ignored. These matters should have gone to arbitration in January 1803 but further

---

[1] PRO, RAIL 803/3, 22 March 1803.
[2] Ibid., 5 July 1803.
[3] PRO, RAIL 803/15, Account of Monies.

disagreements and threats of litigation led only to more delays. The matters in dispute were eventually considered at a meeting between Outram and the committee in July that year, when it was agreed that the final award would be assessed by arbitrators.[1] Outram's solicitor, Mr Lockett of Derby, arranged for James Green of Linton to act as one arbitrator and a Mr Horner was similarly appointed by the company.[2] The proprietors of the company must have felt confident about the outcome of the arbitration, armed as they were with the three valuations assessed by Warner, Parramore and Staveley, all of them lower than the sum of £30,163 already paid out for their contract. However, the arbitrators found in Outram's favour, although the differing valuations for the railways might have been taken into account in assessing the final award. The company did not seem to like the result and, although they sought counsel's opinion, they finally had no option but to pay out £1,300 over the next twelve months for Outram's claims;[3] a final amount of £450 was received by Outram's widow after his death in 1805.

The vexed question of whether payments for the railways were to be according to a fixed-price or an admeasurement contract was never formally answered by the company. Even so, they paid the original price quoted for the works plus agreed, minor variations. Probably the company realised that, taking into account the high rate of inflation and the rising costs of labour of which Outram had frequently reminded them, they had done well to acquire their railways at such a reasonable price compared with what it would have cost had they constructed a canal. Furthermore, notwithstanding the adverse comments of the inspecting engineers, it would be reasonable to claim that, after Outram had completed reparations and all the branches were fully operational, the Ashby Canal Company was provided with a railway system well ahead of its time. The earthworks were neatly balanced, with cuttings and embankments providing easy gradients for the horse-drawn transport. Accommodation bridges, highway level crossings, tunnels, culverted streams, and drainage channels were all finally built to a good standard and those observing the system in use must have been impressed by the scale and quality of the layout. Even the remnants still to be found today suggest that the Ashby railways were indeed a milestone in transport technology and must be seen as a model for the modern railway systems

---

[1] PRO, RAIL 803/3, 9 July 1803.

[2] Ibid., 6 Sept. 1803.

[3] Ibid., 3 and 17 July 1804, 26 March 1805.

which followed thirty years later.

And yet, Outram's satisfaction over this major success in railway engineering must have been marred by this unsatisfactory contract. He probably needed nobody to tell him that he had been unwise to proceed without properly negotiated terms and it must have been galling to find that his trust in the management committee, made up of such distinguished members, was so sadly misplaced. But once he had delivered that first order for five miles of cast-iron plates he was committed and, as time went on, his commitments made it increasingly difficult to extricate himself. Unwise though he might have been in circumventing the contract system in which he was so amply experienced, one can only accept that his motives, as discussed earlier, were reasonable enough. At the end of the day, much of the blame must be laid on the company. The proprietors had the authority of their Act of Parliament and yet they consistently failed to use this to finance their undertaking properly. Their management structure was weak and the collective failure of the proprietors to make decisions allowed them, perhaps unintentionally, to take every advantage of their unsigned agreement. As a result, some of the burden of responsibilty for financing their railways fell on the unfortunate Outram.

Suffice it to say, however, the quarries, limeworks and coal-mines were well-served by the Ashby Canal and its railways in the years that followed, even though they were not an immediate financial success; it was 1828 before the first dividend was paid. Several years of successful endeavour followed but were then overtaken by the Railway Mania from 1845. In 1864 the Midland Railway replaced part of the tramroad by a line from Worthington to Ashby. This followed the original route through the Old Parks tunnel, which was widened and rebuilt, into Ashby. Short lengths of the original railway were retained for local purposes and the branch line to Ticknall was officially closed in 1915.[1]

Today the principal remains can be seen along the Ticknall branch from South Wood. Cuttings and embankments are clearly defined, as well as some drainage works and culverts. There is an accommodation bridge, still in use for farming, and a 50-yards long tunnel in the approach to the covered way which disguises the route of the railway as it passes under the drive to Calke Abbey. Originally described as an 'archway under Sir Henry Harpur's coachroad', this was extended into a 134-yard long brick-lined tunnel in later years and illuminated by four

---

[1] C.R. Clinker and C. Hadfield, 'The Ashby-de-la-Zouch Canal and its Railways', *Trans. Leics. Arch. and Hist. Soc.*, xxxiv (1958), p. 72.

grilled apertures in the soffit, spaced at intervals along the tunnel. At the eastern end of the tunnel the lines diverge; the route to the left passes across a fine arch bridge over the village street in Ticknall to lead into quarries beyond. The other line continues down a steep incline into other quarries and to the site of the former limeworks. The considerable acreage of the quarries gives some indication of the vast quantities of limestone which must have been excavated over many years and which then passed down the railway, *en route* for the canal at Willesley and the markets beyond.

CHAPTER 15

THE MONMOUTHSHIRE CANAL, BRECKNOCK & ABERGAVENNY CANAL AND THEIR RAILWAYS

The Monmouthshire Canal was described in 1831 as 'a useful undertaking, being in the very centre of a country abounding in limestone, coal, iron, tin and lead, which before this work was executed were permitted to remain undisturbed, for want of a conveyance for the produce of the mines'.[1] This could also be said of the Brecknock & Abergavenny Canal which was built as an extension to the Monmouthshire Canal, as well as the many miles of railways (frequently referred to in South Wales as dramroads) which were linked with these waterways (Fig. 28).

A consequence of these new transport facilities (besides those in the Neath and Swansea valleys and elsewhere in South Wales) was the rapid development of a chain of blast furnaces and foundries across a sweep of mountainous country from Hirwain to Blaenavon, some twenty miles long by a mile wide, known as the 'iron belt', where coal and ironstone were found in abundance and limestone for fluxing was close by, to the north. Thus great coal and iron industries flourished whereas before the canal mania years, they had been restricted to the use of pack-horses and waggons for shipment of their products to the ports.

The Monmouthshire Canal Act[2] became law in June 1792. The proprietors were empowered to raise £120,000 and another £60,000 if required, but further Acts had to be obtained in 1797 and 1802 to raise capital up to £275,330 and also to make additional railways.[3] When first built, the main line of the canal ran from the tideway of the River Usk, just below Newport, rising 447 ft by 42 locks in 11 miles to Pontnewynydd. A branch, 11 miles in length, led off from Crindau, rising 358 ft by 32 locks to Crumlin. The maximum dimensions of a vessel using the navigation (and also the Brecknock & Abergavenny Canal) were 64 ft

---

[1] J. Priestley, *Historical Account of the Navigable Rivers, Canals, and Railways, throughout Great Britain* (1831), p. 455.

[2] 32 Geo. III c. cii (1792).

[3] 37 Geo. III c. c (1797); 42 Geo. III c. cxv (1802).

9 in. long by 9 ft 2 in. wide, with draught of 3 ft.[1]

Railways were proposed as feeders to both main-line and branch canals, namely to Blaenavon and Trosnant from Pontnewynydd, and to Beaufort from the Crumlin line with branches leading off to Sirhowy and Nant-y-glo. The first Act also permitted further railways to be built, by either the company or private operators, from within eight miles of the canal or its railways. The main line canal was navigable by early 1796, although the Newport end was further extended by nearly 400 yards in 1798. The Crumlin branch was finished by April 1799. The engineer for these very difficult, heavily locked and expensive routes was Thomas Dadford junior.[2]

The Brecknock & Abergavenny Canal was first proposed during 1792. It was then planned as a separate enterprise to the Monmouthshire Canal in that it was to begin at Newbridge on the River Usk, the line passing by Abergavenny to Llanelly before terminating at Brecon. Not surprisingly, the proprietors of the Monmouthshire Canal were sufficiently alarmed as to invite their neighbours to connect with their own navigation near Pontypool; they also offered to contribute £3,000 to the venture, provided that their own preferred route was accepted.[3] Following agreement with this proposal, Thomas Dadford junior quickly surveyed and estimated for a line from a junction at Pontymoile to Brecon. This passed by way of Mamhilad, Abergavenny and Govilon, where it turned north-east to follow the River Usk to Llanelly, Crickhowell and Brecon. There were three inter-connecting railways: from the canal at Gilwern along the Clydach valley to Waun Ddu near Beaufort; another from Gilwern to Glangrwyney; a third from Llanfoist to Abergavenny. The first two were eventually combined as the Clydach Railway, although the third was never built.[4]

The Act for the Brecknock & Abergavenny Canal was obtained in March 1793 and capital of £100,000 was authorised, with a further £50,000 if required.[5] The canal was 33 miles long and ran on the level for 23 miles from the junction with the Monmouthshire Canal to five locks at Llangynidr and then to another lock at Brynich some eight miles

---

[1] H.R. de Salis, *Bradshaw's Canals and Navigable Rivers of England and Wales* (1904), p. 169.

[2] C. Hadfield, *The Canals of South Wales and the Border* (1967), pp. 127–31.

[3] PRO, RAIL 500/5, 14 Oct. 1794.

[4] P.G. Rattenbury, *The Brecknock and Abergavenny Canal and Railways* (1980), p. 13.

[5] 33 Geo. III c. xcvi (1793).

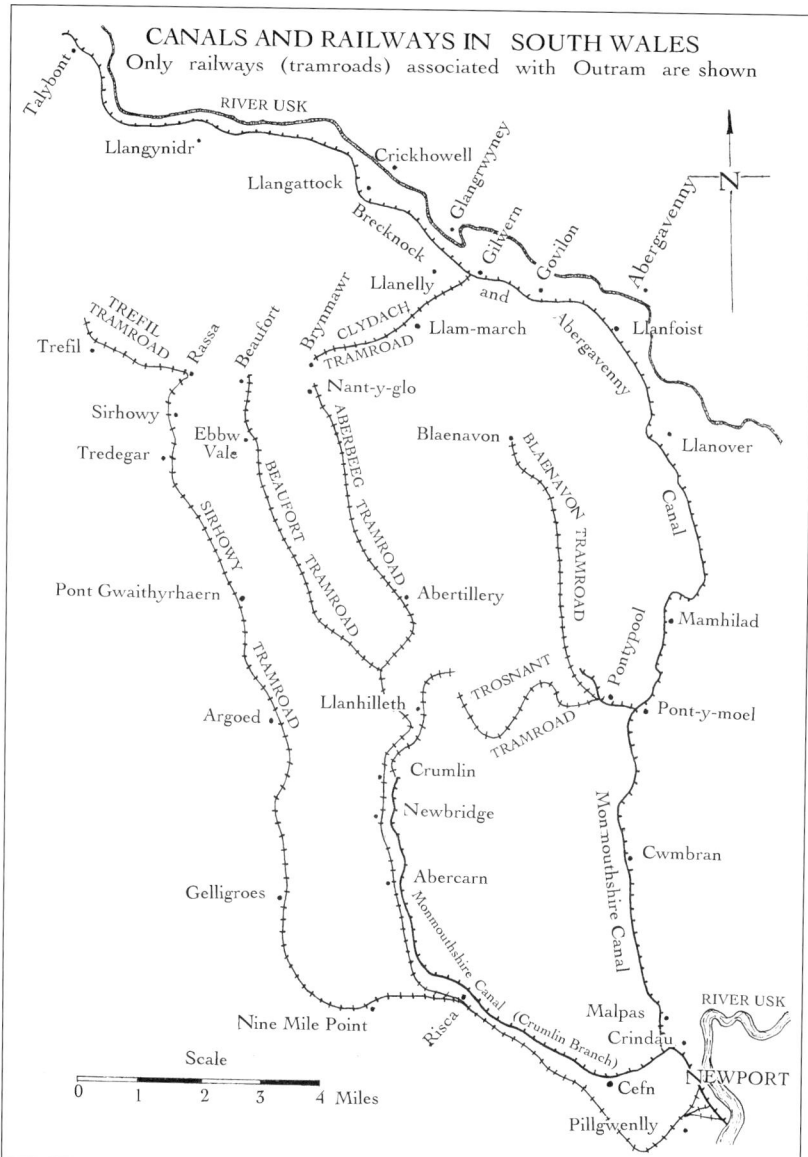

Figure 28 Canals and railways in South Wales.

beyond, thereby rising 68 ft to the terminus at Brecon.[1]

The company and private operators were also empowered by the Act

---

[1] De Salis, *Bradshaw's Canals*, pp. 165, 166.

to construct railways from within eight miles of the canal and indeed the first action of the committee at the inaugural meeting of 14 June 1793 was to appoint John Dadford as engineer and manager for a railway.[1] This ran from the coal mines at Gellifelen in the Clydach valley to Glangrwyney, crossing the Usk at Gilwern by a timber bridge. Iron edge-rails and sleepers were purchased from Samuel Homfray's Penydarren works at Merthyr Tydfil and the line and bridge—the latter built by William Pritchard of Cardiff—were completed twelve months later.[2]

Curiously, the committee made no mention of the need to begin construction of the canal and their preoccupation with railways continued for sixteen months before the Monmouthshire Canal proprietors, having already paid their £3,000, asked politely when construction of the waterway was to begin.[3] They were advised that this would be when the railways were finished and after an engineer could be engaged. Further enquiries some twelve months later were similarly evaded and it was not until the end of 1795 that Thomas Dadford was appointed engineer, probably at the insistence of the Monmouthshire proprietors, but only on a part-time basis. It was not for another three years, when his work on the Monmouthshire Canal was ended, that he was free to take over full-time duties.[4] It is likely that these terms of appointment led to some of the troubles which later afflicted the Brecknock & Abergavenny Canal Company. The piecemeal construction of the canal and the poor quality of workmanship, later noted during the inspection by Benjamin Outram in 1799, showed all the hallmarks of inadequate site supervision by the engineering staff.

Contracts were on offer to canal cutters and stonemasons during March 1796[5] and work on the earthworks, culverts and bridges between Llangattock and Clydach probably began the following month. Later in the year, Benjamin James of Llanover, a stonemason, was awarded the contract for the aqueduct crossing the River Owney at Clydach.[6] In July 1797 a contract for the construction of the approach embankments to the structure, by then complete, was granted to Thomas Powell, a

---

[1] PRO, RAIL 812/3, 14 June 1793.

[2] Ibid., 7 July 1793.

[3] PRO, RAIL 500/5, 14 Oct. 1793.

[4] Ibid., 15 Oct. 1798; PRO, RAIL 812/3, 17 Dec. 1795.

[5] NLW, Maybery 378, handbill.

[6] Ibid., 569, articles of agreement between B&A Canal Co. and B. James (nd).

shopkeeper, and William Watkins, a yeoman.[1] It is surprising that such a substantial and difficult task should be entrusted to such ostensibly ill-qualified persons, but they were only typical of the calibre of contractors engaged by the company. It was indicative of the dearth of experienced men available at the time.

In Thomas Dadford's report of April 1798 it was reported the canal 'from Clydach to Llangunnider, in length eight miles and a quarter is finished and navigable; on this length of canal are twenty-eight stone and three draw-bridges; six aqueducts under the canal'.[2] Much was made of the ancillary works such as feeders, wharfs, coal-yards and limekilns, presumably to impress the shareholders and to divert attention from the fact that work was proceeding very slowly. However, Dadford also noted that the canal from Llangynidr to the north end of Ashford tunnel, nearly three miles in length, was let out; 1½ miles of canal were cut and other sections were in progress. The Clydach aqueduct, 90 yards long, with a single arch of 22 ft span, was then finished and 'the embankment is now making upon it and about one fourth done'. Ashford tunnel, the contract for which had been let out to Thomas James of Merthyr on 29 January of that year, was arched for 80 yards and 'other parts cutting open'. It was due to be finished by 1 November of that year.[3]

On 21 January 1799 tenders were invited for aqueducts at Talybont and Govilon and for the approach cuttings to the Ashford tunnel, besides miscellaneous canal sections.[4] The company, at a general assembly on 18 October 1798, had seemed anxious to accelerate progress and to complete the canal from Brecon to Abergavenny, but at the next assembly on 25 April 1799 the balance of the company's accounts were produced. Dadford then stated that there were insufficient funds to finish the canal to Brecon because the original estimate of £100,000 had been exceeded; he later advised the proprietors that a further sum of £17,430 would be needed to finish the canal from Clydach to Brecon.[5]

Probably in anticipation of this alarming report the Monmouthshire proprietors (the Brecknock & Abergavenny Canal Company had two committees and two treasurers; thus the interests of proprietors from both Breconshire and Monmouthshire were represented by this cumbersome,

---

[1] Ibid., 605, articles of agreement between B&A Canal Co. and Powell & Watkins, 31 July 1797.
[2] Ibid., 382, Engineer's report to B&A Canal Co., 26 April 1798.
[3] PRO, RAIL 812/3, 29 Jan. 1798.
[4] NLW, Maybery 635, handbill.
[5] Ibid., 386, Report of the B&A Canal Co. committee, 25 March 1800.

Plate 22 Ashford tunnel on the Brecknock & Abergavenny Canal.

but apparently workable, arrangement), at a committee meeting on 7 March, had ordered a suspension of all contracts intended to be entered into in their division.[1] They also resolved that an engineer should be appointed to examine the state of the canal and to advise on the possibility of constructing a railway instead of continuing to the Monmouthshire Canal at Pontymoile. Benjamin Outram was recommended as an appropriate choice for that purpose. A meeting of the entire committee was held four days later and that proposal was unanimously accepted, but it was felt proper to delay the decision until after the general assembly on 25 April.[2] However, the committee, increasingly dismayed by the burgeoning costs of their partially completed enterprise, decided at a further meeting on 24 April that their clerk, John Powell, should write to Outram immediately.[3] In the meantime, Dadford was ordered to prepare longitudinal sections for the line between Gilwern to Pontymoile as an aid for Outram during his visit.

Outram appears to have attended the next committee meeting at

---

[1] PRO, RAIL 812/3, 7 March 1799.
[2] Ibid., 11 March 1799.
[3] Ibid., 24 April 1799.

Crickhowell on 20 June.[1] His brief was significantly wider than originally intended; the company was clearly in need of some urgent, sound advice. He was requested to examine the state of the completed works, including the railways, and to advise on any improvements which should be made, and to state the cost. He was also to view the Ashford tunnel, to ascertain the capacity of the various feeders for the canal and to comment on that part of the canal which remained incomplete. Besides requesting an estimate for making the canal between Gilwern and the junction at Pontymoile, the committee also required an estimate for a railway, if this was built along the same route instead, as well as a comparison of the expense of conveying goods by each mode.

This was a tall order but Outram got down to the job and achieved a great deal in the next twelve days, although there is no doubt that he had brought a useful team of engineering staff with him to assist in this task. This included John Hodgkinson, Walter Kenear and David Whitehead (the former apprentice surveyor on the Peak Forest Canal). In his report, completed by 1 July, he noted that the soils of much of the country through which the upper canal was built were decidedly pervious to water and, even though there had been much expenditure on lining and puddling, the line was still seriously affected.[2] There was little alternative but to drain sections of the channel to find the leaks because most of the losses appeared to be in the bottom and off-side banks of the waterway through subterranean strata, rather than through imperfectly consolidated banks. Outram proposed deviations to the intended line to avoid deep cuttings such as those already executed. He recommended setting out the line around the sides of hills where possible and to lessen sloping banks so that 'there be no steep Banks to moulder into the Water' and thus avoid the expense of frequent dredging. For the proposed line from Talybont up to Brecon, Outram altered the position of the aqueduct at Brynich and substituted one lock for the two intended, thereby effectively doubling the length of the summit pound. The depth of the latter was to be increased by 12 inches to act as a reservoir, much as been done on the Cromford and Peak Forest Canals some years previously.

Outram continued his report with remarks about the route of the proposed lower canal to Pontymoile, and in much the same degree of detail. He noted that this 14-mile length was on one level and although

---

[1] Ibid., 20 June 1799.

[2] NLW, Maybery 383, B. Outram, Observations on the B&A Canal and Railways, 1 July 1799.

a railway was feasible and could be profitable in the short term, it should only be built with the ultimate intention of its replacement by a waterway. A railway could be laid at a higher level than the intended canal except for the last three-quarters of a mile to Pontymoile where it would need to be built nearly at top water level. A tunnel would be necessary for the railway but could be partly utilised for a waterway, and earth excavated along the line could also be stockpiled to await canal construction at a later date.

The cost of a canal would be about £50,000, compared with a railway at £22,000, but the land and some works of the latter, which would also serve in the building of a canal, were valued at £9,000. Bearing in mind that the railway might be used for the canal's construction and that the rails would be salvaged eventually, the loss on abandonment of the railway would only be about £4,000. Following a cost comparison for the transhipment of goods by rail and canal, Outram concluded by stating that 'in case a Railway should be adopted, Care should be taken in its Execution to throw no Impediments in the Way of the future Execution of the canal when Extensions beyond Brecon, or other Increase of Trade may render necessary a complete Water Communication'. Clearly he was not unduly biased in favour of railways when considering all the circumstances pertaining to this canal.

During that summer, which Outram declared was an unusually dry one, he found time to gauge the feeder streams along the completed canal and calculated the number of locksfull of water available per 24-hour period, assuming a lock of ten feet fall and 180 tons capacity. Some 189 locksfull were yielded and, although Outram did not comment as to the adequacy or otherwise of the water supplies, he wrote that he had not considered the Usk but if 'it should be found necessary to make a feeding Sluice from that River, Water sufficient for the Supply of many Canals might at any Time be taken from it.'

Outram paid particular attention to the Ashford tunnel, the unsatisfactory construction of which had been of concern to both Dadford and the company for some time past. At Ashford, the canal runs along the west bank of the Usk valley, about 65 feet above river level. The need for a tunnel had been recognised in 1794 when Hugh Henshall, who was assisting Dadford on the early survey work, wrote that 'on the east Side of Ashford I recommend a Tunnel 300 yards in length to be Made to avoid the steep and slippery Bank at the Side of the River along which the Canal cannot be carried'.[1] The tunnel, as built, was a masonry arch structure 370 yards long. It was driven with maximum cover of 26 ft through a spur of glacial drift deposits, just above a stratum of Old Red Sandstone.

Thomas James, the contractor, had experienced difficulties because of constant ground water inflows and flooding of the workings but, even so, his progress was very slow and he fell well behind the contract completion date. James and Dadford had obviously been at loggerheads over the method of driving the tunnel, besides the quality of the finish, and the contract was finally terminated when parts of the finished lining failed. However, six months before Outram's visit, James wrote a letter to the committee to put his own side of the story. He claimed that:

> the Tunnel was exactly carried on conformable to the Engineer's directions. I have sent but two of my workmen who was on the line from the beginning untill I quited the Work when the misfortune happened—Those and several others can prove that I proceeded in every particular by his orders ... If he saw I was wrong it was his place to stop me from proceeding further because I was bound to comply with his injunctions—why should he permit me to go 294 yards and to draw money on account if he did not approve of my proceeding and intervally admeasured my work if it was not done satisfactory to his thinking .... I am willing to stand by any fair determination; but if Mr Dadford will compel me to anything unreasonable I do not mean to do anything contrary to equity.[2]

Outram's comments on the tunnel were brief and entirely in agreement with Dadford's views:

> The Walling and Arching of the Tunnel has been ill done, all those Parts which have failed must be taken down and rebuilt; those Parts which have not failed, do not look well, but perhaps may be sufficiently substantial; a little Time after that Part of the Canal shall be filled with Water, will prove whether they are so.

In fact, newly appointed contractors, Benjamin James and Walter

---

[1] N.A. Thomson, 'The Renovation of the Ashford Tunnel', in *Centenary Conference of the Institution of Civil Engineers, North Western Association* (1986), pp. 21–30.

[2] NLW, Maybery 631, T. James to B&A Canal Co. committee, 27 Dec. 1798.

Walters, had been engaged on 20 June and they went on to complete the works to Outram's more stringent specifications. These required 'an exceedingly good Puddle of eighteen inches over the whole bottom', followed by lime grout and a 12 in. thick inverted stone arch. The side walls were made 3 ft thick with 2 ft of masonry walling backed by 1 ft of pounded earth, so as to ensure a completely watertight structure. Outram also stated that the work should proceed day and night during the dry weather and that two dependable men should be appointed to see that every part of the work was done in the best manner possible. It is likely that the tunnel was rebuilt by 'cut and cover' because otherwise it would have been difficult to comply with the specification.

In recent years, inspection of the tunnel showed that the central section, some 155 yards in length, was rebuilt in accordance with Outram's specifications.[1] It was still in good condition and remained impermeable to groundwater. Outside these limits, however, the tunnel linings were of a much inferior rubble construction, only half the thickness that he recommended, and had been subject to much remedial work over the years. The cause of most of these maintenance problems has been the ingress of groundwater through the west walls and crown. Moreover, quantities of silt have washed through the tunnel linings, leading to the development of voids behind the walls, as well as piping channels deep into the hillside. As a result, there has been some surface subsidence and ponding, causing further inflows of water into the tunnel.

Outram turned his attention to the railways which had, by then, been established and operational for some years. He was immediately critical of these which were 'laid out agreeable to the old System, and expensive to work and troublesome to keep in Repair'. The railways he had seen were typical of the dramroads being developed in South Wales at the time.

Perhaps the closest description of the design of these railways was that published by Archdeacon Coxe in 1801,[2] following a visit to the Blaenavon ironworks and railways, a year or two before Outram's visit. He observed the construction of what may have been sidings, or even sections of the railway leading down to the Monmouthshire Canal. The foundation for the railway was a layer of broken stone in a trench, excavated 6 ft wide by 2 ft deep, and made-up frames of rails and sleepers were placed in line upon this base, then bolted together. The rails were of the edge-rail type, each 4 ft long by 3 in. deep and 1½ in.

---

[1] Thomson, 'Renovation of Ashford Tunnel', p. 21.

[2] W. Coxe, *An Historical Tour of Monmouthshire* (1801), pp. 200–202.

wide, with alternate concave and convex ends so that lines of rails slotted neatly together. The ends of the rails were also pegged down, to a gauge of 3 ft 4 in., on to cast-iron cross-bars, or sleepers. The latter, in turn, were pegged at their extremities to square blocks of timber, known as under-sleepers. In practice it was found that the cast-iron cross-bars often broke and they were eventually replaced by timber sleepers. On completion, the frames were covered with ballast and earth until only the rails protruded, 'over which the wheels of the car glide by means of iron grooved rims, 3½ inches broad'. It is commonly assumed that the tram wheels were a double-flanged type, similar to a pulley. For the most part these railways were constructed to follow the ground profile and little attempt was made to limit the gradients by earth banks and cuttings. The trams weighed up to 3½ tons when loaded and were drawn by a single horse.

Outram suggested that the railways 'might be worked to better Advantage, by using lighter Waggons with smaller Wheels, turning on the axles; but if instead of the present Railroads, others were substituted on the System I have introduced, the Expence of Conveyance might be reduced three fourths, and the Expences of Repairs in Proportion'. In his report he expounded at length on the 'improved plan' and listed desirable gradients. Outram was severely critical of the existing Clydach railway, and recommended abandoning the route on the east side of the valley in favour of an entirely new line on the west side. This would reduce gradients and necessitate the building of a self-acting inclined plane 400 yards long, with a fall of 140 yards. The line would terminate at Gilwern, to the east of the Clydach aqueduct. Outram concluded by claiming that 'On the Plan I now propose, Articles of Merchandise might be carried up Hill at less Expense per Ton than the Coals now cost in bringing down'. He reckoned that the cost of preparing the ground for the new system, making culverts and bridges, preparing the stone sleepers and fixing, but excluding costs of new rails, would be £4,500. The existing edge-rails could be salvaged and sold for £5,000, hence the final costs would not be unduly excessive. Outram also recommended a new line to Abergavenny, a distance of eight miles, for the transport of coal and lime from Gilwern, and concluded with the suggestion that 'each Waggon by the help of a Crane should be set on a Pair of Wheels, on which it might be drawn to any Part of the Town'.

No doubt the canal company recognised the good sense of most of the recommendations in Outram's report but it seems that nothing was ever done to change the existing railways, nor to consider further the replacement of the canal down to Pontymoile by a temporary railway, although his comments on improving the upper waterway to Brecon were accepted and put in hand almost immediately. Their acute financial difficulties discouraged them from Outram's more radical changes, even though the engineer had taken every care to minimise the costs of his proposals.

The canal from Gilwern to Brecon was finished by the end of 1800 but a shortage of capital brought work to a standstill for two years. By 1805 the canal had reached Llanfoist, near Abergavenny, before construction ceased once more. The company then seemed content to let matters rest and to concentrate instead on the generation of business on the completed canal and associated railways. It was not until 1809 that sufficient capital was raised and the company restarted construction. This late development owed much to the Monmouthshire Canal proprietors, who were threatening litigation because of the failure of their neighbours to complete their waterway to Pontymoile. William Crosley was appointed engineer and he and John Hodgkinson re-surveyed the route before work proceeded, to be completed in 1812.[1]

Outram's inspections had not passed unnoticed by the Monmouthshire Canal Company and it was not long before he and his colleagues were invited to examine the state of their canal and railways. His report was submitted on 13 January 1800.[2] Although he had inspected the canal and made recommendations for reservoirs on the Crumlin branch, most of his report was concerned with the railways, all of which he found rather unsatisfactory. He wrote:

> The Railways to both Beaufort and Blaenavon are better laid out than most I have seen on that System. The Trosnant Road should have been on a more gradual Descent but They are all constructed on a Principle far inferior to the System which I have introduced and which after several Years Proof in great Variety of Situations have been found to exceed every other Mode hitherto adopted. The Labour of one Horse on a Railway of this kind being equal to the Labour of four Horses on the best Railways that have been constucted on the Plan of those belonging to the Monmouthshire Canal Company ...

He suggested that all parties interested in the trade on the canal and railways, as well as in the mines and works served by the system, should

---

[1] Hadfield, *Canals of South Wales*, p. 165.

[2] PRO, RAIL 1005/54, B. Outram, Report on the Monmouthshire Canals and its Railways, 13 Jan. 1800.

take immediate steps to reconstruct or modify their railways according to his improved plan, 'as such Works would be attended with less Expense and Inconvenience if done now than at any future Period'. Outram went on to describe his system, which clearly aroused the interests of the proprietors. The latter were only too aware of the costs of the constant breakages of the cast-iron edge-rails adopted for their railways, hence they were soon to convert some of them, notably the Beaufort, Trosnant and Rassa lines, with the 44 lb flanged rails laid to Outram's standard gauge of 4 ft 2 in.

At the time of Outram's visit, the company had been considering adding other railways to those already completed, which by then were carrying much traffic, in spite of the defects outlined in Outram's appraisals. They were willing to construct railways for industrialists provided the latter could subscribe 7½ per cent of the costs, plus the expense of maintenance, and seemed eager to pursue this course of action.[1] For example, in February 1798 the committee had agreed to build a railway to connect the collieries of Sir Richard Salusbury at Llanhilleth (or Llanhiddel, as it was then called) with the company's existing railway above Crumlin.[2] They were determined to act on this matter 'with all speed' and to expend £1,000 out of company funds for the purpose 'on any plan and lines most commodious as Sir Richard thinks fit, under the superintendence of their engineer'. However, articles of agreement were not drawn up until July 1799 and it was at least another twelve months before further consideration was given to the matter. Shortage of capital was probably the root cause but whatever the reason, Outram was offered, and accepted, a commission to survey and to prepare plans and specifications for the Llanhilleth railways, and probably other lines as well, late in 1800.

Outram wrote to the company clerk, Morgan Parry in Newport, from Butterley on 8 December 1800 to note that:

> You must think me tardy in sending you the necessary minutes for setting out the Llanhiddel Railway; but I could not get the plans from Whitehead the Surveyor till a few days ago and as it was so near the time of my Agent's (Mr Hodgkinson the bearer) journey to Wales I thought it better to send by him than the post. He has my plans and a copy of the minutes enclosed together with other

---

[1] PRO, RAIL 500/5, 16 Oct. 1797.

[2] Gwent Record Office, NPT 3518 (M250), Contract for Railroads to Llanhilleth Collieries, 22 July 1799.

necessary instructions to assist in setting out the Railroad if his assistance therein is required. If you have been preparing Rails, Blocks etc you will not have lost much time by the want of my assistance.[1]

Unfortunately, the plans for this railway have not survived but it is known that the line had twin tracks and connected with the Beaufort railway by an inclined plane at Argoed, a short distance above Crumlin. When the railway was completed Outram wrote again, on 5 August 1801, to note that:

> I am sorry to receive Mr Hodgkinson's account of the Pontypool or rather Llanhiddel R[ail] Road. The deviation from the plan I had recommended will I fear be found to lead the Co to great expence & unpleasant circumstances—but more on this subject anon.[2]

The deviation is unknown but it is more than likely that, as a means of reducing costs, the railway was built to follow closely the existing ground levels; Outram's advice on limiting gradients, with emphasis on the associated earthworks, was probably disregarded.

There is no doubt that these canal and railway developments were steadily encouraging major investment in the iron belt. A prime example concerned two ironmasters, Richard Fothergill and Matthew Monkhouse, who managed the Sirhowy ironworks, located at the upper end of the Sirhowy valley, under a lease granted by Charles Burgh of Abercarn in 1778.[3] They also leased substantial acreages down the valley, although the conditions imposed by the lessor prevented mining and other exploitation. Subsequently, these lands became part of the estates of Sir Charles Morgan of Tredegar Park, near Newport, by way of his wife's inheritance, and he was to prove amenable to granting leases which permitted the developments intended by the partners. However, before attempting to negotiate new leases, the latter had wisely joined forces with the successful, and very wealthy, Samuel Homfray of Penydarren ironworks, probably for financial reasons but perhaps also because his wife was the daughter of Sir Charles Morgan. No matter the reasons,

---

[1] PRO, RAIL 1014/17, B. Outram to M. Parry, 8 Dec. 1800.

[2] Private collection of W.J. Skillern, B. Outram to M. Parry, 5 Aug. 1801.

[3] J. Lloyd, *The Early History of the Old South Wales Ironworks, 1760–1840* (1906), p. 137.

leases were signed in March 1800 and the new partnership was granted the rights to exploit all the coals and mines 'of all descriptions' under Bedwelty Common and surrounding lands, an area of some 3,000 acres. They were also granted the liberty 'to make a Tram or other Road down Sirhowy Valley, to join the Canal' and were also permitted to erect works, at a cost of at least £10,000. This became the Tredegar ironworks (in deference to Sir Charles Morgan), situated a short distance south of the Sirhowy works.

An efficient transport system to Newport was an early priority and the partners soon set about constructing a railway from their new works, down the Sirhowy valley to Pill on the river Usk, close by Newport. The line proposed was separate from the Monmouthshire Canal, whose proprietors were naturally alarmed by this move which would avoid their canal from Crumlin and be injurious to their interests. Moreover, they were soon to write to the Sirhowy company complaining that the railway they were making crossed over the canal company's line and that they were making another 'parallel to the Co. Railway which will interfere with the same.' Both these actions were considered illegal[1] and perhaps Homfray and his partners agreed, because they were induced to abandon their independent line all the way to Newport. They consented instead to a railway which terminated at the waterway in Risca, so that all coal and iron from the new works could be exported by canal. An agreement was signed on 18 December 1800.[2]

Outram was aware of these negotiations because the agreement included a paragraph stating that the canal company would, at their own cost, make the railway by the 25 December 1801 and that this would be built according to the plans

> Of Benjamin Outram … from the Tredegar Ironworks to join their canal near Risca Church with all convenient turnouts, and, if required, to construct a double rail; such tramroad should be made of sufficient width for Carts to pass by the side and the road to be open to the public on payment of tolls … .

The company were so keen to secure this new business that they were even agreeable to reimbursing Homfray and his partners with costs already incurred on their partially completed railway. It also agreed to maintain the railway and to keep the canal in good order and free from

---

[1] PRO, RAIL 500/5, 19 March 1800.
[2] Lloyd, *Old South Wales Ironworks*, p. 142.

ice, provided that all coal, iron and other products were sent only by that line. Tolls were to be paid at 3s. per ton for iron, timber and other goods and 2s. per ton for coal. It was a very good deal for Homfray and his colleagues, who must have been only too pleased to accept.

In his letter to Morgan Powell of 8 December, Outram had written that:

> I learn from Mr Homfray that the Monmouthshire Canal Co have agreed terms with the Ironmasters to make a railroad down from Sirhowy Valley to join the Canal. I am glad of it, as I am sure it will, if properly laid out and executed, serve to greatly secure the interests of the Monmouthshire Canal; which should be preserved, the Key of Commerce in that district, but for that purpose some burnishing is necessary to rub off its Rust, and preserve the Polish.[1]

From the tone of his letter he was clearly, but politely, hinting that he would be interested in acquiring any contract which might soon be on offer for constructing the railway. And indeed, within the next few days he was commissioned to prepare plans and estimates, although because of disagreements among the proprietors, his input was delayed for another six months.

The fact that the agreed project was financially flawed from the point of view of the canal company must have been recognised by Outram readily enough. Nevertheless he had secured the design brief for the project, and so he deferred, somewhat uncharacteristically, to the wishes of the proprietors in approving an ill-thought out scheme which required financial input only by the canal company, and would involve much needless and expensive double-handling of bulk materials at Risca. Fortunately, common sense prevailed because on 19 December 1800, the day following the signing of the agreement between Homfray and his partners and the canal company, there was a general assembly of the shareholders at which the one-sided deal was rejected and a more appropriate scheme substituted.[2] Thus it was resolved that a more equitable scheme would be for

> S. Homfray & Co. to make a Tram Road from or near Sirhowy Works to a point at the distance of 8 miles from Pill Gwenlly or

---

[1] PRO, RAIL 1014/17, B. Outram to M. Parry, 8 Dec. 1800.
[2] PRO, RAIL 500/5, 19 Dec. 1800.

the place where the goods are to be shipped at or near on the River Usk. The Company to make a Tram Road on the Plan and under the direction of Mr Outram from such point to Pill Gwenlly or some other convenient place for stopping that shall be determined upon between the Pill and Newport.

This was the basis of a more sensible scheme, still based on Outram's improved plan, to link the new works directly with the harbour at Newport by railway only.

Negotiations then proceeded between Homfray and partners and the canal company for the preparation of the bill to build the railway and to raise funds. Their Act became law on 26 June 1802, when the partnership was incorporated as the Sirhowy Tram Road Company with a capital of £30,000 and powers to raise another £15,000.[1] They were to be responsible for constructing and maintaining that part of the railway (with a carriageway alongside) from the Sirhowy ironworks, nearly 15 miles down the western side of the valley to a place known as Nine Mile Point (a distance of nine miles from Pillgwenlly, near Newport). They would also construct a branch from near the new Tredegar ironworks to the limestone quarries at Trefil, as well as another line from that branch to the Union Furnace at Rhymney. The Monmouthshire Canal Company were to build and maintain a single line from Pillgwenlly up to Nine Mile Point, except for the section crossing Tredegar Park (known as the 'Golden Mile') which was to be made and maintained by Sir Charles Morgan, and also a branch from Risca to the Crumlin line.

Hodgkinson and Kenear surveyed the entire route of the proposed railway[2] and a book of reference and estimates, completed and signed by Hodgkinson, were ready for submission with the bill by 25 September 1801.[3] This was accompanied by a plan of the Sirhowy Railway and its branches which Hodgkinson had set out, but which were surveyed by David Davies.[4] Hodgkinson prefaced his estimate with the statement that it was 'for making a Railway on Outram's improved plan'. It seems certain that Outram had a hand in the design because in a letter to

---

[1] 42 Geo. III c. cxv (1802).

[2] Gwent RO, D749/113, Affidavit of W. Kenear, 3 Jan. 1822.

[3] HLRO, J. Hodgkinson, Estimate for making a Railway on Outram's Improved Plan along the western side of the Sirhowy River ... to Newport on the river Usk and branches, with book of reference, 22 Sept. 1801.

[4] HLRO, Plan of an intended Railway or Tramroad from Sirhowy Furnace to communicate with the Monmouthshire Canal and the River Usk, by David Davies, taken from the line marked out by John Hodgkinson, Sept. 1801.

Morgan Parry of 5 August he had written that 'I enclose a sketch of my Ideas on the Sirhowy Railway'.[1] It would be surprising, knowing something of Outram's temperament, if Hodgkinson had deviated from any part of the plan, except for those minor details which can only be determined on site after work begins, when variations are found to be necessary because of unforeseen ground and other conditions. The estimates were prepared by Hodgkinson and these related only to the sections constructed by the Sirhowy Tram Road Company and the Monmouthshire Canal Company; there was no mention of Sir Charles Morgan's Golden Mile. The section above Nine Mile Point was priced at £25,000 with £5,000 contingencies, and the two branches were given together as £4,000 with £1,000 contingencies. There was no attempt to break down the costs into elements and it seems likely that these rough and ready estimates were given to Hodgkinson by Homfray and his partners. They are quite different to the more methodical estimates prepared for the canal company's southern section which were: £20,666 for the line to the terminus of the canal at Newport, and a further £1,462 for the branch linking the main line near Court-y-Bella Farm with Pillgwenlly. The Crumlin branch was priced at £13,038 10s. and thus the total estimate for the canal company's railways was £35,166 10s.

Although it seems that Outram was the designer of the project, Hodgkinson was the engineer in charge when work began on the railway. He was not engaged by Outram for the appointment, instead he was employed directly by the canal company to whom he had offered his services for twelve guineas per week.[2] He was given responsibility for the entire line, including the section through the Golden Mile and accordingly, the canal and tramroad companies each contributed eight-seventeenths, and Sir Charles Morgan one-seventeenth, to his salary, as well as to the costs of his original surveys. Walter Kenear acted as a resident engineer, judging by a statement that he made some years later, when describing his relationship with Outram and the work of the Sirhowy Railway. He 'was for many years employed by the late Mr Outram of Butterley Hall in the County of Derby, Civil Engineer, and assisted him in the setting out and forming of several tramroads'.[3] He

---

[1] Skillern Collection, Letter from B. Outram.

[2] PRO, RAIL 500/5, 4 Nov. 1802.

[3] Gwent RO, D749/113; PRO, RAIL 500/5, 4 Dec. 1805. Outram does not seem to describe himself as a 'civil engineer' in any documents examined by the author. This is the only evidence found which refers to this professional title. It should also be noted that in contemporary usage railways were described as railroads, tramroads or dramroads and no attempt has been made here to standardise this terminology. Outram, as noted earlier,

came with Mr John Hodgkinson, who was also in the employment of Mr Outram, into Monmouthshire to level and set out the line of the Sirhowy Tram Road which was then about to be made from the Sirhowy Ironworks to a place on the River Usk near Newport.

In describing the line Kenear stated that 'no Tram Road ought to have a less Fall than five Inches in a Chain [i.e. 1 in 160] and that all practical Engineers in setting out Roads endeavour to obtain at least that Fall'. In fact, the gradients from the Sirhowy works to Risca had been set at 8 in. to the chain (1 in 100), except for a two-mile section above Nine Mile Point where it had reduced to six in. (1 in 130), and the last 6 miles into Risca where it was just under 6 in. Below Risca the falls averaged 3 in. to the chain (1 in 265) which, though unavoidable, gave the engineers little cause for satisfaction.

The Act of 1802 committed all three parties to complete their sections of the railway by 29 September 1803, and penalties for failure were severe. If the Sirhowy Tram Road Co. and Sir Charles Morgan failed to complete within the specified time then the Monmouthshire Canal Co. could take over and complete the works and those parts would become vested in them. If, on the other hand, the canal company and Sir Charles failed to meet their deadline then the tramroad company could step in and the sections they completed would be vested in them. These arrangements made for some promising competition but, in the event, none of the parties succeeded in finishing their sections on schedule and accordingly the penalties were never enforced.

A more realistic works programme had to be arranged and this was outlined in an agreement made between all the parties, dated 22 September 1803.[1] After stating that 'from unforseen circumstances it has become impracticable to complete either of the said Railways or Tramroads by the times respectively stipulated', it was agreed that the Sirhowy Tram Road Company would complete, within twelve months, the railway then in course of construction as a narrow tramroad from Sirhowy Furnace down to the Gellygroes valley, and, within eight months, would complete the railway below the Gellygroes valley to Nine Mile Point as a wide tramroad. They would also, on twelve months notice given by a general assembly of the canal company after the 25 March 1804, convert the railway to a wide tramroad from Gellygroes up

---

invariably referred to his 'Railways'.

[1] NLW, Maybery 240, articles of agreement between the Monmouthshire Canal Co. and Sirhowy Tram Road Co., 22 Sept. 1803.

Plate 23 The Long Bridge at Risca on the Sirhowy Railway in 1900.

to Pont Gwaithyrhaearn or another agreed location. The Monmouthshire Canal Company would complete, within six months, their section of railway from Nine Mile Point to the River Usk as a wide tramroad, except for the bridge over the valley at Risca, which would be made ready for use not later than 25 March 1805; and would defray the additional costs incurred by the tramroad company from the time that their railway was complete until the bridge at Risca was opened. In the meantime, the canal company would provide a man and a horse to assist in the carriage of iron and coal across the valley to the railway below Risca.

The agreement did not mention why the Sirhowy Tram Road Company was building a railway on the narrow plan (presumably with edge rails to a gauge of 3 ft 4 in.) or explain the curious arrangement whereby they would agree to convert to a 4 ft 2 in. gauge plateway at a future date, and only then up to Pont Gwaithyrhaearn, some four miles short of Sirhowy. A reason for these confusing arrangements was probably that Homfray & Co. had already laid out several miles of edge-rail to the narrow gauge after acquiring their leases from Sir Charles Morgan during March 1800, and before agreeing to a new plan with the canal company. The clause in the agreement was simply a safeguard in case completion of the tramroad company's section was delayed for unforseen reasons. They no doubt had every intention of converting to the wide-gauge plateway along their entire section by the time the bridge

Plate 24 The Long Bridge during demolition in 1900.

at Risca was completed. In fact the completion of the bridge was further delayed, Hodgkinson only assuring the committee in January that it would be passable by May 1805.

Hodgkinson's estimates reveal that it was initially intended to build a single arch over the River Ebbw at Risca and another across the highway which passed through the town, on the far side of the valley. Presumably it was planned to build an embankment, about 275 yards long, between the two structures but when work began on the railway, minds had been changed and a viaduct, which came to be known as the 'Long Bridge', had been substituted. Although there no certainty that Outram had a hand in the design of the structure, the details must have been submitted to him for approval. It was the outstanding engineering feature on the railway and, although it was dismantled in 1900, it is possible to deduce the approximate dimensions from Ordnance Survey plans and early photographs. The length of this plain but impressive structure, including wing walls, was 930 ft and there were 32 arches, each of 24 ft clear span and an extrados of 12 ft radius. The distance between abutment centres was 28 ft 6 in., the height to track formation from ground level averaged 28 ft and the parapets were 3 ft 8 in. high. The width of deck, excluding parapets, was 16 ft.[1] Each arch was built

---

[1] PRO, RAIL 500/5, 4 Nov. 1802.

square to the longitudinal alignment except the end arches, the voussoirs, spandrel walls and parapets of which were curved in plan.

The contract for the construction was let in November 1802 to Richard Thomas, William Homfray and Walter Walters for £3,600.[1] There may have been some minor alterations to the design because Hodgkinson was instructed by the committee in the following May to 'finish the arches in such a way as to admit of houses being made under them whenever the company thinks proper'.[2] Indeed, permission was granted to Edward Cairns and Thomas Potts, who opened a copper works at the Graig in Risca, to build two 'double houses' under the bridge at the end of 1806.[3] It is questionable whether these these quaint homes were ever built because they do not appear on any of the early lithographs or photographs.

Initially the bridge had no parapets and it was not until June 1806 that a contract for the work was let to a John Hopkins for £400.[4] Hodgkinson had given notice to the committee the previous month that one day per week would be sufficient for his attendance until the entire works under his control on the railway and canal reservoirs were completed. A grateful committee then granted him forty guineas for his work[5] and yet barely two months later, following complaints from users of the line because of slow progress on the parapets, they reprimanded him because they considered 'Mr Hodgkinson's long absence from superintending the Company's works an Act of negligence'.[6]

The Long Bridge was abandoned and disused following conversion of the Sirhowy Railway to standard gauge in 1865 and sold (it is said locally, for £1 per arch). Photographic evidence suggests that when it was demolished at the turn of the twentieth century, work began at the eastern end and progressed, arch by arch, across the valley. The abutments were strutted in case the arches had been designed to be structurally interdependent. This was a precautionary measure which could prevent a progressive, uncontrolled collapse. The attractive stone from the structure was then salvaged and used for building houses nearby.

During the meeting of November 1802, important contracts for the

---

[1] Ibid.
[2] Ibid., 3 May 1803.
[3] Ibid., 27 Nov. 1806.
[4] Ibid., 25 June 1806.
[5] Ibid., 7 May 1806.
[6] Ibid., 30 July 1806.

laying of railways were also let: to Henry Lewis for the building of the section from Court-y-Bella to the park wall of Morgan's estate, including a road bridge, for £1,217; and to the ubiquitous Walter Walters for making that part of the line from Risca valley to Tredegar Park, a distance of 6,380 yards, for £2,500. Presumably these contracts were to follow others awarded for the substantial earthworks which were essential parts of the project. They also excluded the provision of cast-iron rails because tenders were invited that month in the Gloucester and Bristol newspapers for 600 tons of rails, each 45 lb in weight 'of best dark grey metal'. It was unlikely that Outram ever expected to sell rail plates and other iron goods from the Butterley works to either of the companies. In fact, it seems that the tramroad company made the plates for their main-line and for the branches up to the Trefil quarries and the Rhymney furnace. As regards the line from Newport to Nine Mile Point, the Blaenavon works of Hill and Hopkins won an order for plates from the canal company in July 1803, although substantial quantities were probably also provided by Alexander Raby of Llanelli.[1]

The line and its operation was described by T.G. Cumming as he saw it in 1824.[2] It is clear that the railway was successful commercially and was still regarded then as an impressive feat of engineering:

> This rail-way was completed about twenty years ago, also a turnpike-road by the side of it for about 17 miles, the total expense amounting to about £74,000, or about £3,000 per mile. About £40,000 of this sum was expended by the canal company in consequence of building a bridge, and some deep and expensive cutting, whilst the Tredegar iron company completed nearly double the distance at the cost of £30,000. Sir Charles Morgan expended £4,000 upon one mile but he had some deep cutting and a double road to make. Notwithstanding the expense, this road pays the proprietors 30 percent having a considerable trade upon it in coal and iron, which pay the same tonnage as the canal. For the first 9 miles out of Newport, being the parts made by Sir Charles Morgan and the canal company, it is a double road; one for the loaded waggons to come down, and the other for the empty ones to return; and on the Tredegar iron company's part of 15 miles, it is a single road, with frequent turnouts for the teams to pass. The whole line of the road for 24 miles is an inclined plane,

---

[1] Ibid., 20 July 1803.
[2] T.G. Cumming, *Origin and Progress of Rail and Tramroads* (1824), p. 26.

averaging about the eighth of an inch in the yard, or something more; but that part made by the Tredegar iron company is of somewhat greater declivity than the rest. The coal and iron are conveyed upon it in waggons each carrying about forty-five to fifty hundredweight, exclusive of the waggon; and a team of four or five horses will draw about fifteen of these waggons down and take the same number of empty ones back with ease.

It seems improbable that Outram made a great deal of money out of his sojourns into South Wales, except for his professional fees as consulting engineer.[1] As noted earlier, he did not sell any rail plates to either of the canal companies or the tramroad company, but he could hardly have expected anything else in that country of many ironmasters. Perhaps too, he might have been disappointed with the cautious response of some of his clients and their reluctance to carry out all the improvements as he recommended. There can be no doubt, nevertheless, that his ideas aroused a wide interest and, in that respect, his visits were a resounding success.

By way of an example, Outram's report to the Brecknock & Abergavenny Canal Company was received by William Taitt, the partner in charge of policy at the Dowlais ironworks in Merthyr Tydfil, a few days after its publication in July 1799. Taitt, an able but somewhat ruthless man (with an eccentric manner of writing letters), replied to the sender that

> I have yours of the 4th covering an account of a Wonderful Road [of] which I do not believe a Word as to its affects. It seems to be one of the worst kind of Railroads because of its great Width and Weight carried in the Waggons and the Plates are made similar to those on the Inclined Planes in Shropshire. Were Railroads to be continued, the Plates cast in that way would certainly be an improvement ... .[2]

A result of this ambivalent letter was that George Overton, newly appointed to Dowlais as engineer in charge of mine and surface railways (he was later to conduct the first surveys for the Stockton & Darlington Railway, in company with David Davies, in 1818–20) was sent away to

---

[1] PRO, RAIL 812/3, 17 June 1799.

[2] Glamorgan Record Office, Dowlais Letterbook 416, W. Taitt to J. Guest, 7 July 1799.

certain unspecified sites to learn more of Outram's improved 'Road and its effects ... '.

That Overton approved of Outram's methods there can be few doubts. In later years he wrote an influential book[1] in which he stated his marked preference for plates rather than edge-rails and claimed, surprisingly enough, that the resistance to motion was less on the former than the latter. Overton's most famous railway was the 9½ mile-long Penydarren Railway, which was built in 1802 from near the Dowlais works to the Glamorganshire Canal at Abercynon. This was a plateway of 4 ft 2 in. gauge, laid to carefully levelled gradients, and obviously owed much in its design to Outram's improved plan. This railway was used for Trevithick's famous trials with his high-pressure steam locomotive in February 1804, of which stirring event Outram must have been told, but the Sirhowy railway had to wait until 1816 before William Stewart of Newport designed and tried out a locomotive on its lines. Unfortunately there was still a decided lack of interest in these experiments and it was not until 1829 when the *Britannia*, a locomotive purchased from the Stephensons, was delivered to the Tredegar works. It made its uncertain, but successful, first journey to Newport in December of that year.[2]

After Outram's death in 1805, his loyal assistants, Hodgkinson and Kenear, remained in South Wales where they continued for many years in intensive employment as railway and canal engineers, using much the same techniques they had practised under Outram in earlier days.[3]

---

[1] G. Overton, *A Description of the Faults and Dykes of the South Wales Mineral Basin* (1825), i, pp. 40–2; C.F. Dendy Marshall, *A History of British Railways down to the year 1830* (1938), pp. 87, 88.

[2] C.E. Lee, 'The Sirhowy Valley and its Railways. Part II', *Railway Magazine*, lxxxv (Oct. 1939), pp. 262, 263.

[3] Hadfield, *Canals of South Wales*, p. 24.

CHAPTER 16

BENJAMIN OUTRAM'S OTHER CANALS AND RAILWAYS

All Outram's major civil engineering works have now been described but he undertook many minor consultancies in addition to these capital projects, such as reports on the progress of the Melton Mowbray Navigation,[1] the Manchester, Bolton & Bury Canal and the tiny Fletcher's Canal in Lancashire.[2] All these comprised the briefest of tasks and commanded but a mention in company minute books; there are likely to have been many assignments of that kind which went unrecorded. He also prepared imaginative designs for several substantial canal and railway schemes which, for various reasons, were never constructed according to his plans. The canal projects included proposals for the River Dun Navigation Company and the Stratford upon Avon Canal Company, both of which are summarised here.

Whilst Outram, in his later years, never entirely lost his early enthusiasm for canal transport, his experiences had proved that in some circumstances railways were a preferred alternative to canals. He never hesitated to advocate his convictions in this respect and several of his ventures around the turn of the nineteenth century were plainly efforts to promote railways and also to generate business for the Butterley ironworks. As a result he became widely known for this work, particularly in central England and Wales. Some projects, such as the Ashby and Sirhowy railways, were the successes he had promised but others were not blessed with such good fortune and were never brought to fruition. His 'improved plan' was a marked advance in railway technology but unfortunately his beliefs were not always shared by the promoters who controlled the investment. They were often of a conservative bent and were reluctant to embark on a new, and in their minds at least, an unproven system. Thus, Outram's hopes for a wider use of railways were only partially realised during his lifetime. Reference is made here to the railways Outram proposed for the Worcester & Birmingham Canal, the Somersetshire Coal Canal, the Forest of Dean mining interests, the

---

[1] East Sussex Record Office, Diary of Col. J.F. Turner, Aug. 1795.

[2] C. Hadfield and G. Biddle, *The Canals of North West England* (1970), ii, p. 250, 261.

Blisworth Hill Railway on the Grand Junction Canal and other railway contract work. Mention is made of the Liverpool to Manchester railway, a route reputed to have been surveyed by both Outram and Jessop.

Commissions from the River Dun Navigation Company

The Canal Mania had encouraged the River Dun Navigation Company (i.e. the body responsible for the Yorkshire Don) to participate in the construction of new canals to link their river with the Calder and Trent, and also to consider a Sheffield Canal to run from the wharfs at Tinsley into the heart of Sheffield. The latter had excited public interest since 1792 when a proposal also included a branch to Eckington (Derbys.).[1] Further consideration and opposition from landowners delayed progress[2] but the plans were not entirely abandoned and Benjamin Outram, then but 29 years of age, was commissioned the following year to review the proposal. His report and estimate for a canal were submitted to the company on 19 August 1793.[3]

Outram stated that he had examined the country between Eckington and the collieries on the west side of the River Rother, up to Tinsley and Sheffield and 'had tried various Levels to ascertain the best Line for a Navigable Communication between those Places'. His selected route began as a narrow canal at Park Brook, close to the highway linking Beighton and Eckington, then ran north-westwards by the River Rother before turning westwards on the approach to Tinsley. The line passed through Attercliffe, where it opened up into a broad canal, to terminate in Sheffield, a distance of 12¾ miles. A branch, also suitable for river barges, was to connect Attercliffe with the River Dun at Tinsley wharf, the total lockage there being nearly 60 ft. Outram does not seem to have submitted a plan with his report but the intention may have been to read the latter in conjunction with the plan prepared the previous year by William Fairbank, the Sheffield surveyor.[4] On this was plotted the canal

---

[1] C. Hadfield, *The Canals of Yorkshire and North East England* (1973), ii, p. 265; Sheffield Archives, Fairbank Collection, CP 4(44), Account book of W. Fairbank, Sept.–Nov. 1792.

[2] University of Nottingham, Dept of Manuscripts, Manvers of Thoresby MSS, esp. MaB 536/7, 27 and 33.

[3] B. Outram, *Report on the proposed Sheffield Canal, 19 August 1793* (copy in Sheffield Central Library, acc. no. 17925).

[4] Sheffield Archives, Fairbank Collection, W. Fairbank, Plan of an Intended Canal from Sheffield with a Branch to Eckington, 1792.

previously proposed which followed a similar but far longer and more circuitous route than Outram's of sixteen miles. Outram had devised a much improved line by judicious use of embankments and deep cuttings, but also included a tunnel, 770 yards long. Locks were avoided throughout, from Eckington to Sheffield, by these means. He added to his report that a branch could be made to the Chesterfield Canal, beginning about half a mile below Eckington.

The total cost of the project was estimated at £52,215 and revenue was only assessed at £4,000 per annum but, by way of encouraging his clients, Outram went on to claim that 'a Canal through a Country full of Coal and Manufactures, with the Town of Sheffield at its head, cannot fail to have many sources of Commerce that are unforseen, and its Benefits must progressively increase so long as these Maufactures flourish'. Unfortunately these were difficult times financially and the scheme languished yet again. Although there were periodic reviews, an Act of Parliament was not obtained until 1815,[1] and then only for a canal linking Sheffield with the Don at Tinsley wharf.

That was not the end of the involvement of Outram in the affairs of the company. Two years later in 1795, the committee, still apparently of a mind to effect improvements of a sort, invited him to survey the River Don 'from one end to the other', to report on its state and condition, and to recommend 'Repairs, Amendments and Alterations' that seemed necessary.[2] Several of the committee were to accompany him on his inspections, as well as John Copeland, the manager, and George Miller, their new surveyor. Outram sent in his report on 19 November 1795 and Copeland was requested to check it for any obscurities before it was presented to the committee 'on a day to be called'. Athough Outram was paid £40 16s. for his 'long report' in May 1796, this statement was not tabled until August that year.[3] Even then it was resolved that 'time be taken to consider of the same, several of the Proprietors not having read the report'. Thus an excuse was found for putting off the decision once more.

In August the following year the committee invited Robert Whitworth to survey the river in company with them and the next month they at last got around to discussing both his and Outram's reports—or at least they agreed to have them both printed and sent to each proprietor 'as soon as

---

[1] 55 Geo. III c. lxv (1815).
[2] PRO, RAIL 825/1, 10 Sept. 1795.
[3] Ibid., 11 Aug. 1796.

possible'.[1] However, it was not until May 1799 that Copeland was instructed to take steps to carry out improvements in the vicinity of Sprotbrough 'according to the opinion and direction of Mr Whitworth and Mr Outram who have made their report'.[2] And so Benjamin Outram's association with the company ended.[3]

The Worcester & Birmingham Canal railway proposals

The Worcester & Birmingham Canal was sanctioned by Act of Parliament in 1791.[4] Capital expenditure of £180,000 was authorised with a further £70,000 if found necessary. The canal is thirty miles in length and begins at the junction of the Birmingham Canal with the Birmingham & Fazeley Canal near Birmingham. From there its line runs south-west to meet the Dudley Canal at Selly Oak and then continues south-east to the Stratford upon Avon Canal at Kings Norton. It then turns south-west again to Tardebigge, running east of Bromsgrove and Droitwich to terminate at the River Severn, just south of Worcester. The summit level is 14½ miles long and extends from Birmingham to below Tardebigge tunnel, where the first of the 58 narrow locks, which take the canal 428 ft down to the Severn, is located. There are five tunnels: at Edgbaston (103 yards), West Hill (commonly known as Kings Norton) (2,750 yards), Shortwood (608 yards), Tardebigge (568 yards) and Dunhampstead (326 yards). The short Droitwich Junction Canal links with the main line at Hanbury, ten miles north of the Severn terminus.[5]

Work proceeded quickly at first but money ran out during 1797 as a result of very substantial engineering works leading to, and including, the great tunnel at West Hill, near Kings Norton.[6] A further Act in 1798[7] authorised higher tolls and more fund-raising but money was in short supply at the time and the canal was to languish unfinished beyond

---

[1] Ibid., 26–28 Sept. 1797.

[2] Ibid., 27 May 1799.

[3] For the company's later history see Hadfield, *Canals of Yorkshire and North East England* (1973), ii.

[4] 31 Geo. III c. lix (1791).

[5] H.R. de Salis, *Bradshaw's Canals and Navigable Rivers of England and Wales* (1904), pp. 331–6; J. Priestley, *Historical Account of the Navigable Rivers, Canals, and Railways, throughout Great Britain* (1831), pp. 690–5.

[6] C. Hadfield, *The Canals of the West Midlands* (1966), p. 139.

[7] 38 Geo. III c. xxxi (1798).

Hopwood, only nine miles out of Birmingham, for several years.[1]

This state of affairs did not suit everyone and so it was in November 1798 that Joseph Wilkes, that 'judicious and influential' gentleman so admired by Dr Anderson, came to recommend Outram to his colleagues on the committee of the canal company, as a person well-qualified to examine the line of the canal and to consider the possibility of constructing a railway.[2] The latter could be intended 'as a permanent matter or as a temporary expedient'; it would extend from the end of the summit of the canal, south of the Tardebigge tunnel, down to Worcester. The idea was of much interest to the members who were clearly devoid of ideas as to the next step to take. The clerk was instructed to write to Outram, who was quick to respond, his report being submitted to the company in time for the January meeting.[3]

Outram surveyed the route for a railway as close as he could to the Parliamentary line already staked out for the canal, stating in the report that the few deviations he had made were necessary to ensure easier gradients for the railway. Moreover, he pointed out that the waterway should eventually follow that line also, as it would lengthen the shortest pounds between the proposed locks. Naturally, as an enthusiastic proponent of railways, he had 'great satisfaction to find, that a Railway may be constructed to produce speedily, great advantages to the Proprietors, and the Public ... '. Quite apart from the obvious advantages of opening up the line for trade, he stressed that a railway could also be used for canal construction at a later date when funds were available, by providing rapid and cheap transport for high quality stone from Tardebigge for the locks, instead of bricks as had been intended. A source of lime from Dunhamstead Hill could be exploited as well and used for mortar in building every lock and bridge. A railway would also prove advantageous during earthwork construction along the entire waterway.

Outram recommended that the summit level be terminated just south of the proposed Tardebigge tunnel, 700 yards shorter than originally intended, and the railway would begin its descent from there. The tunnel could be built to the original size specified if it was found that good stone could be taken out of it for use on the works, otherwise a tunnel of smaller dimensions would suffice for the railway. Outram claimed that goods could be sent the 15½ miles down to Worcester from Tardebigge

---

[1] Hadfield, *West Midlands*, p. 139.

[2] PRO, RAIL 886/4, 27 Nov. 1798.

[3] BCL, Boulton & Watt Collection, Canal Item 73, B. Outram, Report on the W&B Canal and Railways, 8 Jan. 1799.

in four hours and the return would take about five hours. This could be achieved by arranging three stages: supposing that 20 tons of coal were loaded in 12 or 14 waggons, two horses could convey them down the first stage to Astwood; four other horses could take over the next stage, which was level, to Offerton and the final downhill run into Worcester could be drawn by three horses. These nine horses, with three drivers, could make two journeys each day and thus transport 40 tons to the city and return with upgate loadings of say, one-fifth (8 tons) of the downgate loads. Taking into account the time for navigating along the canal, he reckoned that 20 tons of goods could be taken from Birmingham to Worcester in ten hours. A double railway was recommended although, surprisingly, Outram stated that the rails of one should be sufficiently strong to carry two-ton waggons and the other, for returning waggons, one-fourth lighter. The estimate was:

earthworks, culverts, fencing, blocks of stone, and a horse path (all of which would also serve for the proposed canal):	£10,000
cuttings, fencing, forming and laying the railway (none of which would serve for the canal):	£4,700
cast iron rails (to sell at four-fifths of the original cost):	£20,500
Total	£35,200

Because the railway was a temporary measure the actual cost, following the sale of the salvaged rails, would be about £9,000. Outram also claimed, reasonably enough, that the later savings in shifting earth, stone, lime and other materials for the canal by means of a railway would be nearly equal to such costs. Hence at the end of the day the railway would cost nothing! All this was quite apart from the gains in making the company's system fully productive two years sooner than expected. This plan latter assumed that the money was immediately available to effect a speedy completion and this was certainly not the case at that time. Should funds be insufficient for a dual track, then Outram priced a single-track railway with passing places as:

earthworks etc, useful when making the canal:	£8,700
works applicable to a railway alone:	£3,300
cast-iron rails (to sell at four-fifths of original cost):	£13,000
Total	£25,000

Taking into account the eventual sale of salvaged rails, actual costs of a railway would be less than £6,000 and, again, this would be easily recoverable through savings on the incarriage of materials and earthmoving during the canal's construction.

Outram added that a communication could be made with Droitwich along the line of the intended canal to that town at a cost of £2,200 for a single-track railway and £3,100 for a double. Moreover, he assessed the costs of carrying coals from the summit, say, would be 8d. and 1s. per ton to Droitwich and Worcester respectively, although other costs would differ, depending on the nature of the merchandise. Outram thus made out a strong case for a temporary railway and although the proprietors did show an initial flurry of interest in his report, it was not until the May meeting that it was published and distributed.[1] Outram attended the next meeting in July 1799 to offer further advice and to collect his fee and expenses for £57 15s. before leaving.[2] He probably suspected by then that nothing would be done.

And yet six months later the committee resurrected the scheme, resolving that 'no profit from the Canal can be expected therefrom to the Proprietors, until the canal is extended'. They were right about that, of course, hence they proposed to complete the waterway to a point south of Tardebigge tunnel as recommended by Outram and to engage him, 'an engineer of the first abilities', to build the railway.[3] They reckoned this would bring in an income of £4,000 per annum at least but to pay for the works they would have to ask shareholders for £20,000 by subscription, or £12 10s. per share. Even so, more delays ensued before another resolution of the meeting of July 1800 confirmed that a railway would be adopted, because even a 'moderate calculation' now showed that an income of £25,000 per annum would be the result of completing the line to Worcester.[4] It was also disclosed that it would cost £19,127 to finish the canal up the northern portal of the Tardebigge tunnel (the added costs of driving the tunnel and completing the waterway to the head of the railway were not stated) and 'the sum of £5,000 only being subscribed ... ', the committee would have to devise other means of persuading shareholders to part with sufficient cash.

The company's financial difficulties continued and it was not until March 1807 that the waterway was finished to Tardebigge Old Wharf

---

[1] PRO, RAIL 886/4, 8 May 1799.
[2] Ibid., 3 July 1799.
[3] Ibid., 30 Dec. 1799.
[4] Ibid., 2 July 1800.

(13¼ miles from Birmingham). In fact, the entire navigation was opened only in late 1815.[1] Regrettably the lifeline of a railway project never materialised but acute cash shortages probably prevented that worthwhile step. One can only speculate on what Outram's unspoken views might have been about his commission but it seems fairly evident that that, from the outset, a permanent railway from Birmingham to Worcester would have been a better solution than that heavily locked canal; its total costs would have been but a fraction of the monies expended, even as early as 1797, on the unfinished waterway. Not that such a view would ever have appealed to the committee members: in common with many canal companies they were rather conservative, preferring to keep to well-established practices in which they had more confidence and understanding.

At least one of the proprietors realised at an early date that the railway was something of a pipe-dream. John Wall, who was also a committee member of the Stratford upon Avon Canal Company, wrote to Mathew Boulton at the Soho Works, Birmingham, in December 1799, enclosing a copy of Outram's report with the remark that, as regards the railway, ' ... I don't think it will ever be put into Execution'.[2] Nevertheless, he thought well enough of Outram to advise Boulton, who was then contemplating building a railway from the Soho Works up a steep incline to the Birmingham Canal,[3] that he might be advised to contact ' ... Mr Outram, a Person much noted for laying down Railways [who] will be at the White Lion at Stratford upon Avon on the evening of the 28th inst. He is desired by the Committee of the Stratford Canal to examine the unfinished line of that Concern between Hockley House and the Junction with the Warwick Canal ... '.

There is no evidence that Outram was ever invited to visit Soho to offer advice to the able cohort of engineers there, but he certainly went to look at the Stratford Canal.

The business with the Stratford upon Avon Canal Company

The Stratford Canal begins at Kings Norton on the Worcester & Birmingham Canal, and runs east for three miles to Yardley Wood before

---

[1] Hadfield, *West Midlands*, p. 142.

[2] BCL, Boulton & Watt Collection, Canal Item 36, J. Walls to M. Boulton, 21 Dec. 1799.

[3] Ibid., Plan and Estimate for Railroad at Soho (nd).

turning south-east to Hockley Heath and Kingswood, where a short branch connects with the Warwick & Birmingham Canal. The Stratford Canal then turns south and follows a circuitous route via Preston Bagot and Wilmcote to Stratford upon Avon, a total distance of 25½ miles.[1]

An Act was passed in March 1793 to empower the proprietors to raise £120,000 for the building of their narrow canal and, should that prove insufficient, an additional sum of £60,000.[2] Work began from Kings Norton late that year under the engineer Josiah Clowes but a second Act was deemed necessary in 1795 in order to sanction a two-mile junction canal at Lapworth to join the Warwick & Birmingham Canal.[3] A further expenditure of £10,000 was authorised but exhaustion of these funds brought work to a halt at Hockley Heath in mid-1796 and by then a mere 9¾ miles of waterway had been finished.[4] The Dudley Canal was completed in 1798 and this, coupled with good progress on the cutting of the Warwick canals, generated new enthusiasm among the Stratford proprietors. Another Act was obtained in June 1799[5] to allow a deviation in the unfinished section, such that the proposed link with the Warwick & Birmingham Canal at Kingswood was reduced in length to one furlong. The right to raise another £35,000 was also granted.

By this time Clowes had died and although he was replaced by his assistant, Samuel Porter, it was to Benjamin Outram that the proprietors turned in November 1799 with an invitation to examine the proposed line and to report on its feasibility.[6] Outram presented his report on 28 December but he was in a hurry to continue a journey into Wales, hence he took away with him a few points for consideration and further clarification for another meeting on 8 January.[7] In the meantime, Porter was ordered to survey sites pointed out by Outram as suitable locations for reservoirs and to have plans finished by his return. Only three copies of Outram's report were prepared for distribution to the committee and unfortunately none of these has survived. However, it seems that he recommended, among other things, deeper and fewer locks than had been proposed, although this advice was later rejected by the committee,

---

[1] De Salis, *Bradshaw's Canals*, pp. 159–61.

[2] 33 Geo. III c. cxii (1793).

[3] 35 Geo. III c. lxxii (1795).

[4] C. Hadfield, *The Canals of the East Midlands (including part of London)* (1966), p. 179.

[5] 39 Geo. III c. lx (1799).

[6] PRO, RAIL 875/3, 20 Nov. 1799.

[7] Ibid., 8 Jan. 1800.

probably on the advice of Porter.[1] Tenders were invited for cutting the canal from Hockley Heath to Lapworth in March 1800 and the contract was awarded to Thomas Cartwright, who was also successful in winning the contract for the junction canal in December that year.[2] That phase of the works was opened in May 1802, although it was not until June 1816 that the entire line was finished to Stratford and joined there with the river Avon.[3] Outram's total contribution to the scheme is unclear, although it is likely that his commission was to check on and also recommend improvements to the scheme projected by both Porter and Cartwright.

On Blisworth Hill

The Grand Junction Canal, described as a 'stupendous and most useful line of navigation', begins at a junction with the Oxford Canal at Braunston in Northamptonshire and runs by Weedon, Blisworth, Stoke Bruerne and Cosgrove, and then Wolverton, Tring and Uxbridge, before it terminates in the river Thames at Brentford. The mainline is a 93¾ miles long barge canal with 101 locks and there are major tunnels at Braunston (2,042 yards) and Blisworth (3,056 yards). There were over 40 miles of branch canals, several of them with narrow locks. James Barnes, formerly engineer to the Oxford Canal Company, was commissioned by the Marquess of Buckingham and others to conduct the preliminary surveys, after which the line was examined and approved by William Jessop. The Act was obtained in 1793, with an authorised capital of £600,000.[4] Barnes was appointed the full-time engineer responsible for all the works, his onerous duties extending to the supervision of contractors and direct labour gangs. Jessop was described as chief engineer with overall responsibility for the engineering design of the entire canal.[5]

All the tunneling works were beset with difficulties, some of which were due to the inexperience and incompetence of contractors, notably at Blisworth where driving commenced on a different line than that established by Barnes and Jessop. Neither engineer had approved nor had

---

[1] Ibid., 5 Feb. 1800.
[2] Ibid., 31 Dec. 1800.
[3] Hadfield, *East Midlands*, p. 180.
[4] Priestley, *Navigable Rivers*, pp. 297–8.
[5] PRO, RAIL 830/39, 3 June 1793.

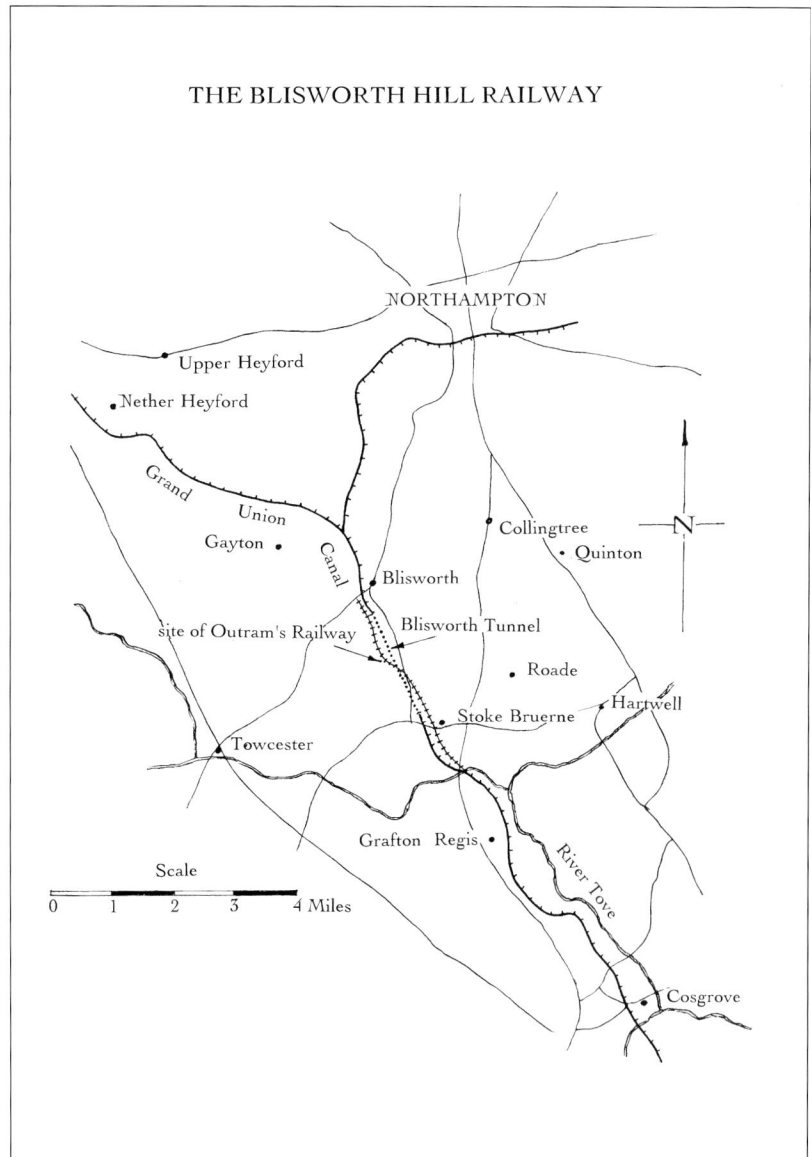

Figure 29 The Blisworth Hill railway.

any prior knowledge of this change, which seems to have been unofficially agreed between the local resident engineering staff and the contractors concerned. Barnes discovered and disclosed this matter to the committee only about eighteen months after works began, which did not say much for his capabilities as a supervisor. Substantial sections of the brick arching were also condemned or had collapsed and this, coupled with troublesome rock strata and flooding of the workings, brought the driving to a halt in November 1795. Jessop was for abandoning the tunnel altogether and proposed building locks over the hill instead, the water supply being provided by pumps powered by a steam engine. However, Barnes still favoured driving a tunnel on the original line and the committee approved of his scheme in mid-1796, but only after consulting Robert Whitworth and John Rennie on the merits of their engineers' reports. Jessop's plans were rejected and Barnes was awarded a contract to complete the tunnel within three years, for a sum of £48,000.[1] He was soon at work on both ends of the tunnel but was ordered to suspend operations in March 1797, because of a temporary shortage of funds. He was ordered instead to concentrate on the southern sections of the canal (Fig. 29).[2]

Barnes then constructed a toll road over Blisworth Hill to by-pass the abandoned tunnel so as to encourage some commercial activity and generate much-needed revenue. This was finished in August 1797 and it linked the northern section of the canal, then terminating in Blisworth village, with the turnpike road to the south. The measure was only partially successful and; it must have been obvious that more appropriate action would need be taken were traffic to increase. It was none other than Joseph Wilkes, also a shareholder in this company, who, at a special meeting in February 1799, urged the committee to review their policy.[3] It seems that he had a temporary railway in mind as the only satisfactory solution to the bottleneck at Blisworth and indeed, had already attempted to interest the committee in a railway on Outram's improved plan some twelve months previously, to link the canal at Blisworth with Northampton. Wilkes had later proposed that Outram might accompany Barnes on an inspection of the canal, although it is unclear what was the exact purpose of this proposal. The minutes state that it was simply 'to view the works thereof' but probably Wilkes intended it as a means of getting Outram's opinion on the practicalities of laying a railway over Blisworth Hill.[4] Neither of Wilkes's proposals excited the interest of the committee at the time but, at that special meeting in February, they at least agreed to an inspection of the canal. Jessop was to be invited to do this in company with Barnes.

---

[1] C. Hadfield and A.W. Skempton, *William Jessop, Engineer* (1979), pp. 115–19.
[2] PRO, RAIL 830/39, 15 March 1797.
[3] PRO, RAIL 830/40, 22 Feb. 1799.
[4] Ibid., 12 Sept. 1798.

One month later Jessop, who no doubt had by then conferred with both Wilkes and Outram, wrote to recommend the building of a temporary railway from the Blisworth wharfs, to climb the hill and then to run on down to the junction of the main line with the Buckingham branch near Cosgrove, some nine miles away.[1] The main-line waterway was still incomplete there but construction was proceeding rapidly from the south and Jessop expected that the railway and canal construction would meet near Cosgrove at much the same time, when traffic could begin immediately. On this occasion the committee listened and agreed with Jessop; it was said that these proposals were 'most expedient' and he was requested to survey the best route for a railway, in company with Barnes, and to prepare sections and estimates in time for the next meeting. The work was done, copies of Jessop's report were printed for distribution to shareholders, and the committee was authorised to proceed with a railway the next month.[2]

Members of the committee were unfamiliar with the new form of railway but Wilkes once more stepped in with an offer to demonstate the effectiveness of the system at his collieries in the Midlands.[3] The offer was accepted and the chairman, William Praed, and other members of the committee, accompanied by Wilkes, Barnes and other company officers, journeyed to Leicestershire and Nottinghamshire to see the operation of the railways at the collieries near Measham and Brinsley. Several demonstrations had been arranged for them, using horse-drawn waggons loaded with coal and timber on lines of different gradients. The resulting data were listed in the minutes[4] and shortly afterwards they were also reported at length by Wilkes in a paper printed in the *Repertory of Arts and Manufactures* for 1800 (these investigations were also mentioned by Dr Anderson's article referred to earlier).[5] A typical reference was that a single horse 'of £20 value' drew 21 waggons laden with coals and timber, amounting to 35 tons weight, down an incline of 1 in 115 and returned, drawing 5 tons up with ease. Another horse 'of £30 value' drew 21 loaded waggons, totalling 43 tons 8 cwt, down an incline of 1 in 108, and 7 tons on the return journey. Wilkes's railways were all built on Outram's principles because he stated that the rails were

[1] Ibid., 13 March 1799.
[2] Ibid., 10 April 1799.
[3] Ibid., 29 July 1799.
[4] Ibid., 21 Aug. 1799.
[5] C.F. Dendy Marshall, *The History of British Railways down to the Year 1830* (1938), pp. 41–2; above, pp. 203–4.

3 ft in length, weighed 33 lb and the gauge was 4 ft 2 in. Some lines were laid on stone but others differed from the improved plan because they were fixed to timber sleepers which, in turn, were set on a bed of slack coals. He costed a single line, built in favourable conditions but excluding major earthworks, bridges and culverts, at £900 to £1,000 per mile.

After some debate with Jessop and Barnes regarding the time it would take to finish the tunnel, the committee ordered that the railway should be built without delay and the clerk was requested to write to Outram, who was presumably invited at the suggestion of both Jessop and Wilkes, asking him to inspect the site and to design and estimate for a railway. By then it had been decided to terminate the railway below the bottom lock at Stoke Bruerne, fairly close to the tunnel exit, instead of at Cosgrove as Jessop had at first suggested. The committee also ordered work to begin immediately on the canal at Stoke Bruerne to meet the waterway now steadily approaching from the south.

In an effort to form slacker gradients for the railway, Jessop and Barnes had selected a route over the hill which deviated from the line of the tunnel and in places was outside the limits sanctioned by the Act of 1793. As a result, the company had to approach the landowner, the Duke of Grafton, to seek his agreement for this proposed change of alignment. The duke did not take very kindly to the suggestion.[1] In a long, rambling, grumbling, letter he claimed that he knew little enough about the proposals (even though Praed had already visited him to explain the nature and purpose of the railway) and voiced disapproval of the supposed inconveniences that his tenants might suffer. He also asked:

> Are waggon Drivers to be brought during the use of a Railway under due recognition and to have an establishment so as to lodge and board them and prevent them from becoming a nuisance to the quiet Inhabitants of the neighbouring Villages who will receive no benefit from these Measures and whom I consider under my protection?

There was perhaps some justification for this unease about the character and behaviour of the drivers because the canal builders had proved to be an unruly, drunken lot (and Outram's railway workers were to prove equally as bad) who had sometimes attracted the attention of the

---

[1] PRO, RAIL 830/40, 4 Oct. 1799.

magistrates.[1] The suspicion remains, however, that the duke's lack of enthusiasm was because he did not wish to see railways, or any other industrial development for that matter, anywhere near his estates if he could possibly help it. He would prefer them to go elsewhere, a common enough attitude of those whose properties were affected by industrial expansion then, it seems, as it is today.

Praed was not in the least abashed by Grafton's response and, in an equally lengthy letter,[2] he reassured the duke that the company's policy was always to co-operate with landowners affected by the canal's construction. As regards concerns about the waggoners, he was told that these men would always be carefully supervised. In any case, compared with haulage by road, far fewer would need be employed as drivers on the railway (in fact only one-tenth as many). Moreover, Praed stated:

> after the Railway is finished it will relieve the Inhabitants of Stoke Bruerne, Blisworth and other villages from the repair, to which they will be compelled by the Public, of the ancient Roads which will be destroyed every Month by immense accumulation of Waggons upon them.

Grafton was not to be placated; his opposition to the scheme remained and so Praed wrote no more. The company was thus forced to restrict the railway to the parliamentary line but fortunately, in the opinion of their engineers, this did not seem to matter very much. Outram and Barnes reported that although the line would be shorter and far steeper in places than previously planned, it would be manageable, in spite of the fact that the summit of the hill was 136 ft above the canal in Blisworth and 192 ft above the level at Stoke Bruerne wharf. One advantage, which they felt should not be overlooked, was that the new alignment would greatly facilitate the construction of the tunnel because it was so close to the line of the construction shafts, up which much of the spoil would be hoisted for disposal.

Outram speedily completed his specifications and estimates for a double railway over the hill and along the parliamentary line. The earthworks were formed over an 8 yards-wide strip, 6 yards of which were covered with a 12-in. layer of stones or broken brickbats. The track was in accordance with his standard specification with rails of 37 lb;

---

[1] V.A. Hatley, 'The Blisworth Hill Railway, 1800–1805', *Reports and Papers of the Northants. Antiquarian Soc.* (1962–63), p. 5.

[2] Ibid., p. 3; PRO, RAIL 830/40, 4 Oct. 1799.

waggons were to carry 30 or 40 cwt and never more than 2½ tons. There were no special arrangements for assisting the horse-drawn traffic up and down the steep inclines, even though average uphill gradients were as much as 1 in 35. The estimate was in two parts; Outram estimated for 4,600 yards of railway from Blisworth to Stoke Bruerne at £6,604, the costliest item being for 310 tons of cast-iron rails at £4,185. An extension of the main line by 1,400 yards from Stoke Bruerne village to the River Tove was priced at £1,921 which included for 95 tons of rails priced at £1,282 10s. The location of the terminus was so described before the waterway was built but in fact it was eventually located at the basin below the bottom lock at Stoke Bruerne, close to the crossing of the river. To these sums Outram added £800 for 80 waggons and four cranes, £250 for 200 yards of canal wharf construction and £175 for attendance and expenses; a grand total of £9,750. The cost per mile for the double-tracked railway was £2,500, which appears to be reasonable enough compared with that quoted by Wilkes in his article. Outram was offered and accepted a contract from the company and began work during October 1799.

As Outram got on with his task, Jessop was asked for his views on the practicalities of constructing another railway from Blisworth to Northampton.[1] He priced a single-track railway from the canal below Gayton, leading to the western end of the town, at £9,000, but did not recommend it wholeheartedly to the committee because it would cross meadows susceptible to floods. In the event, however, a twin-track railway was built in 1805 using materials recovered from the then abandoned Blisworth Hill railway. This was eventually replaced by a canal branch in 1815.[2]

Meanwhile, Outram had contracted to finish his works by 1 August 1800 and the committee were anxious to hold him to this because canal construction was, by then, fast approaching from the south. He wrote seeking an extension until 1 September, the date when the canal was expected to reach Stoke Bruerne, and the committee, reluctantly agreeing, replied that they hoped 'he will not consider the progress of the canal as a guide for his exertions' because even if the canal was not finished by the planned date, the company would still derive much benefit from the handover of a completed railway.[3] By 29 September Outram was still unable to confirm completion and another extension to

---

[1] PRO, RAIL 830/40, 15 April 1800.
[2] Hadfield, *East Midlands*, p. 114.
[3] PRO, RAIL 830/40, 19 Aug. 1800.

25 October was agreed by an increasingly impatient committee.[1] He was required to enter into a bond for £500 if he failed to meet this new deadline, or pay double the loss incurred by the company, whichever was the greater. It seems that the canal was finished by then but the railway was not. Outram, in an effort to placate his restive employers, offered to convey coal and other bulk commodities across the hill and to crane it into boats, at 1s. per ton. He was to write to the committee yet again on 3 November; work was still unfinished because it was recorded that 'when the Railroad be completed the Company shall direct Mr Barnes to examine and value extra works due and performed for which he [Outram] is to be paid by valuation'.[2] The minutes do not record the payment of any penalties by Outram and indeed it is unlikely he was ever required to do so because Barnes recommended that he to be paid a further £1,000 for work extra over contract. The committee, probably by then well pleased with their railway, paid out without demur.[3]

To some extent their benevolence was a result of the help that both Jessop and Outram had rendered to the company's staff responsible for operating the railway. Assistance had been given in devising procedures to avoid confusion and conflict between traders using the line and besides this, staff had visited the Butterley Ironworks to observe practices on that company's railways. They had even been to the Peak Forest Canal railways and also to the Bridgewater Canal in Manchester to inspect the wharfs and to learn about the packet boat system extensively used there.[4]

With the completion of the railway, the canal was open for business throughout its entire length and there is no doubt that the company thrived from that time onwards. The canal tunnel was finally opened in March 1805[5] and the railway, which had proved to be a most successful temporary measure, then fell into disuse.

The Somersetshire Coal Canal and railways

> This canal is of great importance in the export of coal, with which the neighbourhood of Paulton and Radstock abound. That useful

---

[1] Ibid., 29 Sept.1800.
[2] Ibid., 3 Nov. 1800.
[3] Ibid., 21 Feb. 1801.
[4] Ibid., 16 Dec. 1800.
[5] Hadfield, *East Midlands*, p. 111.

article is thus forwarded eastward to the Kennet and Avon and Wilts and Berks Canals, by which it is supplied to places on their lines, and also to others on the borders of the River Thames; besides entirely supplying the City of Bath and a part of the neighbourhood of Bristol.[1]

So wrote Joseph Priestley in 1831 of a canal and railway system which served a little known but prosperous coalfield.

The canal and railways were sanctioned by an Act of 1794 which allowed the raising of £80,000, plus another £40,000 if required, although a further enactment was necessary in 1796 to authorise deviations to the alignment of both waterway and railways.[2] Yet another Act in 1802 sanctioned more deviations, as well as the raising of funds for the construction of locks.[3] The canal began in the Kennet & Avon Canal, west of the Dundas aqueduct near Limpley Stoke, and ran for two miles parallel to Midford Brook as far as Midford Mill, where it divided. One branch continued for another 8½ miles by Combe Hay, Dunkerton and Timsbury to Paulton, and another followed the Cam Brook for some six miles to Radstock, by Wellow and Writhlington; there were several short railways which joined the upper pounds of both lines to nearby collieries (Fig. 30).[4]

Both canal branches had to surmount a sharp rise in the terrain to the west of Midford up to their summit pounds and for several years the company toyed with some innovative ideas to link the levels, in preference to conventional pound locks. Eventually they reverted to the latter and 138 ft of lockage was completed on the Paulton line between Combe Hay and Midford. A shortage of funds then forced the company to adopt a short railway, instead of locks, between the upper pound leading to Radstock and the Midford junction. In later years the Radstock branch was entirely converted to a railway.

John Rennie planned and estimated for the original scheme submitted for the Act of 1794. At the time he was also the engineer for the Kennet & Avon Canal. Shortly after the Act was obtained, however, Rennie had further thoughts and doubts about aspects of his designs but these matters were resolved following a re-survey by him and William Jessop. This

---

[1] Priestley, *Navigable Rivers*, p. 582.

[2] 34 Geo. III c. lxxxvi (1794); 36 Geo. III c. xlviii (1796).

[3] 42 Geo. III c. xxxv (1802).

[4] For a general account see K. Clew, *The Somersetshire Coal Canal and Railways* (1970).

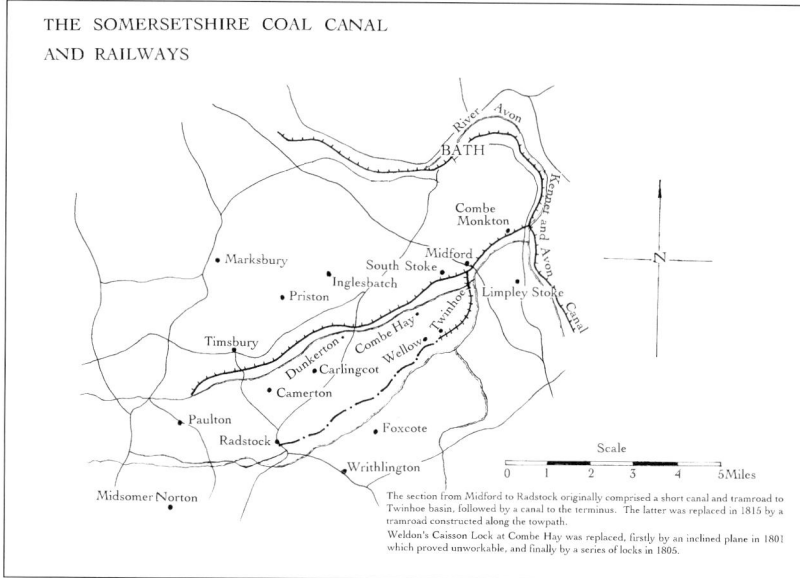

Figure 30 The Somersetshire Coal Canal and railways.

resulted in variations of the alignments and levels of the branches, subsequently authorised by the Act of 1796, and John Sutcliffe of Halifax was then appointed engineer for what was to be a narrow waterway. The canal was of a conventional design which should have presented no serious engineering difficulties, and yet for a long time the committee appeared to be unsure of the steps to take in building it. In particular they could not make up their minds as to the best method of linking the upper and lower pounds with the main line. Money was always in short supply in those years of national financial crises but the committee wasted much time (and markedly increased their costs) by their constant dithering whilst seeking a cheap solution to the lockage problem. Several engineers and businessmen were approached, some of whom were motivated more by personal interests rather than those of the company, and in most cases they were unable to offer sound solutions to what was, after all, a straightforward problem. In the event it took several years before the committee decided to use conventional pound locks after considering, and even attempting, various untried and dubious designs.

The first of these methods was the caisson lock, a form of boat lift, invented by Robert Weldon, an engineer. This device had been demonstrated in experimental form at Oakengates, on the Shropshire Canal,

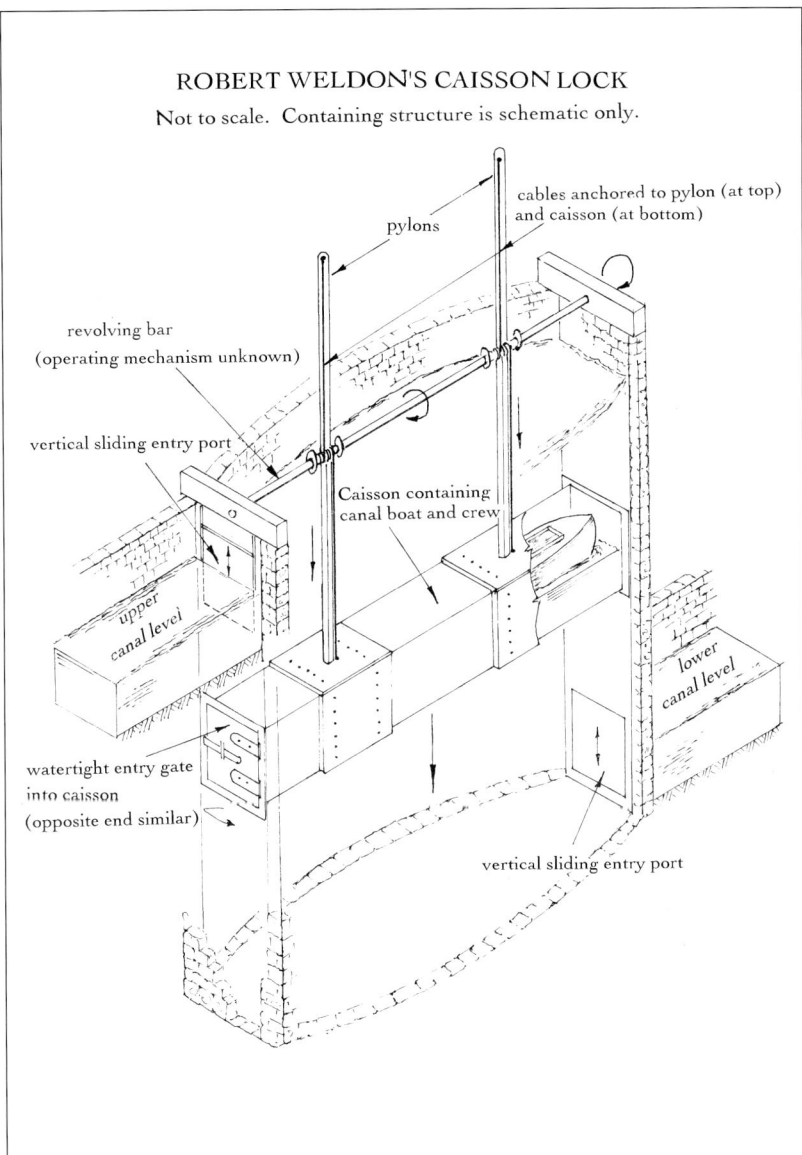

Figure 31 Robert Weldon's caisson lock.

during 1795[1] and advertised widely as a time- and water-saving alternative to pound locks. Both the Kennet & Avon and Somersetshire

---

[1] 'Canals' in *Cyclopaedia* (ed. A. Rees) (1819), vol. vi.

Coal Canal companies became interested and agreed to share the costs of the inventor for building a trial lock located close to the line of the canal below Combe Hay. The system built for these companies consisted of a masonry-walled chamber, 88 ft in both length and depth (to foundations), situated between canal pounds and filled with water to the top level. Short tunnels led from the pounds into entry ports at top and bottom of the chamber respectively, and these could be sealed off by sliding gates. The caisson was a watertight, rectilinear timber box, fitted with gated ends to admit a floating narrow boat (and its crew) via the entry ports from the canal pounds. This structure could move 46 ft vertically between upper and lower levels of the canal controlled by racks and pinions, and was just long enough to abut the entry ports at either level. The vertical movement down to the lower level was aided by the admission of water into the caisson so that its total weight was just greater than the upwards buoyancy force exerted by immersion. When the boatman inside the caisson pumped water out, it rose back to the upper level once more. Consequently very little external force was needed to move the caisson but it seems that the movement, especially the connecting of the caisson with the entry ports, was a tricky operation. It was claimed that the system could replace one or more conventional pound locks and that only a small quantity of water was lost from the chamber during use (Fig. 31).

Weldon's caisson lock must have been seen as an astonishing advance in technological invention, much in keeping with the remarkable revolution the general public had witnessed in the nation's transport system during the canal mania years. From today's standpoint, however, it seems strange that the potential hazards and dubious engineering design features of this apparatus went unrecognised by many who observed its operation, and by a few engineers who, even at that time, should have known better. The caisson was invented at a time when British engineers possessed but a modicum of theoretical knowledge and understanding of materials science and structural design practice and Weldon would have had very little but rule of thumb and common sense to assist him in his task. Fortunately the trials did not result in catastrophic collapse and loss of life, although they came close to that on more than one occasion. It is recorded that several proprietors, brave enough to lock down during a trial, were fortunate to escape asphyxiation when the mechanism jammed and they were left marooned below water level for some time before they could be released. Sometimes the machinery worked quite well but there were frequent failures which, coupled with the ever-increasing costs, eventually caused the committee to lose enthusiasm. They abandoned the scheme early in 1800, possibly after hearing the advice of Benjamin Outram. More than £5,000 and six years of delay had been wasted on this passing fancy.

Outram was invited to visit the canal late in 1799, shortly after the proprietors learned that there were insufficient funds to complete the navigation. At the time, construction was well advanced but the upper pounds were finished from their western limits only as far as Dunkerton and Wellow respectively. Thus the principal works which remained unfinished were the completion of these pounds and the linking of them with the canal near Midford junction.

Outram stated in his report of February 1800 that, having surveyed all the canal and railways, his intention was to ascertain the most advantageous methods of communication, 'either temporary or permanently', from the points of view of coal-owners, canal company and traders, from the source of the canal down to the city of Bath.[1] He was aware of the financial constraints on the canal company yet, quite properly, he offered no cheap solutions. His principal recommendations were that the canal should be completed quickly by linking the pounds with railways, and these might be regarded as temporary until such time as the company could afford more permanent installations, such as pound locks. In this respect they were not dissimilar to his proposals for Blisworth Hill. Outram's basic aims were to get the company operating so as to generate much-needed revenue. He also recommended a method for speeding up the export of coal, as well as some improvements to the already completed works, some of which he found wanting. His several ideas made good engineering and economic sense but, not surprisingly, they did not go down well with the resident engineering staff.

The caisson lock was cautiously criticised. Outram was never one to reject new ideas out of hand but he seemed to have recognised the dangers inherent in this system, quite apart from the weaknesses in the design. He felt that:

> the attempts in this respect have been unfortunate, partly owing to the difficulties inseparable from the execution of new inventions but principally owing to the very imperfect manner in which the Mason Work of the Caisson have been executed. I entertain no doubt of the practicabillity of passing boats from one level to another on Mr Weldon's principle but think it would be absolutely necessary to simplify his Apparatus materially.

---

[1] SRO, DD/MY 30, B. Outram, Report to the Company of Proprietors of the Somersetshire Coal Canal, 6 Feb. 1800.

He estimated that it would need another £16,000 (the original estimate was £1,200) to finish Weldon's work and that the cost of maintenance of the required number of caisson locks would not be less than that of the equivalent pound locks (13 in this case) 'and with the Caissons, the risks are certainly greater than with Locks'. He also stressed the point that the trade on the canal 'would be confined to Vessels adapted to Caissons'. What Outram said to the proprietors in private is unrecorded but, as noted above, Weldon's caisson lock was abandoned shortly after his visit.

Outram recommended that the lower level could be widened so as to accommodate broad vessels from the connecting Kennet & Avon Canal. In fact he thought it would be desirable to extend such a canal about three-quarters of a mile along each branch from Midford to terminate in basins with wharfs and depots, where large quantities of coal could be stored. Common pound locks could be avoided if railways were adopted to connect the upper with the lower pounds. Thus, on the Paulton line a dual-tracked railway, 2,200 yards long, could be built from the extremity of the canal at Dunkerton to run downhill at a gradient of 1 in 55 to a level some 12 ft above and 25 yards distant from the lower canal, from where the coals could be discharged into the depot. It could be advantageous if a 200 yards-long branch ran around the basin to the opposite side. A similar dual-tracked railway (the length was not given) could begin from the eastern extremity of the Radstock arm with a falling gradient of 1 in 70, reducing to 1 in 150 on the approach to the canal and terminating at wharfs similar to those already described. The total costs of his railways (these were not inclined planes), plus the extended waterway branches from Midford, would be £11,500 (£9,700 for the railways and £1,800 for the canal branches) compared with Weldon's costs of £34,000. An important clause in his report stressed that if at a later date the company wished to replace the railways by a 'navigable communication', then this could readily be achieved along either line of canal without any disruption of trade.

Outram was critical of the company's railways which ran from the several collieries down to the upper pounds of the canal. They were 'of a construction very inferior to those which I have introduced into extensive practice'. Naturally he felt that these should be relaid on the improved plan, claiming that costs would not be prohibitive. The iron rails *in situ* were twice the weight he normally recommended for heavy traffic (i.e. 37 lb) and no doubt they could be sold 'on the spot' for £5 per ton and replaced with lighter plates. Likewise, the timber sleepers could be sold off for use in the collieries at the original price paid by the company, because there had been a considerable increase since they were first purchased. Outram estimated that the 4,700 yards of colliery lines could be relaid for 2s. 6d. per yard, a total cost of £600. However:

> the descents of such roads are very irregular and in many cases so steep as to occasion much trouble to prevent laden Waggons running too swiftly downhill and to require great labour in drawing back the empty Waggons. It would be much for the interests of the Owners of the Collieries to alter the directions of such roads.

In these circumstances, inclined planes should be located so as to ease the gradients of the lines to the head, and from the foot, of the incline. Outram was unimpressed by the haphazard way in which coal was exported and dumped by the canal at the end of the railways leading from each coal pit. A more methodical practice was suggested. Coal from each pithead could be transported in gangs of twelve 2-ton capacity waggons, drawn by two horses, down to the canal. Waggons with detachable bodies were preferred, such that these could be lifted directly into 24-ton capacity boats, which would then float down to the railhead where the bodies would be craned on to sets of axle frames and wheels and run down to the depots. Coal destined for Bath might be retained in the bodies and shipped directly to the city where they could be hoisted on to sets of cart wheels for distribution to points along the highway. He thought it unlikely that there would be any upgate transport of goods beyond the lower canal levels but he nevertheless asserted that a single horse would draw 2-ton loads up the railway to Dunkerton in half an hour and on the other branch a horse would draw 3 tons up the line to Wellow in an hour. Outram did not go into any details when outlining a plan for the transport system that he felt might be adopted for the entire coalfield. He preferred only to illustrate how such a system might operate by describing the plant required for dealing with the export of the daily output from the Camerton Colliery, which was at the far end of the Dunkerton branch canal. He calculated that this would need the services of five men, 15 waggons, three horses and two 'long Boats' to export about 7,000 tons annually from that one pit.

There was nothing new about his recommendations, which were clearly based on personal experience. The methods outlined had been applied and had proved satisfactory on the Derby and Peak Forest canal railways already. Perhaps as a means of emphasising the soundness of his proposals, Outram saw fit to mention 'the Railways I am now constructing for the Grand Junction and the Ashby de la Zouch Companies'. These could be inspected should further proof of his abilities be needed.

Essentially, Outram's report was a radical review of the company's system. However, had these several wide-ranging improvements been implemented immediately, they would have greatly added to the company's capital costs, no matter how desirable they were for the future of the canal. The proprietors probably thought as much and this may explain why they seemed only too eager to listen to the several critics who crowded in to reject Outram's ideas in preference to their own. In the circumstances, Outram might have worded his report more carefully. He should have stressed far more that the railways might be regarded only as short-term substitutes for locks, and that all the additional works discussed were possible future developments, only to be implemented when funds allowed. Considering the short lengths of canal involved and the unavoidable double-handling of goods necessitated by the dual transport system Outram had recommended, the collieries might have been served better and cheaper from the outset by railways alone. Certainly the terrain suggested as much and no doubt Outram recognised that fact immediately, even though he never mentioned it. Instead, he deferred to his clients by devising a makeshift system which was not entirely satisfactory. Once again, it was a case of his having to make the best of a bad job.

Outram's report was considered at a general meeting of the proprietors which was attended by two businessmen, Charles Norton and William Whitmore. They were very critical of the report and, although the substance of their criticism has since been lost, Outram's indignant letter to the committee in reply suggests that they had scorned his ideas and had claimed that they could do the work far cheaper:

> I shall not retaliate for the misrepresentations Messrs. W. & N. have bestowed on my Report; ... With respect to your observation that 'offers have been made to do the Works that I have proposed, for £2,000 less than my estimates'—I have only to say, that, either a part of these Works were forgotten by those who made such offer, or they mean to use inferior materials, or that they misconceive the method of executing the Works. My Report and Estimate are founded on experience acquired by the execution of more than 100 miles of the most difficult Canals in the Kingdom, by seven years of great and successful attention to the improvement of Railways, and by establishing and conducting of Collieries, Iron Works, and Lime Works for self and partners in which more than 500 men are employed:and must say that I will not undertake your Works for one farthing less than before proposed. But were I to

do them, I should make them well worth the money and I think, the Somersetshire Coal Canal Committee have seen enough of defective execution and cheap contracts ... .[1]

Norton and Whitmore were not to be disposed of so easily and they returned to the fray with further comments and estimates, this time supported by William Underhill, resident engineer for the Dudley Canal.[2] They wrote:

> To the notes we have made upon and annexed to Mr Outram's report we request the Company will direct their attention by which they will perceive the great difficulties they have to encounter, the several expences they must incur and the almost total impracticability of his Plan: it is certainly new, but it is most erroneous in many particulars.

The real purpose of their criticism of Outram's scheme then came to light with the remarks that

> We have long been aware of the opposition to our proposals for the erection of geometrical locks,[3] and we are not disappointed in what we have experienced, though satisfied ourselves that on every account, it is the safest and least expensive mode yet discovered for the conveyance of boats in water from a higher to a lower level, yet we shall cease to impress them on the Company by any other observations further than to state that we are certain the conveyance of coals by water will exceed and be superior to any mode of land carriage ...

They went on to state that if the committee were not disposed to accept their plan, then

> We recommend (as next in value to our geometrical lock) the erection of inclined planes at our own expense to bring vessels from 8 to 10 tons burthen from the higher to the lower level at 2d. per ton and we will also convey all the upgate trade being half the

---

[1] Ibid., Outram to Canal Committee, later appended to report of 6 Feb. 1800.

[2] Ibid., Report and Estimates by C. Norton, W. Whitmore and W. Underhill, 29 March 1800.

[3] A form of boat lift, or balance lock, which was made in model form but never in a full-size working version.

burthen at the same price ... .

And finally, if the committee

> gave a preference to Rail-Roads, which after the very plausible Report of Mr Outram is not improbable, we propose to make and erect all the Rail Roads necessary for the trade of the Company (and which Mr Outram estimates at £9,700 and his Commission £500—total £10,200) in a proper and workmanlike manner for the sum of £8,200.

Norton and his colleagues then saw fit to prepare a revised estimate for all Outram's work, even adding substantial sums for the extensions proposed from the Kennet & Avon Canal and for the number of compartment vessels they considered were necessary on the upper pounds. Their total estimate was £30,595, plus annual extra charges of £1,732. Having grossly inflated Outram's estimates they then offered to complete the works for £17,300 except for the geometric locks which they proposed to install, operate and work with 20-ton boats at their own expense, the company to be charged at 4d. per ton on 90,000 tons of coal annually. Alternatively, they would sell the locks to the company for £15,000. In spite of their claims and their obvious determination to win a contract, it seems clear that their estimates would exceed those of Outram; it was hardly surprising that the company showed no great interest in their proposals.

Barely a week later, yet another contender for the contract appeared. He was a 'Mr Dodd' (presumably the engineer Ralph Dodd) who appeared briefly and hurriedly before the committee to demonstrate a working model of an inclined plane with which he proposed to connect the upper with the lower levels, The committee members were tentatively interested because they had already considered inclined planes for this purpose. Dodd said that a plane could be built for each branch at a cost of £4,000, with annual running costs of £250. He promised to submit a full report within a few days but failed to do so and made no further approaches to the committee.

At that time, the engineer for the Somersetshire Coal Canal was William Bennet, who was formerly engaged in a similar capacity on the nearby Dorset & Somerset Canal (the same man who had prepared the surveys for the Ashton Canal). Bennet was invited to report to the committee on all the schemes put to them and, in so doing, he took the opportunity to put forward one of his own.[1]

It is evident from Outram's report that he intended the vessels used on the upper pounds to be narrow boats, suitably modified to take boxes of coal. It does not seem however, that he had thought through this aspect of his plan from the point of view of the dimensions of the vessels and problems of navigation along those narrow, winding, waterways. Moreover, he had only described how coal from one pit could be exported and had not fully assessed the total numbers of boats needed for operations on the upper pounds. His basic ideas were sound enough but his lack of attention to detail was immediately seized upon by his critics. All of them pointed out that the aim was to export some 90,000 tons of coal annually (curiously, Outram did not mention this figure in his report) and all of them found reasons why Outram's methods would fail to deal with this quantity.

An incorrect assumption made by Norton and company was that Outram proposed to use rafts, and they arbitrarily decided that 28 of these craft, each 9 ft in width (the length was not given), would be needed on the upper levels. Bennet also assumed that rafts were to be used but he calculated that only 17 such vessels would be necessary and that each of these would be 90 ft long. However, he claimed that it would be necessary to widen the waterways, as well as to provide passing places, if Outram's scheme were to be adopted, because 80 ft was the maximum length of vessel which could conveniently negotiate the several sharp bends on the upper canals. Bennet also worked up an estimate for Outram's project, based on the expected coal output of 90,000 tons annually; this came to £22,035, plus £6,540 for 'floats or barges', with a further £2,065 for annual running costs. He seemed to disagree with almost everything in Outram's report and yet he admitted that on first reading it he was enthusiastic about the plan, only changing his mind after he got down to the details of its operation.

About that time, an experimental balance lock (similar in operation to the, as yet untried, geometrical lock) was in course of construction by James Fussell, on the nearby Dorset & Somerset Canal. Bennet priced the scheme proposed by Norton and his colleagues, assuming that the geometrical lock would cost about the same as Fussell's lock, the estimates for which were available to him. His calculations came to £24,065, plus annual maintenance charges of £260. This was compared with their own estimate of £32,300 (i.e. £17,300, plus £15,000 for the geometrical lock), which Bennet pointedly noted would provide them

---

[1] SRO, DD/MY 30, W. Bennet, Report and Estimates, 2 April 1800.

with a very generous profit indeed. Fussell's lock was eventually completed and was demonstrated successfully during October 1800 by lifting a 10-ton vessel, and simultaneously lowering another of the same weight vertically through 20 ft. Unfortunately this was too late for the proprietors of the Somersetshire Coal Canal to show much interest.

Bennet finally assessed the cost of inclined planes built to his own design. That for the Combe Hay line would be £7,804, and for the Wellow line £11,938, with annual operating charges of £400 at each site. Clearly his proposals were cheaper than any of his rivals but when his appraisals were submitted to the committee, some of the gentlemen present queried why his estimate for Outram's scheme included such high appended charges, whereas the other schemes did not. Bennet explained that his costing of Outram's plan included for an additional 8d. per ton charged on coal collected from the depots. This covered the expense of delivery from the various collieries, whereas for all the other schemes it was assumed that traders would collect their coal directly from the collieries at their own expense, using their own boats.

The committee must have been bewildered by this mass of detailed information on engineering methods and costings but they still could not make up their minds about which scheme was best. Finally, they decided to invite John Sutcliffe of Halifax to advise them. Sutcliffe was known to the members, of course, because he had been the first engineer to be engaged in 1794. He had resigned late in 1795 and was replaced shortly afterwards by Bennet. Sutcliffe came to inspect the site and reported in May 1800. In his long-winded report he did not take kindly to Outram's plan, probably because most of its recommended changes related to works which he himself had constructed during his term of office as engineer.[1] The recommendation that the existing colliery railways be changed to the improved plan was dismissed abruptly, as was every aspect of that system: 'The present Railways will answer every purpose that can be wanted for the Coal Owner and the Rails will be good and substantial, when those made on Mr Outram's improved Plan are worth nothing except for old Metal'. He was opposed to the 'heavy and unnecessary Expence' of making inclined planes because 'no man of sound mechanical Knowledge will ever recommend them where they can possibly be avoided'. Sutcliffe was not in favour of 2-ton waggons, preferring those of 10 to 12 cwt capacity, which were easier to operate on steep slopes, and cheaper to repair: 'In Yorkshire the Coal Owners

---

[1] Ibid., J. Sutcliffe, Report to the Proprietors of the Somersetshire Coal Canal Co., 26 May 1800.

have reduced their Waggons from two tons burthen to one; and others have contracted them to 10 cwt and have found great benefit from it'.

He then turned to the boats. He mistakenly assumed, as had Norton and Bennet before him, that Outram had recommended rafts, commenting that: 'These Rafts, I understand are proposed to be 90 feet long and 9 feet wide'. Sutcliffe agreed with Bennet that such vessels would not sail easily along the narrow canal already built. They would be affected by wind when loaded with waggons and much work would be needed to ease several bends and rebuild bridges, thereby interfering with existing trade. Sutcliffe could not have read Outram's report with any understanding because he also recorded, again incorrectly, that

> I am aware that 12 Waggons are proposed to be placed on the Raft together, though Mr Outram has not mentioned in what manner it is to be done, nor what the expence will be of making convenience to keep the Rafts steady when the Horses enter upon them. That the latter Plan may also be effected I have no doubt, though it will be extremely difficult to bring the ends of 24 Railways [that is, 12 on the Rafts and 12 on the Wharf] parallel to each other and great power will be requisite to draw the Waggons along such short curves, and if they are not so it will require a Raft almost as long as a First Rate Man of War.

Sutcliffe made several other uninformed criticisms of Outram's scheme and he even saw fit to comment adversely, and irrelevantly, on the Peak Forest Canal railways in order to illustrate his remarks. He stated that, because large waggons were used there, 'the improved Railway has so many times broke in pieces'. Moreover, 'as for the inclined Plane,[1] which has been, and now is attended with so much Expence, there was certainly no necessity for it, for the Limestone might have been got from another Rock, much preferable to that which is now working, far cheaper, upon a Railway had it been judiciously laid'. As regards Outram's letter in reply to Norton and company's comments, Sutcliffe's statements turned into more personal attacks when he felt

> at a loss to discover where these 100 Miles of the most difficult Canals in the Kingdom are that Mr Outram has executed; and whether the tunnel for the Huddersfield Canal be included in them, which will be full 3 Miles long, which he has often pledged

---

[1] i.e. the one at Chapel en le Frith; above, pp. 138–40.

> himself to finish it and the Canal in 5 years, and for his estimate. The Company have I think been at work on that Tunnel above 4½ years and there is not one Mile of it finished, and but about half the number of Pits sunk that was proposed and the Funds have long since been exhausted and still there is a considerable part to execute besides the Tunnel and perhaps near one-half the Proposed reservoirs to make ... The deficiency due to the Engineer will now discover on the Capital will be upwards of £9,000 per Mile (though only a narrow Canal) the particulars I shall give in a subsequent publication.[1]

He then went on churlishly to sneer at the designs for the Cromford Canal and also the 'extraordinary Plan and Report for improving the Worcester and Birmingham Canal'.

The reason for these unsavoury outbursts is perhaps not hard to find. Sutcliffe had presented a report on the state of the Monmouthshire Canal and railways shortly before Outram had been invited to do the same, earlier that year.[2] Apart from acknowledging Sutcliffe's recommendations in the minutes, no further action seems to have been taken by that committee. On the other hand, Outram's report and proposals were accepted wholeheartedly, which could hardly have pleased the Yorkshire engineer.

Sutcliffe felt that the Whitmore and Norton geometrical lock, a model of which 'Mr Whitmore very politely shewed me', was preferable to the caisson lock. But he felt that the committee should proceed with caution before adopting any untried system such as this, or even Fussell's balance lock. He considered that the best way of effecting communication between the levels was by conventional pound locks. These he recommended to the committee as the only logical solution to their problems:

> I am fully satisfied that when every circumstance and conveniency are impartially examined that will attend making it by common Locks, it will far exceed any other Plan that can be devised, and will be found most advantageous to the Company, and beneficial to the Public, not withstanding the first expence may far exceed that of the balance Lock or even that of inclined Planes. But with

---

[1] Sutcliffe was referring to his *Treatise on Canals and Reservoirs* (1816).

[2] PRO, RAIL 500/5, 28 April 1800, which refers to Sutcliffe's report on the state of the canal and railways. It does not seem that his advice was acted upon. His account for £66 was settled on 20 Oct. 1800.

regard to executing Mr Outram's Plan, it is impossible to say what it would cost, as the Canals would in general require making over again before the Rafts could conveniently navigate upon them, but he appears grossly ignorant, or strangely negligent, as not one shilling is allowed in his estimate for it, nor one word mentioned of any inconvenience that would attend navigating the Rafts upon the Canals in their present state.

Sutcliffe's estimate for locks was £22,323 for the Dunkerton branch and £18,980 for the Radstock branch, sums which included for steam engines and pumps on both canals for maintaining the water levels in the upper pounds. In case there was concern over the costs of running and operating steam engines, he pointed out that the Birmingham Canal used them and yet 'was there no Canal in this Kingdom that ever paid the Proprietors like that of the Birmingham Canal. If this be a fact, why should the Committee hesitate one moment in what manner these Canals should be finished?' Although Sutcliffe was correct in his assertion that locks were the only appropriate solution to the problem in the long run (reading between the lines of Outram's report, he thought so too), he had overlooked the fact that there was an acute financial crisis at the time and large sums of money such as he suggested were simply not available to the company. He made no attempts in that rambling and arrogant report to solve the problem by temporary means as Outram had tried to do, and for which the latter engineer seemed to receive no credit from any quarter.

During the period when Sutcliffe was preparing his report it seemed that Outram had called on two committee members namely, Samborne Palmer and Brudenell Barter, in order to discuss aspects of his scheme.[1] These proprietors then invited Outram to supervise the construction of parts of his scheme, as an experiment, in order to demonstrate its practicality, but he was unable to accept because of the numerous projects on which he was engaged in several distant places. In any case, he thought it was unnecessary to subject the proprietors to the costs of such experiments whilst a similar system which they could visit was then in course of construction in the north of England (he was no doubt referring to the temporary railway down the site of the Marple locks on the Peak Forest Canal).

Outram told Palmer and Barter that the system which he had recommended for transporting coal from Camerton Colliery could be adopted

---

[1] The date of the meeting was about 20 May 1800.

immediately at all the collieries using the existing railways and 'common boats might be used, no new Expence of any Kind [would] be incurred for altering the Canal or present Railways'. If railways were built between the levels as he recommended then

> the Expence of Repairs of the Rail Roads on both Lines will not amount to more than the Maintenance of two common Labourers; and above all, that in no case can any Accident happen in the transfer of Commodities on the Rail Roads that may not be repaired in the Course of an Hour;—whereas in Locks of any description, inclined Planes, or any other Mode of Descent, Days, or perhaps Weeks may elapse before the Trade can be recommenced. That with proper Zeal and Activity, the whole Rail Road, uniting the two Levels on the Dunkerton Line may be completed in three Months, when the lower Level will be ready to receive the Trade.

Outram stated that the coal-owners should not believe that the cost of detachable boxes would exceed their benefits, and that this proposal was not a theoretical speculation but a well-proven practice. However, if any coal-owner preferred not to use boxes there was no reason why he could not send coal loose in the body of the boats. Finally, he again stressed that the railway system recommended could be substituted by locks or other means at a later date, when the fortunes of the company permitted it. In that case, trade could still be maintained without interruption and of course, all the rails could be salvaged and sold off, so reducing the first cost of the project.

Outram's last recorded words to Palmer and Barter (both of whom seemed to have some sympathy for his views) were that

> he wished he had been spared the Necessity of Observing, that the printed words on his Plan made by different Persons, and sent to the Subscribers, had not been conducted with that Equity and Fairness, to which he thought himself entitled, and which Nothing but Justice to himself would have induced him to mention.[1]

With that rejoinder Outram departed and apparently had no further dealings with the company. That appointment took place before Sutcliffe submitted his report, hence Outram would have no inkling of that

---

[1] SRO, DD/MY 30, Bennet, Report and Estimates, 2 April 1800.

gentleman's vindictive observations. It is not difficult to imagine what his reponse might have been had he been aware of them.

The committee did not accept Sutcliffe's report and eventually turned to the inclined plane recommended by Bennet, probably because this was the cheapest mode on offer. The plane for the Dunkerton branch was in service by late 1801 but shortage of capital prevented the company from also finishing the Wellow branch by this means. A railway, very much on the lines of Outram's design, was laid instead. The plane operated by gravity but, unlike that on the Peak Forest railway, it had an endless chain to which the waggons were lashed. Some or all of the coal-owners had by then adopted detachable boxes and these were hoisted on to trams, just as Outram had recommended, prior to running on to the incline. Unfortunately the plane was neither an engineering nor an economic success and, sufficient to say, the company was forced to replace it by locks. Shortage of capital once more prevented them from finishing both branch connections by these means and only the Dunkerton line was completed. The Wellow railway was extended, and the branch canal abandoned along its entire length, several years after Outram's death. No doubt it would have interested him to know that the engineer for that scheme was John Hodgkinson.

The Forest of Dean railways

The Forest of Dean is situated in the north-west of Gloucestershire, where it is bounded to the west and north by Monmouthshire and Herefordshire respectively. The River Severn skirts the eastern boundary and the River Wye the western and northern limits. In the early nineteenth century it was said that the region contained an 'inexhaustible quantity of Coals, Building Stone, Lime Stone, Tile Stone and Clay, besides other products of Great Value' but unfortunately it was also 'considerably elevated by bold and sharp Precipices above the surrounding Country; which renders the access to the Forest and the travelling on it, difficult for Waggons and Carriages of Burden'.[1]

There are some 26,000 acres of woodland in the Forest of Dean, inside which lies a coalfield which forms a more perfect basin than any other in England. Within the basin were eight seams over two feet in thickness and the total area of coal measures within the outcrop of the lowest seam, known as the Trenchard Delf, was about 19,000 acres. A

---

[1] GRO, D421/E49/2, Memorial to Treasury, 1807.

group of sandstones about 140 yards thick divides the coal measures into two series and these have been extensively quarried over the years for the excellent stone yielded. Both woodland and minerals belong to the Crown and the nature of the tenure under which coal was formerly worked was unique. Mining in the forest dates back to 1282 and licences were granted by the Crown only to so-called free-miners, who were males above the age of 21, had been born within the hundred of St Briavels and had worked for a year and a day in a coal or iron mine. A qualified free-miner had the right to demand of the Crown Gaveller a 'gale', this being a spot of ground chosen by himself for sinking a mine. In the late nineteenth century a gale might be anything between 100 and 2,000 acres in extent but free-miners of the previous century seldom had sufficient capital to work other than a small crop gale. This explains why there were so many bell and ladder pits near the outcrops at that time, compared with more permanent drift and deep mines.[1]

Detailed production records for the numerous Forest mines of the late eighteenth century are not available, nor is it likely that they were ever made, hence it is impossible to determine with any accuracy the total output of coal from the region. Surviving evidence suggests that output was low; one source noted that only 30 tons of coal were delivered to the Severn daily and another claimed that 'the Produce, it cannot be stated with much certainty—the Miners at present send to Purton on an average only 300 Tons per Week'.[2] By 1800 there was a growing demand for coal from the industrial regions in the vicinity of the Stroud and the Thames & Severn canals, as well as from Herefordshire and Gloucestershire generally. These local markets were being steadily undermined by cheap coal imported from Staffordshire and Shropshire and also from Newport, where the construction of the canals and railways of Monmouthshire had rendered coal exports highly competitive.[3]

The principal reason for this low output was the inefficient transport system used by the miners of the Forest. Coal could only be sent down to the Severn and Wye along primitive, unmetalled tracks either by pack-horses, or by one-ton capacity waggons drawn by as many as eight horses.[4] As a result, transport costs were so high that, although prices

---

[1] C. Pamely, *The Colliery Manager's Handbook* (1896), pp. 52–4.

[2] GRO, D421/E48/4, J. Moggeridge, Arguments in favour of railways, 3 Nov. 1801; D421/E48/8, T. Tovey to C. Bragge MP, 13 June 1801.

[3] GRO, D421/E48/8.

[4] GRO, D421/E48/4.

at the pit-heads were invariably cheaper than those in South Wales, the latter could still undercut Forest coal delivered to places only fifteen miles from its source. To illustrate, it was stated that coal at one pit-head in the Forest was 6s. a ton but the cost of conveying it some four miles to the Severn was a further 8s. It was small wonder that the local pits were at a standstill.[1] The distance of any coal-pit from the Severn or Wye was obviously critical and this discouraged the easy exploitation of some rich and shallow workings in more remote parts of the region, whereas some deep mines with far more expensive produce, but nearer to the rivers, could survive. A less obvious effect of this predicament was that timber suitable for pit props was fast running out in the Forest and the costs of importing, even from areas just beyond its limits, were prohibitive; this prevented the development of mining everywhere. Moreover, other products such as cheap slack coal for lime-burning, limestone and paving stones, could not be exported at competitive rates. The net result was that investment in mining development was minimal. There was also much unemployment among the mining community and indeed many were leaving at that time to seek work in the mines of Monmouthshire and elsewhere.

From 1799 onwards, several meetings had been held under the chairmanship of the mayor of Hereford (where coal during winter had risen to as much as 35s. a ton) but it was only at a meeting early in 1801 at which a wider representation of citizens attended, including a strong group of gentlemen from Gloucester, that it was resolved that 'general Railways from the Collieries in Dean Forest to the Wye and Severn would be highly advantageous to the Cities of Hereford and Gloucester and the neighbouring Country and would be the means of reducing the price of Coal'. A committee was established and shortly afterwards Benjamin Outram was invited to visit the Forest to recommend the best way forward.[2]

Outram went to the Forest in the autumn of that year and soon selected a promising route for a railway which connected the mining districts with both Wye and Severn.[3] John Hodgkinson, who had accompanied him to the site, took longitudinal level sections and Henry Price, a local man, surveyed the route and plotted the plans which were

---

[1] GRO, D421/E49/2.

[2] GRO, D421/E48/1, Extracts from the Forest Journal kept by the Gentlemen of Herefordshire (1807).

[3] GRO, D421/E48/TRS 21, B. Outram, Report and Estimate for the proposed Railways from the Collieries in the Forest of Dean to the Rivers Severn and Wye, 5 Sept. 1801.

presented with the report on 5 September 1801.[1] Outram had 'caused Surveys and Levels to be taken in various directions'. He noted, but not very accurately, that the richest part of the coalfield, 'a district of the Forest nearly central', extended eastwards three miles from Lydbrook, and southwards some six miles from the road between Coleford and Ruardean. He mentioned in passing that several steam engines were then in operation for draining and working certain mines; there were also two furnaces for iron manufacture at Cinderford and Parkend.

Outram's preferred route was from the summit of the valley which runs from the north-west corner of the above district and along its western edge, then south to the Severn below Lydney. It also included the northern extermity of this valley which descends from the watershed towards Lydbrook and the Wye. His railway was to run along the eastern slopes of this valley to communicate with both rivers. Branches could then link up with most of the collieries in the coalfield, except for a small number close to the south-east corner of the district. A branch skirting the northern side of Surridge Hill, to meet the main line at the summit would serve the many pits located on the outcrop, besides encouraging the opening of others 'that would serve the demand of many years'. Deeper mines existed on the southern slopes of the hill and branches from these could run downhill to either the Wye or Severn, although he thought that a terminus at the former would prove more economical. The line north followed the valley along the sidelong ground on the eastern side, dipping steadily at twelve inches in the chain (1 in 66). This slope could be maintained along a circuitous route until it fell 150 feet to the river Wye by a self-acting inclined plane, 180 yards long. This route was preferred to the very steep passage along the narrow valley bottom which, in any case, was fully occupied by the river, turnpike road and much industry. These left very little space for a railway, particularly one which was eventually expected to have dual tracks.

The main line south was also to fall at 1 in 66 for the first mile around Surridge Hill, reducing first to 10 in. and then to 8 in. to the chain (1 in 80 to 1 in 100) as it passed the Parkend Furnace. For much of this section Outram avoided the valley bottom. He selected a line for the railway which ran over sidelong ground, crossing several rills by culverts and shallow embankments, gradually reducing in slope to 4 in. to the chain (1 in 200) as it passed Lydney Upper Forge. The track then crossed over to the western side of the valley before passing over the

---

[1] GRO, Q/Rum 5, H. Price, Plan of the Intended Railway through the Forest of Dean to the Rivers Severn and Wye, 1801.

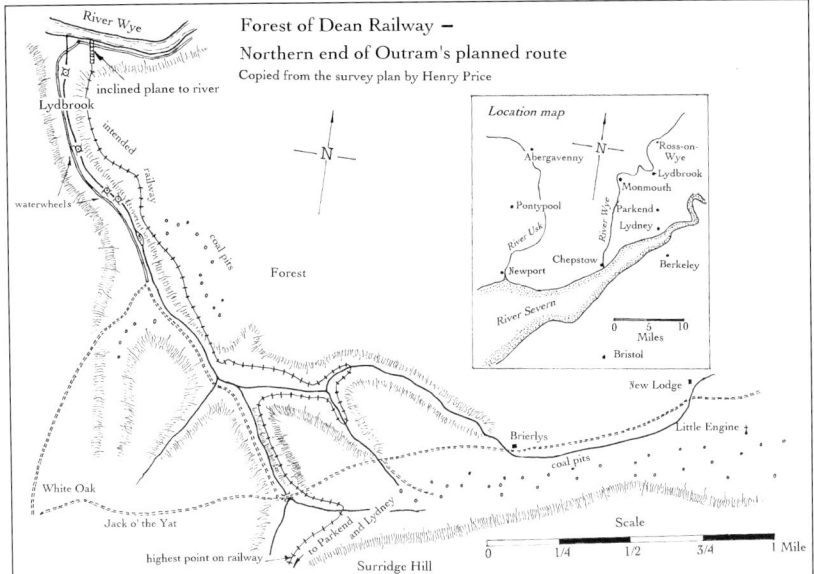

Figure 32 Forest of Dean Railway.

turnpike road at Lydney to the Lower Forge. It continued on the level to the wharf at Jack's Pill. Outram concluded by discussing the possibility of reducing gradients in places at the expence of increasing the length of the line, but he did not think that the cost of these modifications, which could be adopted during the construction stage, would materially affect his estimate of £21,500 for the line as described (Fig. 32).

The length of the line from the Severn to the Wye was 13¾ miles and as usual Outram specified a railway built according to his improved plan. He recommended a single line (although a formation of six yards in width would be prepared to allow for future dualling of the track) with passing places at 300 yards intervals. However, 300 yards of twin tracks would be necessary at the Pill below Lydney and also near the inclined plane at Lydbrook.

Outram declined at that stage to estimate for branch lines linking all parts of the coalfield with the main line but he nevertheless offered some useful forecasts regarding the operation of his system. If, for example, coal was brought from outlying pits down to the summit of the main line west of Surridge Hill, then one horse could take six or eight waggons loaded with ten to twelve tons to the head of the inclined plane at Lydbrook and return with them empty in two hours. Five such journeys could be completed each day by the same horse 'with ease', a statement which said much about contemporary attitudes to animal welfare. The same could be achieved on the southern route, with times proportional to distance, except for the short level section between Lydney and the Pill. Along that length, either the train would have to be divided, or else horses would need to be kept at Lydney to provide additional haulage. Outram reckoned that the cost of conveying coals from the collieries down to the rivers would not exceed a halfpenny per ton per mile excluding tonnages (tolls), loading and unloading, and repairs to the waggons. His estimate for carrying from a colliery situated six miles from either river was 18d. per ton and he predicted a trade of 100,000 tons annually. This quantity, carried on the average of four miles, would pay £10 per cent of the cost of the main line estimate, allowing £350 per annum for repairs and other expenses. Outram's speculations on the export of coal by railway also included for the possibilities of upgate transport of merchandise from both rivers, all of which must have greatly interested his clients.

Finally, Outram did not miss the opportunity of diverting attention of his clients to the state of the River Wye as a navigable waterway. He felt that 'It is a great Pity, that so fine a River should have remained, to so late a Period without any material Efforts of Art being exerted, in aid of what Nature has so liberally done, towards making it a fine Navigation'. Some indication of his high regard for his friend William Jessop may be gleaned from his subsequent remark that 'If Mr Jessop's Advice could be obtained on this Subject, his very great Abilities, and long Experience in River Navigation, would point out the most certain and economical Means of effecting an Object of so much consequence to the Country'.

His report, brief and sketchy by today's standards, was a competent enough document for the time and it was clearly written by an engineer who knew his business. It was read at a meeting held at Ross three weeks after its submission and the first of the resolutions adopted there stated that 'Mr Price's Survey of the intended line for the Railways from the Collieries in the Forest of Dean to the Rivers Severn and Wye, together with Mr Outram's Report and Estimate be approved of'.[1] The second resolution was that 'application be made to Parliament the next Session for powers to carry the same into effect'. Outram seems to have offered to take on the contract himself at estimate price for which he was requested to provide a breakdown. At £1,550 per mile of railway, the committee would have been provided with an excellent railway at reasonable cost because it crossed some very difficult terrain. Unfortunately the committee were soon to be thwarted in their efforts to progress

---

[1] GRO, D421/E48/10, Resolutions at a Meeting held at Ross on Wye, 22 Sept. 1801.

the business.

Initially there were disagreements with some landowners, whose estates were traversed by the railway, over the allotment of profits. These matters were satisfactorily resolved, only to be followed by the emergence of a second party of gentlemen who intended to promote an alternative scheme for a railway from Lydney, to terminate only at Parkend. Fortunately, following much discussion at a meeting in November 1801, this project was united with Outram's general scheme.[1] All then looked set for an early petition to Parliament for leave to present a bill but the entire project was soon to collapse with the withdrawal of the Gloucester party. Their support had been mainly for the section leading down to the Severn but they lost interest in this on the sudden stoppage of construction of the Gloucester & Berkeley Canal due to acute financial difficulties. It seems that an additional factor in the demise of Outram's scheme was the apparent opposition of the verderers of the Forest and certain Crown officials.[2] Both these parties had shown a growing hostility to the development of railways and the extensive branches which were bound to follow. They felt sure that these would interfere with the well-being of the woodlands and also obstruct the haulage of naval timber, a very significant factor during the war with France. The subsequent endeavours and the eventual establishment of an efficient transport mode for the Forest of Dean mining and other industries is another story.[3] These matters were, surprisingly in view of its crucial importance to the miners, unresolved until some years after Outram's death.

Contract work for John Rennie

Prior to 1800 the port of London had no commercial docks on the Thames, and the practice was to load or unload ships moored in the river by lighters. An increase in traffic and overcrowding of the river necessitated the construction of several docks during the early years of the new century and both Rennie and Jessop were engaged on such major constructions.[4] Whilst Outram was not directly interested in work

---

[1] GRO, D421/E48/12, Resolutions at a meeting held at Ross, 12 Nov. 1801; E48/9, T. Tovey to T. Bathurst, 14 Nov. 1801.

[2] GRO, D421/E48/1.

[3] C. Hadfield, *The Canals of South Wales and the Border* (2nd ed., 1967), pp. 209–17.

[4] A.W. Skempton, 'Engineering in the port of London, 1789–1808', *Trans. Newcomen Soc.*, 1 (1978–9). This paper gives a comprehensive review of the work of these engineers.

of this category, his strong advocacy of railways and the need to increase sales of Butterley products led to his seeking contract work both as a supplier of railway components and as a general contractor. Accordingly he was successful in winning an unusual railway contract for John Rennie's project at the London Docks, at Wapping on the north side of the Thames.[1] This was an interesting example of the versatility of railways when utilised temporarily for heavy construction works.

Rennie's task was the construction of a 20-acre dock, 20 ft in depth, which was approached from the river through an entrance lock 180 ft long by 45 ft wide into a three-acre basin (later known as the Wapping basin).[2] A communication lock 160 ft long then led into the splendid dock, known later as the Western Dock (closed only in 1969).[3] He also proposed to build a range of massive warehouses on the northern edge of the basin totalling 1,400 ft in length and 180 ft in width. In his report of July 1801, shortly before work commenced, Rennie explained that upwards of 200,000 cubic yards of earth would have to be excavated but nearly five times that quantity was eventually shifted. Some 40 per cent of this was spread on site to lift up the general level and to fill in behind the dock walls, but enormous quantities were also disposed of, some of it on to adjacent land and the remainder taken by ships down river. Preparatory work also entailed a great area of site clearance of domestic and other premises, which had to be completed before any building could begin. The Butterley contract was for the provision of railways for the removal of spoil down to the jetties and to adjacent landfill sites, as well as for site transport and distribution of bricks, stone and other materials.

Rennie recommended

> Messrs Outram's Estimate Of Railways and Waggons for carrying Brick and Stone to different parts of the Premises, as well as Railways and Waggons for conveying Earth. ... The price of the rails appears high, but if they are made of the best Metal, and on the best construction, I apprehend they will not be found much to exceed the Real Value and I submit to the Directors the propriety of ordering the Rails immediately & about 60 Brick and 20 Stone Waggons....I think it will be best to contract for the furnishing the rails & laying and forming the Road.

---

[1] Institution of Civil Engineers, Rennie Notebooks, Report to the Directors of the London Dock Co., 7 July 1801.

[2] Skempton, 'Port of London', p. 95.

[3] S.K. Al Naib, *London Docklands: Past, Present and Future* (1980), p. 3.

Three railways were planned and priced by Benjamin Outram. The first was from a jetty to the side of the entrance basin, comprising a twin railway linking into a 'fourfold' railway on the jetty. The second ran from another jetty, leading to the large basin as a twin railway with single-line branches leading across the site, with 'sixfold' railways on the jetty. The third line was another twin railway running from the large basin to a landfill site. Initially this was located north of the dock but when works began the site selected was to the east, at Fawdon Fields. The rails were standard plates of 3 ft in length and 44 lb weight although, because they were for temporary use during construction of the docks, they were laid on wooden sleepers for ease of relaying as work progressed. The lines were, nevertheless, to be laid to controlled gradients on prepared earthworks and reinforced where public streets were crossed. Even short temporary tunnels were to be built where highways passed over the lines. The total cost of the scheme, which was carefully listed, specified and priced, came to £3,703. In addition, waggons capable of carrying 2½ tons for spoil and trucks carrying 3 tons of building materials were costed at £12 each.

As regards the contract terms (which seemed to suggest that Outram & Co. did not wish to be responsible for any later relaying of rails) Joseph Outram jun., who signed the documents on behalf of the company, sought to impress Rennie.[1] He stated that they

> would prefer only the laying down of the Railways & completing them if the Dock Compy would form the Ground and Gravel the Road, & even would prefer only serving them with the Castings but from regard to their Reputation for making Railways on the best and most approved System, and that these may be a Specimen of their Utility in a Country where they are so little known, and having paid rigid attention to that Quality of metal which they manufacture and which is proportioned to the Burthen which it has to carry, induces them, if they are favoured with the order, to lay the Railways and to see them Executed in a proper & workman-like Manner.

In spite of these bold claims, Rennie reported only in the following April that 'Messrs Outram & Co are now proceeding with more expedition with their deliveries'.[2] Presumably this was because Outram & Co.'s

---

[1] ICE, Rennie Notebooks.
[2] Ibid., Report to Directors of the London Dock Co., 27 April 1802.

London agents, Oswald & Anderson, had just noted that 950 gang rails and 14 trucks were on their way by sea from Gainsborough.[1] In fact Aydon & Co. of Bradford were then also supplying rails, so that Butterley must have been having some difficulty in maintaining supplies. The use of temporary railways for earthmoving and for general site transport was then something of an innovation and the method proved to be a great success. In 15 months up to the completion of earth-moving in October 1804, the rate of excavation (all by pick and shovel) was 8,500 cubic yards per week, and all this spoil was transported by railway.[2]

Contract work for William Jessop

William Jessop was no less enthusiastic than Outram in promoting railways built on the improved plan and pride of place must be given to his Surrey Iron Railway, which is generally acknowledged to be the world's first independent public railway. Interest in this project began in the 1790s during the war with France, when there was a constant threat to British shipping sailing up the Channel and entering the port of London. Consideration was then given to improving overland links between London and Portsmouth by canal or railway.[3] Eventually, less ambitious notions focused on a canal to Croydon from Wandsworth and Jessop was commissioned to examine possible routes. He afterwards persuaded the subscribers to abandon their intentions of building a canal in favour of a railway, claiming that a canal was impracticable because of the difficulties and costs of providing an adequate water supply. In his report he asserted that railways 'have been brought to the degree of perfection, which now recommends them as substitute for Canals; and in many cases they are much more eligible and useful'.[4]

On the strength of his advice, an Act was obtained in 1801 for a railway which started at a basin connected to the Thames at Wandsworth. From there it ran south to Mitcham (where there was a branch to Hackbridge, near Carshalton) and then on to Pitlake in Croydon, a distance of nearly nine miles.[5] The railway was double-track along the

---

[1] Ibid., Oswald & Anderson to Rennie, 6 April 1802.

[2] Skempton, 'Port of London', p. 96.

[3] C.E. Lee, 'Early Railways in Surrey', *Trans. Newcomen Soc.*, xxi (1940–1), p. 50.

[4] Ibid., W. Jessop, Report on the Proposition for making a Navigable Canal from the River Thames at Wandsworth to Croydon, 9 Dec. 1799, pp. 50–1; see also Hadfield and Skempton, *William Jessop*, p. 175. Neither cites a location for the original report.

[5] 41 Geo. III c. xxxiii (1801).

whole length and nowhere did the gradient exceed 1 in 120.[1] The plates were 38 lb in weight and the gauge was 4 ft 2 in. The estimated cost was £33,000 and the Act authorised a maximum expenditure of £50,000. A further Act was necessary in 1805 to permit the raising of another £10,000 to cover the cost of improvements to the initial design, particularly to the basin.[2] The latter was a quarter of a mile long and could accommodate more than thirty vessels. The railway to Croydon was in operation by July 1803 and the branch was probably finished twelve months later.[3]

A bid was made by Benjamin Outram & Co. under the hand of Joseph Outram junior for the railway, which would be 'executed in as Compleat and perfect a manner as any of Benjamin Outram's improved Railways have been heretofore done', for £29,200.[4] The bid was unsuccessful, which says something for the integrity of Jessop in that he obviously did not collude with his colleagues at Butterley over their tender. His estimate, nevertheless, was public knowledge before tenders were received and this must have influenced submissions by contractors.

An extension of the line quickly followed, again with Jessop as engineer, and an Act for the Croydon, Merstham & Godstone Railway was obtained in May 1803 to the same specifications as before, the line joining the Surrey Iron Railway to the west of Croydon, then running south to Merstham and on to Reigate.[5] The Godstone branch began at Merstham, turning south-east through Chevington to terminate at Godstone Green. From Croydon to Merstham the length was nearly 8¾ miles and a further 3¾ miles to Reigate; the branch to Godstone was 3¼ miles long. Jessop's estimate was £52,347[6] and this time the tender of Benjamin Outram & Co. was successful with a bid of £36,350. The company took on the job as a general contractor responsible for all aspects of the work, included earthworks and bridges, until completion. It seems that Benjamin Outram took personal charge of the work, although he died before the contract was finished in August 1805. The extensions to Godstone and Reigate were never completed.[7]

---

[1] Priestley, *Navigable Rivers*, p. 609.

[2] 45 Geo. III c. v (1805).

[3] Lee, 'Early Railways', p. 54.

[4] DRO, D503/58/1. Draft letter, 1 July 1801.

[5] Surrey History Service, Z/306, Surrey Iron Railway: report of meetings to discuss a proposed extension (1802); 43 Geo. III c. xxxv (1803).

[6] Priestley, *Navigable Rivers*, pp. 181, 182.

[7] Hadfield and Skempton, *William Jessop*, pp. 179–81.

The Manchester to Liverpool railway

The story has long persisted that both Jessop and Outram were commissioned to survey for a railway between Manchester and Liverpool, Jessop in 1797 and Outram the following year. These endeavours have been mentioned by several writers[1] and, so far as can be determined, the originator of the reports was Clement Stretton, writing in 1901.[2] The suspicion remains that this was but another of that author's historical falsehoods but the story is apparently confirmed by Margaret Outram's biographer and great grand-daughter Frances in 1932.[3]

Frances Outram stated that during the early 1830s Margaret took her first journey by train from Manchester to Liverpool on her way to Ireland and wrote that 'years before, both her husband and Jessop had surveyed alternative routes for a horse-drawn railway' between the two towns. Accordingly, in spite of the absence of any engineering reports or other details, there apears to be some grounds for leaving an open verdict on the subject. Outram had invested heavily in land in the vicinity of the Ashton Canal terminus in the Piccadilly district of Manchester during the latter years of the eighteenth century. This was at a time when it seemed right to compete with the transport monopoly long established by the Duke of Bridgewater's Canal and a horse-drawn railway might have provided that challenge. As noted earlier, Outram had written to his brother Joseph in October 1798 that he could not leave Manchester for a while because to do so 'may be the greatest Loss in my private Affairs'.[4] This could have had much to do with a meeting with the proprietors of the Ashton Canal but the strong emphasis of his note may suggest that something much more important was afoot. It would not be surprising if a railway between the cities was then under consideration and Outram would naturally wish to be involved in any deliberations leading to the award of contracts. Future research may provide the confirmation for this notion, although it has eluded investigations to date.

---

[1] R.E. Carlson, *The Manchester & Liverpool Railway Project, 1821–1831* (1969), p. 37, referring to G.S. Veitch, *The Struggle for the Liverpool and Manchester Railway* (1830), pp. 24, 25, who does not cite a source.

[2] C.E. Stretton, *The History of the Liverpool and Manchester Railway* (1901) (copy in Manchester Central Library).

[3] M.F. Outram, *Margaret Outram (1778–1863). Mother of the Bayard of India* (1932), p. 233.

[4] Above, p. 155.

CHAPTER 17

APPRAISAL

Very little is known about the physical stature, appearance and personal characteristics of Benjamin Outram. Unlike many of his contemporary engineers, a portrait of him does not seem to have been painted, nor was a comprehensive obituary ever written to perpetuate his memory and fame. Little survives, apart from the sketchy reminiscences of his wife written many years after his death, his own formal business letters and reports, and the few personal comments recorded by members of his family and his professional peers.

Margaret Outram wrote of her past,[1] apparently for the benefit of her second son James, and in those memoirs brief reference was made to her life at Butterley. Of Benjamin's demeanour she wrote 'My husband, like so many other men of great talent and comprehensive and generous mind, was hasty in his temper, feeling his own superiority over others. Accustomed to command he had little toleration for stupidity and slowness, and none for meanness or littleness of any kind'. This tribute complements the persona projected by Benjamin Outram's correspondence in particular. During his work on the promotion and construction of the Cromford Canal he appears to have been a brisk, able, cheery (and sometimes cheeky) young man. By the turn of the century, however, maturity had produced an engineer who was serious in manner, highly motivated, somewhat impatient of others and with a shortness of temper to match. The latter characteristic had probably been exacerbated by his protracted illness of earlier years.

It must have become obvious to Outram, some time after 1802, that his promotional work on railways was not yielding much work for his own private practice. Effort had not been spared to persuade industrialists to invest but his imaginative schemes had often been planned in vain. There was, nevertheless, plenty of work for him to do in developing and extending his coal and iron interests at Butterley and he directed most of his energy to these activities during the few remaining years of his life.

Some part of this time, as well as considerable sums of money, were devoted to refurbishing Butterley Hall for his young wife and growing

---

[1] M.F. Outram, *Margaret Outram (1778–1863). Mother of the Bayard of India* (1932). Quotations and other reference to Margaret in this chapter are from this book except where otherwise noted.

family. Until his marriage, Outram's style of living at the Hall, as a bachelor living at the Hall, seems to have been modest enough and much in keeping with the homely standards with which he was familiar in his father's house in nearby Alfreton. All this changed when Margaret came to live at the Hall. She was a vivacious lady of determined character and with ideas somewhat above her station in life. In later years, when reflecting self-critically on the vanity of her youth, she wrote that it was a time when she had been treated 'by the surrounding country as a superior being, looked up to by the ladies as the leader of fashion and manners, and by the gentlemen with admiration'. Somewhat regretfully she also noted that 'for to my other misfortunes I was considered to have a handsome person, and vivacity that attracted too much attention. I was a child of nature, with talents but no regulation of mind'. She affirmed that the Hall had been 'improved by him [Outram] at a great expense' and 'fitted up with every elegance', and that £3,000 had been expended during that period on 'the most expensive fixtures of every kind, patent laundry, baths, boilers etc.'. The outbuildings, stables and gardens were likewise improved.[1]

Margaret was presumably responsible for designing these self-indulgent and costly renovations, yet it would be unreasonable to blame her entirely for such extravagances. Benjamin Outram, desirous for social status though he certainly was, should have exercised greater caution because these costs had to be borne by his annual salary from the ironworks, plus the limited sales of minerals from the Butterley Hall estate and the fees from his occasional practice as a civil engineer. These modest funds should also have financed, if only in part, his substantial debts to John Wright, although Outram probably reckoned in any case that these would gradually diminish with the increasing profitability of the company. The evidence of the company accounts, limited though these are, suggests otherwise.[2] Outram was spending well above his means, his better judgment clouded because, according to his wife, he harboured 'such expectations of realising a splendid return' from his business interests. Unhappily for Margaret this was not to happen; for her the return was anything but splendid.

The days of reckoning followed closer than anyone thought likely. Benjamin Outram left home during May 1805, bound for London on business. Shortly afterwards, on the very day that he was expected home,

---

[1] DRO, D503/42/1, Furnace Ledger, 1800–22, p. 63. Entries for 25 May and 14 Nov. 1801 are typical and appear to relate to work at the Hall.

[2] Ibid., p. 35 and later pages relating to Outram's cash account.

Margaret's brother, James Anderson, came down from the city to tell her that her husband had died on 15 May of a 'brain fever'. Margaret, devastated by grief, immediately left by chaise on a journey of 130 miles to London, accompanied by her brother and stepmother and also by her daughter Anna, then but three years of age. Benjamin, who was staying in the parish of St Olave's, Southwark, at the time of his death, was buried in the churchyard of St Paul's Cathedral on May 18.[1]

Outram's death was widely reported, notably in the *Gentleman's Magazine* in which it was written that:

> The strength of his understanding, the generosity of his heart, and a spirit of enterprise and activity almost unequalled, enabled him to surmount with ease the most formidable obstacles to such public improvements as he recommended. His death, which has cast an unusual gloom over the neighbourhood of Derby and Manchester, will long be regarded as an irreparable loss, not only to his family and friends but to his Country.[2]

Somewhat closer to home Alexander, Margaret's eldest brother, wrote to George Goodwin from London on the day following Outram's death.[3] For some time his own firm had been acting as the London agents for Benjamin Outram & Co., hence he had more than family interests to concern him over the affairs at Butterley. Goodwin was regarded as Outram's deputy at the time and Anderson gave him his place when he wrote.

> As the dreadful calamity which has befallen us has taken away the manager of all the affairs of Butterley it will be necessary for the surviving partners to make some speedy arrangement for conducting the business and though I have not yet recovered the shock occasioned by this most distressing event, I cannot help turning in my mind the evils which may in the meantime result from it in those extensive works. It is fortunate that you are on the spot and are so well acquainted with everything that relates to the business which will enable you to give great assistance in keeping matters in their proper train and I am sure you have every

---

[1] *St Paul's Cathedral Registers* (Harleian Soc., Parish Register Series, xxvi, 1899), p. 186.

[2] *Gentleman's Magazine*, lxxv (2) (May 1805), p. 581.

[3] DRO, D503/103, A. Anderson to G. Goodwin, 16 May 1805.

disposition to give all the assistance in your power. I have seen Mr Jessop and urged him to appoint a speedy meeting with the other partners ... when some plan will be adopted for carrying on the business. ... I cannot doubt that you and all the other gentlemen about the works will use your utmost endeavours.

Obviously troubled, he wrote that he would 'be with you after having paid the last sad duty to my dear friend, both to comfort my sister in her distress and to give any little assistance I may be capable of arranging her affairs'.

Anderson also noted that 'Mr Outram told me before he was taken ill that this is about the annual period at which the books of the partnership are balanced and I hope you will get it done as soon as possible that it may be seen exactly how matters stand'. It was shortly after this letter arrived that a thorough examination of the books did take place and the deficiences left by Outram were then exposed. He was heavily in debt to the partnership and it was found soon afterwards that he had died intestate and, moreover, had never appointed a successor as was permitted under the partnership deed. Margaret Outram had to procure letters of administration to act as her late husband's legal representative, a task which brought her much anguish and trouble over the next eight years. In the meantime and in spite of the demise of the managing partner, the company continued with business much as before. This, at least, spoke well for the managerial abilities of Outram and the efficient organisation that he had established over the years, which had been achieved without much direct input of any of his partners.

The four partners were fully aware of each other's investment in the company because these were annually summarised in their private ledger. Outram had subscribed £2,500 in cash by April 1796 but as, the capital of the company was steadily increased to £80,000 by April 1802, his cash investments, which were mostly paid in by John Wright on Outram's behalf, had increased by a further £11,340.[1] This latter sum was subject to interest but it was reduced simultaneously through Wright's receipts of the profits. The actual sum owed to the partnership by Outram on the 20 March preceding his death was eventually agreed to be £8,556.[2]

The partnership deed stated that a meeting of the proprietors would be held annually on the second Monday in April, when inventories and

---

[1] DRO, D503/28/1, Private Ledger, 1790–1856.
[2] DRO, D503/1/3A, Chancery Master's Report, 8 Dec. 1813.

valuations would be made of all the property, and accounts of all cash, debts, notes, securities and bonds belonging to the concern would be examined. Although the partners met annually, they did not audit all the books as would be the practice today, otherwise they would have noted, with some anxiety, another substantial debt of their managing partner of which apparently, they had no previous knowledge. Outram had opened a personal account in one of the sales ledgers in May 1801,[1] although reference to a sum brought forward from 'the old ledger' suggests that he might have commenced borrowing from the company shortly after his marriage took place. This practice of mixing personal with company accounts was common enough during the eighteenth century, even in large concerns,[2] and Outram's account was evidently permitted by his partners. Indeed William Jessop also opened such an account after 1805, although his spending was more carefully controlled than was Outram's.[3] It is evident nonetheless that, perhaps out of courtesy as well as trust in their colleague, they did not examine Outram's account until after his death, otherwise they would have found that his borrowings from the company were reaching intolerable proportions.

In November 1801 Outram's debit in the furnace ledger amounted to £641 and included costs of such items as window frames, gates and the services of several workmen, all of which might be assumed to have been for the purpose of renovating Butterley Hall. This had increased to £1,476 by 25 March 1802 but was balanced exactly on that occasion by a £500 loan, sanctioned and granted by his partners, plus rents due to the estate and even by foregoing his annual salary of £500. An indication of the parlous state of Outram's finances at that time is suggested by the inclusion in the credits of only £2 'by cash' and the return to the company of a fire-grate valued at 6s. By 26 February 1803 the debit had increased to £1,906, against which he could only set credits of £27. Even this small sum was made up of his own valuations for 'old wood, old metal, lead and deal' which he preferred in lieu of cash, although it is difficult to understand how such items could be profitably used by the company.

The situation was deteriorating further by October that year, when Outram's debit stood at £3,641, with credits of only £1,601. In fact, the latter included a loan of £500 from Richard Arkwright (Sir Richard Arkwright's son) on the company's account which, it later transpired,

---

[1] DRO, D503/42/1.
[2] S. Pollard, *Genesis of Modern Management* (1965), p. 265.
[3] DRO, D503/42/1, p. 473.

must have been arranged without his partners' knowledge;[1] thus the debt then stood at £2,040. Six months later the debt had risen to £2,780, mainly as a result of several cash withdrawals, and on 4 September following Outram's death it had grown to £4,553.

Margaret Outram soon found that the partners spared her scant sympathy on their discovery of Benjamin's debts. She left Butterley Hall shortly after returning from the funeral and William Jessop his family moved into the premises, refusing to pay anything more than a nominal rental of £52 10s. a year and declining to purchase any part of the fittings which had been so expensively installed. The younger William Jessop, then only 22 years of age, took over the task as manager of the business which in 1807 was renamed the Butterley Company. His father continued with his practice as a civil engineer and his association with the company remained much as before. No written evidence remains regarding the elder Jessop's views at this distressing time but it seems from the events which took place in the years ahead that he was unwilling to forgive his friend, who could be said to have betrayed the trust of his partners whilst in pursuit of a pretentious lifestyle.

Later examination of the ledgers by the partners resulted in a further increase in Outram's debts; by February 1806 this had reached an alarming £10,646 and included an 'advance as settled by Mr Outram 25 March 1805'. This sum was neither summarised nor explained in the ledger but it may have incorporated part of the capital due to the partnership.[2] For several years thereafter Margaret Outram was frequently, but unsuccessfully, pressed to settle her husband's debts. This stalemate was most unsatisfactory for everyone concerned and the surviving partners, William Jessop and John Wright, frustrated and anxious to clear up the dispute, eventually resorted to litigation. The case was heard before the Master of the Rolls in the Court of Chancery on 16 May 1811.[3]

In submissions to the court, counsel for the plaintiffs, Jessop and Wright, claimed that Outram was indebted to the partnership in sums of money greater than the value of his share; moreover on his death he was possessed of considerable personal estate, all of which had passed to Margaret Outram as his administrix. Frequent applications had been made to her for payment of the monies due to them but she had consistently refused to ascertain the balance of the amount due and to

---

[1] Ibid., p. 188.
[2] Ibid., p. 321.
[3] DRO, D503/1/3A.

settle to their satisfaction. The wrangling in court continued, with the plaintiffs alleging that Outram had borrowed money from several persons on the credit and in the name of the partnership without the knowledge of his partners and had converted these to his own use. Additionally it was asserted that he had received debts due to the company from various persons which were unaccounted for at the time of his death, and that he was also indebted for goods manufactured by the company for his own purposes and which he had purchased on credit. Finally, the plaintiffs' counsel declared that the partnership had ceased on the death of Outram and that accounts and valuations should be assessed only up to that time.

Margaret Outram's counsel argued that, on the contrary, the partnership was indebted to Outram to an extent which exceeded the partners' claims. His client insisted that the original covenant was still valid and that she was entitled to one quarter of the value of the partnership as it stood on 2 April 1805, which should include the cash, bills of exchange, bonds and securities in hand at the time of Outram's death. Because the value of such a share had never been paid out and had continued in the partnership, a proportionate part of the profits accruing from these funds should also be paid to her from that time onward.

The Master of the Rolls finally declared that the defendant was entitled to a share of the profits arising from the partnership and he reserved judgement on the case until one of the masters of the court, J.S. Harvey, could take account of all dealings up to that time and report as to whether it was to the benefit of Benjamin Outram's estate and his infant children that the partnership should continue in accordance with the indenture of 1792.

Harvey submitted his report to the court on 8 December 1813.[1] He affirmed that by 2 April 1802 the capital of the undertaking amounted to £46,801 6s. 8d. Of this sum, William Jessop, John Wright and the late Francis Beresford had each paid £13,839 6s. 8d. and Benjamin Outram only £5,283 6s. 8d. Accordingly, £8,556 was due from the latter on 20 March preceding his death. George Goodwin, described as the accountant to the partnership, had sworn an affidavit to the effect that he was acquainted with Benjamin Outram's affairs and he confirmed that Outram was entitled to a one-quarter share of the partnership. Goodwin had sold off sundry articles of furniture and other effects, on behalf of the administrix, for £1,284. He believed that Outram's personal account stood at £600 and that was the amount of his estate, apart from a few canal and turnpike shares of small value. He revealed that Outram was

---

[1] Ibid.

also indebted to several people to the amount of about £1,000, exclusive of his funeral expenses and the cost of obtaining letters of administration. It was hardly surprising that Harvey then noted that Margaret Outram 'was not of ability to make good the said sum of £8,556 out of the assets of the said Benjamin Outram'.

Harvey went on to consider if it was for the benefit of the estate of Outram and his infant children that the partnership should continue, which led to his ordering a valuation of the entire undertaking. For this purpose he engaged Joseph Butler of Killamarsh who claimed, in his affidavit, that for 28 years past he had been in the business as an ironmaster and as such 'was well acquainted with the value of Iron Founderies'. Butler's valuation of the furnaces, works, collieries, ironstone mines, lime quarries and kilns, lands, wharfs and buildings and premises of the partnership, was £55,840. His detailed listing also provides a good account of the extraordinary progress of the company in the 21 years of its existence. Moreover, Goodwin and John Cressy Hall, an attorney, had sworn a joint affidavit relating to their evaluation of all the books, accounts and papers belonging to the partnership, besides stock in trade, good and bad debts. Stock was valued at £31,116, good debts £33,054, doubtful debts £700 and bad debts £7,063. All these assessments, including those of Joseph Butler, totalled £127,773 on 25 March 1813.

The evaluations also revealed that Outram's cash account in the furnace ledger amounted, with interest, to a debt of £5,059. This was added to the above figures and, since debts due from the company were £62,998, the entire property of the partnership was worth £69,834. Harvey pointed out that this was less than the amount of the sums advanced for capital and interest from 5 April 1802 (the private ledger showed this was £115,000, based on an assumed dividend of 5 per cent). In conclusion, he stated that all the evidence indicated that the assets of Outram were insufficient to make good what was owed and that the partnership should not continue according to the articles of December 1792.

Finally, as a result of Harvey's report, the Master of the Rolls ordered, in May 1814, that the partnership be dissolved with effect from 25 March 1813. Margaret Outram was to be released from all her indebtedness to the partnership on payment of £1,290.[1] The intricacies of the calculations which resulted in this sum have not survived.

---

[1] Abstracts of conveyances of 6 May 1814 and 30 May 1815 in the 'Butterley Bible' (in possession of Butterley Brick; information from Philip Riden).

Outram's quarter share, as well as all real estate vested in him in trust for the partnership, was assigned to Jessop and Wright in the proportions of one third to Jessop and two thirds to Wright. This arbitrary award probably did only justice to Wright as the major investor in the company and as Outram's creditor. These transactions were thus formalised by a deed of 30 May 1815 which finally ended the association of the Outram family with the Butterley Company. Before this happened William Jessop died on 18 November 1814,[1] when his share in the company passed to his son William, who was the nominated successor.

Margaret Outram, whose frank and headstrong ways, and occasional petulance, had not endeared her to the partners during the years of discord, must have been dismayed by the verdict. She nevertheless shrugged off whatever disappointments she harboured and proved to be resourceful in widowhood. After her husband's death she left Butterley Hall and lived with her young family, closer to penury than her previous existence, in a small house in Worksop, sustained by an annual sum of £200 provided by her seven surviving brothers. Somewhat surprisingly, but to their credit, the Butterley Company partners also granted her £50 a year until Francis, her eldest son, attained his majority. Nor was she neglected by the Outram family, particularly by Edmund who was to remain a staunch friend all his life, and also by her concerned and sympathetic father-in-law, Joseph. Her modest income was later augmented by a pension provided by the government in recognition of some work of national importance undertaken by her father in earlier years.

Benjamin Outram had never made a settlement on his wife, although that had been a measure proposed by her brothers before her marriage. Margaret had brushed aside such suggestions as 'most unjust and mercenary' because she knew that what little money Benjamin Outram possessed was invested in the ironworks, and she said that there was nothing to spare for her. She never remarried and remained self-reliant and determined all her life, although she was never short of considerate and influential friends, which says much for her high intelligence and gracious manner. Eventually she returned to live in Scotland, first to Aberdeen and finally to Edinburgh where she died in comfortable circumstances in 1863,[2] a dignified and elegant lady, respected in society as the mother of Major-General Sir James Outram, the baronet

---

[1] C. Hadfield and A.W. Skempton, *William Jessop, Engineer* (1979), p. 269.

[2] National Archives of Scotland, SC 70/1/115, pp. 520–6, 12 Mar. 1863, inventory of Margaret Outram's estate.

known as the Bayard of India for his part in quelling the Indian Mutiny.

Margaret was to write of her husband that 'Could he have foreseen what was to happen, or had warnings of it, he would have been most miserable on my account' but not once in her memoirs did she reproach him over his financial difficulties, in spite of the consequences these had for her and the children. Doubtless her heartfelt remarks were true but in the absence of any recorded comments, Outram's thoughts on his precarious finances must remain a baffling mystery. He was an honourable man in his professional life but sadly, some of the accusations of Outram's partners regarding misuse of the funds ring true, although such issues were not apparently debated in court. In Outram's cash account, so meticulously prepared for Harvey's investigations,[1] were several sums of money received from the company's London agents Anderson & Eades,[2] as well as others. Some amounts must have been received in settlement of accounts with the partnership, although others may well have been loans borrowed on the credit and without the knowledge of Outram's partners. Certain items, notably an account of £200 from John Hodgkinson, were probably personal debts relating to Outram's civil engineering work on canals and railways which he saw fit to pay out on the credit of the company.

Outram had carelessly mixed his personal debts with the finances of the company and as a consequence there would have been no questions in the minds of his partners as to his obligations in respect of his arrears. In the partnership indenture, one of the clauses states that each partner would 'indemnify the others against his own separate debts' and, rather more ominously, another states that 'if either shall upon a distinct account from this concern become bankrupt or any Extent or other Execution be put against him whereby the partnership may be affected. Then such Partner shall be excluded from being any longer a Partner as if he were dead ...'.[3] For these reasons, the response of the three surviving partners to Benjamin Outram's apparent misuse of the company's funds seems to have been a harsh, unrelenting but unjustified harassment of Margaret Outram who, no matter how defensive and uncooperative she seems to have been thereafter, was quite innocent of any misdeeds.

It was reprehensible that the source of Outram's funds was principally the company with which he was entrusted as managing partner and,

---

[1] DRO, D503/1/3A, Master's Report, 25 March 1813.
[2] Ibid.
[3] DRO, D503/4, Articles of Partnership.

furthermore, that he had failed to notify the partners of his actions. Moreover, it is difficult to believe that Outram could ever have justified his excessive borrowings, although it is certain that he made no efforts to hide the figures from his partners. After all, these were openly recorded in the furnace ledger for everyone to see and if his colleagues had taken it on themselves to examine all the books annually (as required of them in the partnership deed), then these mounting debts could have been exposed at any time during the five years they occurred. No doubt he did not consider that he was dishonest and most certainly he would have been deeply offended (and probably greatly angered) by any such suggestions. The company's letterbook shows, in particular, how he always reacted strongly to any accusations of improper behaviour, no matter who was his accuser.[1]

In a sense Outram, in his handling of the company's funds, was emulating the casual practices of engineers which were common enough on many canal projects at the time and with which he was certainly familiar. Outram and other engineers of the late eighteenth century, including the eminent William Jessop, often devised time and cost estimates for construction projects which proved grossly inaccurate. Very little thought was given to the planning of work and the use of accountancy for management control. It so often seemed that, so long as the money was available for payment of goods and services, then all was well. If, on the other hand, costs exceeded estimates, as they frequently did, then it was the task of the proprietors to generate fresh funds from the shareholders or other sources. This was no more apparent than during the construction of the Cromford Canal where estimates of time and costs invariably proved to be wrong.

Outram applied this cavalier attitude to cash-flow during his management of the Butterley finances, at least in so far as it concerned his own debts. He would peruse the ledgers frequently to pick out any delayed payments and was quite prepared to admonish debtors to the company, although he did not seem to think that rapid repayments of his own debts were obligatory too. He was unwise and reckless and he must stand condemned on that score but, as noted earlier, he optimistically looked forward to a splendid return from the company in the years ahead and perhaps he felt there was little to worry about. He needed no one to tell him that he was an honest and respected man; business was booming, his personal expenditure would soon be curtailed and his debts would quickly diminish. Nobody would be any the wiser by these mis-

---

[1] DRO, D503/12/1.

deeds, even if he thought of them as such. He would not be the first, nor will he be the last person to think of his actions in this manner, but unfortunately, he reckoned without a thought for the consequences of his early demise.

In spite of this distressing episode there remains a lingering sympathy for Outram. For fifteen years he had managed the company almost single-handed to establish a major industrial enterprise from nothing. Certainly his partners, particularly Wright and Beresford, had provided the money when needed but Outram's outstanding engineering and managerial skills had alone been applied with great success to such diverse and specialised operations as mining, quarrying, railway and canal engineering, as well as the ironworks and the foundry. He had worked tirelessly on behalf of his partners and also found time from these labours to generate sales and income from the promotion of railways in distant parts of the country. It is regrettable, therefore, that all his achievements should have been set aside by his partners in their panic to safeguard their investments which, as soon became obvious, were quite secure. They had subscribed their monies and had been rewarded with a splendid and thriving company which was to enrich them in the years ahead but, instead of making the best of a situation, they elected to spend several years badgering an impecunious widow for funds which she obviously did not possess. There seems no reason why a more sympathertic solution could have been found to the problem without recourse to the courts. Benjamin Outram deserved better, in spite of the errors of his ways, than the shabby treatment accorded to his memory by his partners.

That Outram was highly regarded within the engineering profession there can be no doubt and it would be sad were his reputation tarnished by the consequences of this single incident. Margaret claimed that

> I have heard my father and many other clever men say, that he united in an extraordinary degree, great genius with equal judgement, and with a spirit of enterprise that seemed to surmount every difficulty. Messrs Rennie and Telford, the first engineers in England, yielded the palm to him in talents and genius.

Curiously, Outram never became a member of the Society of Civil Engineers, of which his friend and colleague William Jessop was secretary for sixteen years. This important association was founded in 1771 by John Smeaton, the leading engineer of his generation, and reconstituted 22 years later.[1] It was the first professional engineering association of its kind in the world and, considering Outram's standing, there seems no obvious reason why he never became a member. There were advantages to be gained from joining the society of which he must have been well aware, hence his omission from the membership remains a mystery.

Benjamin Outram was, without question, one of the outstanding civil engineers of his generation as this exposition of his works will have shown. Of his ability as an engineer, however, it has to be said that, in spite of his flair and the courageous applications of his skills on many complex projects, he was not an innovator of exciting and original engineering designs. Rather he had another quality, equally rare, which can be seen in so many of his works; he could recognise a good idea when he saw one and could select the time and circumstances at which to apply it with advantage. His breadth of knowledge and experience of engineering were second to none and his pioneering work on railways in particular was of great importance and influence nationally, poised as it was at the beginning of the steam era. It is only possible to speculate on the contributions he could have made to the development of the modern railway which began a mere 25 years after his tragically early death, but had he lived it is certain that they would have been more than considerable.

Let Benjamin Outram's father, a perceptive man as ever there was, have the last word. When writing to a client of his son's engineering abilities, he stated that 'I will lay partiality aside, he is the first man I know in this business'.[2]

---

[1] C.M. Norrie, *Bridging the Years* (1956), Appendix B, pp. 195–9.
[2] JRUML, Bagshawe Muniments, 8/4/2881, J. Outram to J. Bagshawe, 28 Sept. 1798.

INDEX

Abbott, — 185
Abercarn (Mon.) 262
Abercynon (Glam.) 273
Aberdeen 327
Abergavenny (Mon.) 250, 259
Agricola, Georgius 198
Alfreton (Derbys.) 1, 2, 5, 7, 196
   Greenhill Lane wharf in 213
   Hermitage Colliery in 183, 212-18
   Somercotes in 197
   Swanwick in 1, 195, 196, 213
Altrincham (Cheshire) 76, 119
Amber aqueduct 37-8, 44
Anderson, Alexander, brother of Margaret Outram 321-2
   James, brother of Margaret Outram 321
   James, father of Margaret Outram 143, 191, 202-4, 207, 209-10, 220, 230, 278, 286
   Margaret: *see* Outram, Margaret
   William 131, 142
Anderson & Eades 328
   see also Oswald & Anderson
Arden, Mrs and Miss 191
Argoed (Mon.) 262
Arkwright, Richard, son of Sir Richard 146, 323
   Sir Richard 10-11, 13-14, 16-19, 28, 30-2, 42-3
Armfield, Jane 1
Ashbourne (Derbys.) 10
Ashby de la Zouch Canal 221-48
   railways 203-4, 221-48, 274, 297
Ashby de la Zouch (Leics.) 110, 139
Ashby Wolds (Leics.) 221, 222, 224
Ashford tunnel 253-7
Ashover (Derbys.) 1
Ashton under Lyne (Lancs.) 104
Ashton under Lyne Canal
   design and promotion 118-28
   completion 149-56
   mentioned 78, 96, 107, 171, 197, 300, 318
Astley (Lancs.) 158
Aydon & Co. 316

Bagshawe, John of Barnoldswick 157-76
   John, guardian of John Bagshawe of Barnoldswick 158
   Revd William 176
   William, guardian of John Bagshawe of Barnoldswick 158

family 3
Bailiffe, William 83, 84, 105-7
Bakewell (Derbys.) 8
Banks, Edward 106, 197, 220
Barber & Walker 183, 203, 211-12
Barnes, James 283-90
Barnoldswick (Yorks.) 157-76
Barter, Brudenell 305-6
Bath (Somerset) 199, 207, 291, 295, 297
Battye, John 81
Beaufort (Brecs.), railway 250, 260-2
Bedwellty (Glam.) 263
Beeston (Notts.) 46
Beggarlee (Notts.), colliery 203, 211
Bellhouse, David 150
Belper (Derbys.) 23, 24, 27, 69, 196
Benjamin Outram & Co. 45, 54, 74, 137, 141, 143, 177-97, 219-20, 314-18, 321-30
   see also Butterley Co.
Bennet, William 119-21, 300-2, 303, 307
Beresford, Francis 10, 13, 17, 19, 30, 178-81, 197, 325, 331
   John 196, 197
Bilborough (Notts.) 211
Birmingham (Warwicks.) 184, 277-81
   St Philip's parish in 4
   Soho works in 187, 281
Birmingham Canal 305
Black Moss reservoir 114
Blaenavon (Mon.), ironworks 258, 271
   railway 250, 260
Blisworth (Northants.), railway 203, 283-90, 295
Bloxam, — 19-20
Bolton, John 163, 164-7, 171-2, 174-5
Booth, John 109, 110
   Joseph 142
Borrowash (Derbys.) 63
Boulton, Matthew 281
Boulton & Watt 90-1, 187
Bradford (Yorks.) 316
Braunston (Northants.) 183
Brecknock & Abergavenny Canal 202, 249-73
Brecon (Brecs.) 250-1, 255, 259-60
Breedon on the Hill (Leics.) 60, 222-4
Bridgewater, Duke of: *see* Egerton, Francis 200
Bridgewater Canal 118, 119, 149, 290, 318
Brindley, James 21, 57, 150

Brinsley (Notts.), colliery 183, 219-20
    railway 286
Bristol 291
Broadhead, William 131, 141, 142
Broseley (Salop) 198
Brown, Nicholas 76, 81-4, 90, 101, 109, 112-13
    Thomas 122-3, 125-9, 131, 135, 141, 143-8
Brynich (Brecs.) 250, 255
Buckingham, Marquess of: *see* Grenville, George
Bugsworth (Derbys.) 122, 126
Bull Bridge: *see* Crich, Bull Bridge in
Burgh, Charles 262
Burton upon Trent (Staffs.) 186
Butler, Joseph 72-4, 199, 326
Butterley Carr (Derbys.), colliery in 182, 218-19
Butterley Co. 210, 212, 324, 327
    see also Benjamin Outram & Co.
Butterley (Derbys.), reservoir 47, 49-50
    tunnel 32-5, 37-8, 43, 47, 54, 95, 179, 182, 183, 213, 218, 220
Butterley Hall (Derbys.) 178, 181, 189, 191, 193, 197, 218, 319-20, 323-4, 327
Butterley Ironworks (Derbys.) 66-7, 98, 108, 203, 213-18, 220, 230, 236, 271, 274, 290
Butterley Park (Derbys.) 9, 197
Buxton (Derbys.) 126, 190, 191
Buxworth: *see* Bugsworth

Cairns, Edward 270
Calder & Hebble Navigation 118
Caldon Canal 127
Caledonian Canal 183
Calke Abbey (Derbys.) 226, 247
Cambridge, University, St John's College in 4
Camerton (Somerset), colliery 297, 305-6
Canada 5
Carshalton (Surrey) 316-17
Cartwright, Thomas 283
Case (*later* Case-Morewood), Henry 213-18
Castleton (Derbys.) 5
Cavendish, Lord George 16
    William, 5th Duke of Devonshire 13, 15-16
Chambre, A. 173-4
Chapel en le Frith (Derbys.) 137-9
    Loads Knowle in 122, 126
Chapel Milton (Derbys.) 122, 125-8
Chapman, William 133-4, 151-4
Chesterfield (Derbys.) 90
    Griffin Ironworks in 180, 184
    ironworks in 179
    see also Sheffield & Chesterfield Turnpike Trust
Chesterfield Canal 21, 57, 276
Cinderford (Gloucs.), ironworks 310
Clarke, George Hyde 122, 129, 136, 137
Clayton, Major 160
Clydach (Brecs.), aqueduct 253, 259
    railway 250, 252, 259
Clyde Ironworks 4, 192
Coalbrookdale Co. 184
Coalbrookdale (Salop) 131
    railways at 199
Codnor Park (Derbys.) 9, 48-9, 193-5
Coke, Revd D'Ewes 7, 10, 212, 220
Coleford (Gloucs.) 310
Coleorton (Leics.) 221, 226
Coleridge, Samuel 130
Colne (Lancs.) 159, 164, 171
Combe Hay (Somerset) 294, 302
Combs reservoir 128, 144, 145-6
Copeland, John 276-7
Corbridge (Northumberland) 32
Cort & Co. 184
Cosgrove (Northants.) 286, 287
Cotes Hall, in Barnoldswick (Yorks.) 158-76
Coventry Canal 221, 242
Cowes, Josiah 282
Coxbench (Derbys.) 69, 74
Coxe, William 258
Crich (Derbys.) 10, 20, 39, 179, 181, 183
    Bull Bridge in 37, 196, 210
    railway 203, 210-11, 218, 220, 235
Crickhowell (Brecs.) 250, 254-5
Crindau (Mon.) 249
Cromford & High Peak Railway 8
Cromford (Derbys.) 6-29
    Willersley Castle in 17, 42-3
Cromford Canal
    promotion 6-29
    construction of 30-45
    and Benjamin Outram & Co. 177-97
    mentioned 5, 46-9, 57, 123, 219, 220, 255, 304, 319, 329
Crosley, William 260
Croydon (Surrey) 316-17
Croydon, Merstham & Godstone Railway 206, 317-18
Crumlin (Mon.) 249-50, 260, 261, 262, 265, 266
Cumming, T.G. 271
Curr, John 188, 200-1
Curzon, Nathaniel, 1st Baron Scarsdale 29

Dadford, John 252
    Thomas 32, 35, 42, 178, 252, 253, 254, 256, 257
    Thomas jun. 250
Dale Abbey (Derbys.), ironworks at 188
Dalton, John 130

INDEX

Darley Abbey (Derbys.) 24
Dartmouth, Earl: *see* Legge, William
Davies, David 265, 273
Dean, Forest of (Gloucs.) 307-13
Denby (Derbys.) 59, 74, 196
　Denby Hall in 69
　Smithy Houses in 58, 60, 69, 73
Denton (Lancs.) 120
Derby 11
　Corporation 13, 24
　mills in 24
Derby Canal
　construction 57-75
　mentioned 97, 123, 132, 186, 223
　railways 137, 140, 199, 201, 203, 205-7, 220, 235, 297
Derwent aqueduct 39-42, 44, 54
Derwent Navigation 13, 27, 57, 59, 74-5
Derwent, River (Derbys.), water supply for Cromford Canal 10-29
Dethick, Lea & Holloway (Derbys.), Lea Hall in 14
Devonshire, Duke of: *see* Cavendish, William
Dickinson, Maillett & Shore 184-5
Diggle (Yorks.) 78, 85, 89, 91, 94, 97
　aqueduct 96
　reservoir 104, 114
Diggle Moss (Yorks.), reservoir 114-15
Disley (Cheshire) 122, 126
Dodd, Ralph 300
Dodsley, Revd Richard 55-6
Don, River (Yorks.) 275-7
Dorset & Somerset Canal 121, 300-1
Dove Holes (Derbys.) 122
Dowlais (Glam.), ironworks at 272-3
Droitwich (Worcs.) 280
Druce, Charles 54
Drury family 74
Ducie, Lord: *see* Reynolds-Moreton, Francis
Dudley Canal 282, 299
Duffield (Derbys.) 64
　Makeney in 24
　New Mills in 24
Dukinfield (Cheshire) 122, 123
Dun Navigation Co. 275-7
Dunhampstead Hill (Worcs.) 278
Dunkerton (Somerset) 295, 296, 297, 307

Eades: *see* Anderson & Eades 329
Eckington (Derbys.) 275, 276
Eddystone lighthouse 70-1, 199
Edinburgh 327
Egerton, Francis, Duke of Bridgewater 23, 130
Ellesmere Canal 134
Elvaston (Derbys.) 5
Erewash Canal 7-10, 12-13, 21, 23, 30, 43, 46-52, 58, 59, 61, 63, 211
Evans, Francis 10, 17, 30
　George 94
　family of Darley Abbey 24
Eyre family 10

Fairbank, William 275
　family 3
Farey, John 64, 65, 89-90, 98, 100, 136, 138, 139, 179, 206, 210, 212, 218, 219
Farrand, Major 160-2, 175, 176
Ferrers, Earl: *see* Shirley, Robert
Ferrybridge (Yorks.) 1
Fiennes-Clinton (*afterwards* Pelham-Clinton), Henry, 2nd Duke of Newcastle 13, 15
Fillingham family 10
FitzRoy, Augustus Henry, 3rd Duke of Grafton 287-8
Fletcher, Edward Green 11
　Samuel 157
Fletcher's Canal 274
Fothergill, Richard 262
Francis, John 209
Franklin, President Benjamin 1-2
Fritchley (Derbys.) 210
Frith, — 126
Fulton, Robert 129-34
Furness, Bethel 131, 142
Fussell, James 301-2, 304

Gainsborough (Lincs.) 1, 316
Gaskell, — 126
Gayton (Northants.) 289
Gell, John 10, 13, 15-19, 21-3, 25-7, 29, 31-2
　Philip 10, 13-14, 16, 18-20, 29-31, 35, 177
　family 3
Gilwern (Brecs.) 250, 252, 254-5, 259, 260
Glamorganshire Canal 35, 273
Glangrwyney (Brecs.), railway 250, 252
Glasgow 4
　Clyde Ironworks in 192
Gloucester 309
Gloucester & Berkeley Canal 130, 313
Godstone (Surrey) 317
Goodwin, George 190, 321-2, 325-6
Gorton & Thompson 184
Govilon (Mon.) 250, 253
Grafton, Duke of: *see* FitzRoy, A.H.
Grand Canal (Ireland) 151, 154
Grand Junction Canal 118-19, 183, 203-4
　Blisworth Hill railway 283-90, 297
Grand Trunk Canal: *see* Trent & Mersey Canal
Greasley (Notts.) 5
Greaves, — 230, 233, 240-1
Green, James 46, 49-50, 246

Timothy 188
Grenville (*afterwards* Nugent-Temple-Grenville), George, 1st Marquess of Buckingham 283
Grey, George Harry, 5th Earl of Stamford 222-3, 228, 232-3
Griffin Ironworks, Chesterfield 180, 184

Hackbridge (Surrey) 316-17
Halifax (Yorks.) 157, 292, 302
Hall, John Cressy 326
Hallam (Yorks.) 5
Hankey, Thomas 54
Hardy, — 160, 167-71, 173-4
Harecastle tunnel 108n.
Harpur, Sir Henry 226, 228, 247
Harrington, Lord: *see* Stanhope, Charles
Harrison, John 157
Hartley, Peter 159
Harvey, J.S. 325-6, 328
Hastings, Francis, 27th Earl of Huntingdon 223
    General 228
Hathersage (Derbys.) 5
Hawley's Pond reservoir 55
Heage (Derbys.) 196
    see also Morley Park
Heanor (Derbys.) 7
    Langley Bridge/Mill in 7-9, 43, 46-7, 211
Henmoor (Derbys.) 69
Henshall, Hugh 256
Hereford 309
Hermitage Colliery Railway, in Alfeton 212-18
High Peak Junction Canal 127-8
Hill & Hopkins 271
Hinckley, David 5
Hinckley (Leics.) 221
Hodgkinson, Edmund 1
    Elizabeth 1
    John 54-6, 234-5, 238, 240, 244, 255, 260, 261-2, 265-7, 269, 270, 273, 307, 309-10, 328
    Samuel 5
    Thomas 7
Holden, — 188
Holmes aqueduct, in Derby 65-9, 74-5, 97-9, 132
Holt, John 81
Homfray, Samuel 252, 262-5, 266
    William 270
Homfray & Co. 268
Hoof: *see* Pritchard & Hoof
Hopkins, John 270
    see also Hill & Hopkins
Hopton (Derbys.) 3, 10
Hornblower, J.C. 199
Horne, E.T.W. 178

Horner, — 246
Horsley (Derbys.) 69
Howard, Charles, 16th Duke of Norfolk 188, 200
Howe, Richard, Earl Howe 16
Huddersfield Broad Canal 78
Huddersfield Narrow Canal
    construction 76-117
    mentioned 119, 123, 142, 144, 149, 157, 183, 197, 218, 303-4
Huntingdon, Earl of: *see* Hastings, Francis
Hurt, Charles 16-17
    family 10
Hyde (Cheshire) 122, 136
    Hyde Bank quarries in 135
    Hyde Bank tunnel in 142
    Hyde Hall in 129

Ingleby (Derbys.) 223
Isleworth (Middlesex) 191

Jack o'Darley's bridge 70-1, 73
James, Benjamin 252, 258
    Thomas 253, 257
Jessop, Josias, father of William 71
    Josias, son of William 8, 192
    William, son of Josias
        and Ashby de la Zouch Canal 221-3
        and Blisworth Hill railway 283-90
        and building of the Cromford Canal 30-45
        and the Butterley Co. 177-8, 180-3, 192, 196-7, 322-5, 327
        and Cromford Canal 8-29, 57
        and Huddersfield Narrow Canal 108
        and the London Docks 313-16
        and Nottingham Canal 46-52
        and Nutbrook Canal 52-4
        and proposed Manchester–Liverpool railway 318
        and railway construction 203, 206-7
        and railways in Surrey 316-18
        and Standedge tunnel 94-5
        mentioned 134, 147, 154, 292, 312, 331
    William, son of William 192, 197, 324
Jewsbury, Thomas 224
Jodrell, John Bowyer 122
Jolliffe & Banks 10n.

Kell, John 115
Kenear, Walter 234-5, 244, 255, 265-7, 273
Kennet & Avon Canal 291, 293, 296, 300
Kenyon, Lloyd, 1st Baron Kenyon 174
Kilburn (Derbys.) 69
Kildare Canal (Ireland) 151
Killamarsh (Derbys.) 326
Kilmarnock & Troon Railway 206

INDEX

Kirk Hallam (Derbys.) 56
Kirkland, William 26-7

Lamb, Peniston, 1st Viscount Melbourne 223
Langley Bridge/Mill: see Heanor, Langley Bridge/Mill in
Lawson, James 90-1
Lawton (Cheshire) 127
Lea Hall: see Dethick, Lea & Holloway, Lea Hall in
Lea Navigation 22
Lee, Thomas 91-2
Leeds, Duke of: see Osborne, Francis Godolphin
Leeds & Liverpool Canal 5, 76
  dispute at Barnoldswick 157-76
Leek (Staffs.) 127
Lees, John 131
Legge, William, 2nd Earl of Dartmouth 102
Leicester 184
Leicester Canal 13
Leicester Navigation 206, 245
Lenton (Notts.) 46
Lewis, Henry 271
Lichfield (Staffs.), St John's Hospital in 4
Limpley Stoke (Somerset) 291
Lingard, John 158, 160, 171
Lister, — 39
Little Eaton (Derbys.), railway 59, 60, 63-4, 69-74, 73, 75, 201
Liverpool (Lancs.) 184
  proposed railway to Manchester 318
  see also Leeds & Liverpool Canal Co.
Llanelli (Carms.), ironworks at 271
Llanelly (Brecs.) 250
Llanfoist (Mon.) 250, 260
Llangattock (Mon.) 252
Llangynidr (Brecs.) 250, 253
Llanhilleth (Mon.), railway 261
Llanover (Mon.) 252
Lockett, — 246
London 207-8
  Blackfriars Bridge in 22
  Docks 207, 313-16
London Bridge Waterworks Co. 183
London Dock Co. 183
Longbotham, John 157, 158
Longden, William 26
Longdon Chambers & Co. 189
Longdon upon Tern (Salop), aqueduct 66, 100
Longroyd aqueduct 98
Loughborough & Nanpantan Railway 206
Loughborough Navigation 7, 12, 13
Lydbrook (Gloucs.) 310, 311
Lydney (Gloucs.) 310-12, 313

Macclesfield (Cheshire) 127

McNiven, Charles 130
Maillett: see Dickinson, Maillett & Shore
Makeney: see Duffield, Makeney in
Mallalieu, John 155
Mamhilad (Mon.) 250
Manchester (Lancs.) 78, 80, 107, 118
  Piccadilly in 318
  and Ashton Canal 149-56
  proposed railway to Liverpool 318
Manchester, Bolton & Bury Canal 274
Manchester Grammar School 4
Manchester to Ashton under Lyne & Oldham Canal 118
Mansfield (Notts.), foundry at 185
March Haigh reservoir 101
Market Bosworth (Leics.) 221
Marple (Cheshire) 122, 123, 126, 127
  aqueduct 122-3, 125, 129, 131-6, 140-2
  railway 305
Marsden (Yorks.) 78, 79, 83, 85, 89, 104, 105, 115
  aqueduct 97, 98
Marsland, Samuel 109, 129
Masson Mill: see Matlock, Masson Mill in
Matlock (Derbys.) 3, 58
  Masson Mill in 17, 32, 43
Meadows, James 121, 139, 148
Measham (Leics.), railway 203, 219, 220, 230, 286
Melbourne, Viscount: see Lamb, Peniston
Melton Mowbray Navigation 274
Merstham (Surrey) 317
Merthyr Tydfil (Glam.) 253
  Dowlais Ironworks at 272-3
  ironworks at 252
Micklethwaite, Anne 1
Middleton, Lord: see Willoughby, Henry
Midford (Somerset) 296
Miller, George 276
Milnes family 10
Mitcham (Surrey) 316-17
Moira, Earl of: see Rawdon-Hastings, Francis
Moira (Leics.) 227
Monkhouse, Matthew 262
Monmouthshire Canal 249-73
  railways 304
Montagu, John, 4th Earl of Sandwich 29
Morewood, George 3, 212
  Helen 212-18
  family 183, 185
Morgan, Sir Charles 262-3, 265-6, 267-9, 271
Morley Park (Derbys.), ironworks at 179
Mylne, Robert 22-3, 25, 27-8, 47-9
Mythom Mill aqueduct 98

Nant-y-glo (Mon.) 250

New Mills: *see* Duffield, New Mills in
Newark (Notts.) 8, 20
Newbold, Thomas 224
Newbridge (Mon.) 250
Newcastle, Duke of: *see* Fiennes-Clinton, Henry
Newport (Mon.) 249, 263-5, 266, 273, 308
   Tredegar Park in 262, 265
Nightingale, Peter 14
Nine Mile Point (Mon.) 265-8
Norbury (Cheshire) 126, 127
Norfolk, Duke of: *see* Howard, Charles
Northampton, railway from Grand Junction Canal 289
Norton, Charles 298-301, 303-4
Norton (Derbys.), Oaks in 3, 158
Nottingham 10
Nottingham Canal 43, 46-52, 211
Nova Scotia 5
Nuneaton (Warwicks.) 221
Nutbrook Canal 52-6
Nuttall, George 3
   John 3, 52, 58

Oakengate (Salop) 292
Oaks: *see* Norton, Oaks in
Oddie, Matthew 160, 162, 163, 165, 168
Oldham (Lancs.) 119, 120
Oldknow, Samuel 122, 137, 146-7
Ollerton (Notts.) 184
Osborne, Francis Godolphin, 5th Duke of Leeds 3
Osmaston (Derbys.) 63
Oswald & Anderson 316
Outram, Anna, daughter of Benjamin Outram (1764–1805) 192, 321
   Anne, half-sister of BO 1, 5
   Benjamin, son of Joseph (d. 1732) 1
   Benjamin (1764–1805): *see next entry*
   Edmund, brother of BO 1, 4, 327
   Eliza, daughter of BO 192
   Frances, great grand-daughter of BO 318
   Francis, son of BO 192, 327
   George, nephew of BO 4-5
   Major-General Sir James, Bt 192, 319, 327-8
   John, half-brother of BO 1, 5
   Joseph, brother of BO 1, 4-5, 155
     and Benjamin Outram & Co. 184-6, 188-92, 194, 196, 211-18, 315-18
   Joseph, father of BO 1-5, 7, 11, 35, 155, 177, 190-1, 327, 332
     and John Bagshawe of Barnoldswick 160, 162-4, 166-7, 171-2
   Joseph (d. 1732) 1
   Margaret, daughter of BO 192
   Margaret, wife of BO 4, 191-2, 318, 319-31
   Sally, sister of BO 1, 5
   Thomas 1
   William 1
Outram, Benjamin (1764–1805)
   ancestry 1-3
   early life 3-5
   his opinion of Sir Richard Arkwright 19
   others' opinion of him 77-8, 164, 319, 321, 330-1
   assessment of his canals 116-17, 128, 135, 144, 331
   assessment of his railways 208-9, 246-7, 271-3, 274, 298, 331
   health 189-91
   marriage and family 191-2, 319-20
   death and aftermath 320-8
Overton, George 272-3
Owen, Robert 130
Owston Ferry (Lincs.) 1

Packington (Leics.) 226
Paddock aqueduct 98
Palmer, Samborne 305-6
Parkend (Gloucs.) 313
   ironworks at 310
Parker, — 162
Parramore, George 242-5, 246
Parry, Morgan 261, 266
Paulton (Somerset) 291, 296
Peak Forest Canal
   design and promotion 118-28
   construction 129-48
   railways 203, 290, 297, 303, 305, 307
   mentioned 8, 95, 106, 109, 149, 154, 183, 255
Pelham-Clinton, Henry: *see* Fiennes-Clinton
Pentrich (Derbys.) 195, 216, 220
Penydarren (Glam.), ironworks at 252, 262-3
   railway 273
Piddocke, — 233
Pike, James 54
Pillgwenlly (Mon.) 265
Pinxton (Derbys.) 7-9, 10, 220
Pont Gwaithyrhaearn (Mon.) 268
Pontcysyllte (Denbighs.), aqueduct 134
Pontefract (Yorks.) 1
Pontey, William 76-81, 83, 116
Pontnewynydd (Mon.) 249-50
Pontymoile (Mon.) 250, 254-6, 259
Pontypool (Mon.) 250
Porter, Samuel 282-3
Potts, Thomas 270
Powell, John 254
   Morgan 264
   Thomas 252-3
Poynton (Cheshire) 126, 127
Praed, William 286, 287-8
Price, Henry 309-10, 312

INDEX

Priestley, Joseph 163, 167, 168, 291
Pritchard, — 107
   William 252
Pritchard & Hoof 108n.

Raby, Alexander 54, 271
Raby & Co. 188
Radcliffe-on-Trent (Notts.) 20
Radford, John 61, 69
Radstock (Somerset) 291, 296
Ramsden, Sir John: see Sir John Ramsden's Canal
Rassa (Mon.), railway 261
Ratcliffe: see Radcliffe-on-Trent
Rawdon, Lord: see Rawdon-Hastings, Francis
Rawdon-Hastings, Francis, 2nd Earl of Moira, Baron Rawdon 13, 29, 228, 232-4
Reigate (Surrey) 317
Rennie, John 183, 285, 291-2, 313-16, 332
Reynolds, Richard 199
Reynolds-Moreton, Francis, 3rd Baron Ducie of Tortworth 155
Rhymney (Mon.), ironworks at 271
   Union Furnace at 265
Ripley (Derbys.) 3, 196, 216
   Hartshay in 220
   Knowts Hall estate in 197
   see also Butterley
Risca (Mon.) 263-71
Rochdale Canal 76, 81, 118, 119, 149, 154, 157
Romiley (Cheshire) 122, 142
Rooke, Mr Justice 169
Rooth, John 84, 107, 108, 109, 110, 112, 114
Ross on Wye (Herefs.) 312
Royal George aqueduct 98
Ruardeen (Gloucs.) 310

Saddleworth (Yorks.) 76, 94
Salterswood (Derbys.) 69
Salusbury, Sir Richard 261
Sandiacre (Derbys.) 59, 63, 74
Sandwich, Earl of: see Montagu, John
Sapperton (Wilts.), tunnel 87n.
Sawley (Derbys.) 7
Scarbottom aqueduct 98
Scarsdale, Lord: see Curzon, Nathaniel
Scout Mill tunnel 78
Shardlow (Derbys.) 58, 107
Sheasby, Thomas 32, 35, 42, 178
Sheffield (Yorks.) 3, 128, 188, 189, 200-1, 275-6
Sheffield & Chesterfield Turnpike Trust 5
Sheffield Canal 275
Shelton (Derbys.) 63

Shipley (Derbys.) 52, 53, 55, 56
Shirley, Robert, 7th Earl Ferrers 228
Shobnall (Staffs.) 186
Shore: see Dickinson, Maillett & Shore
Shropshire Canal 66, 131, 292-3
Sir John Ramsden's Canal 78
Sirhowy (Mon.), ironworks 263-5, 267-8
   railway 250, 265-7, 270, 274
Sirhowy Tram Road Co. 265-6, 267-8
Skyerbottom aqueduct 98
Slaithwaite reservoir 101-5, 114
Smalley (Derbys.) 59, 60, 61, 69
Smeaton, John 70-1, 199, 330-1
Smeatonian Society of Civil Engineers: see Society of Civil Engineers
Smiles, Samuel 201
Smith, — 240
   Ebenezer & Co. 90-1, 180
   George 169
   John 47-8, 90
   William 177
Smithy Houses: see Denby, Smithy Houses in
Society of Civil Engineers 22, 330-1
Somercotes: see Alfreton, Somercotes in
Somersetshire Coal Canal and railways 290-307
Sorocold, George 57
Sowerby Bridge (Yorks.) 118
Spark, George 115
Sparth reservoir 114
Sprotborough (Yorks.) 277
Stakes aqueduct 97, 98, 99, 106
Staley Mill reservoir 104
Stalybridge (Cheshire) 78, 97
   aqueduct: see Stakes aqueduct
Stamford, Earl of: see Gray, George Harry
Standedge tunnel 78-9, 80, 83, 85-93, 94, 103-5, 108-9, 112-13, 116, 218, 303-4
Stanhope, Charles, 3rd Earl of Harrington 3, 5, 27
   Charles, 3rd Earl Stanhope 29, 52, 54, 130
Stanhope (Durham) 115
Stanton (Derbys.) 52, 54
Starkie, John 169
Staunton Harold (Leics.) 221, 226
Staveley, Christopher 206, 245, 246
Stephenson, G. & R. 273
Stewart, William 273
Stockport (Cheshire) 8, 120, 127
Stockport Canal 127
Stockton & Darlington Railway 273
Stodhart tunnel 124-5, 137
Stoke Bruerne (Northants.) 287-9
Store Street aqueduct 96, 151-4
Stratford upon Avon Canal 281-3
Strelley (Notts.) 198

Stretton, Clement E. 201, 318
Stroud Canal 308
Strutt family 24
Surrey Iron Railway 206, 316-18
Sutcliffe, John 90, 94, 103, 116, 162, 292, 302-7
Swanwick: see Alfreton, Swanwick in
Swarkestone (Derbys.) 58, 59, 60, 63, 64, 74
Swarkestone Canal 61
Swellands reservoir 115
Swinshore Common reservoir 105
Sykes, Thomas 44

Taitt, William 272
Talybont (Brecs.) 253, 255
Tame aqueduct 129
Tamworth (Staffs.) 32
Tardebigge (Worcs.) 278-81
Telford, Thomas 66, 100, 127, 134, 332
  and Huddersfield Narrow Canal 110-15
Thames & Severn Canal 87n., 308
Thomas, Richard 270
Thompson: see Gorton & Thompson
Thornber, — 159, 175
Thornewells, Messrs 186n.
Thorniley, John 142
Ticknall (Derbys.) 20, 221, 223, 226, 242
Tinsley (Yorks.) 275-6
Todds Brook reservoir 128
Tredegar (Mon.), ironworks at 263, 265, 271-3
Tredegar Park: see Newport, Tredegar Park in
Tredgold, Thomas 202
Trefil (Mon.) 265, 271
Trent & Mersey Canal 7, 9, 57-8, 60, 63, 108n., 118-19, 127, 183, 186, 223
Trent Canal 58, 59-60, 61
Trent Navigation 7, 9, 46-51, 57-8, 59, 60, 222-3
Trevithick, Richard 273
Trosnant (Mon.), railway 250, 260, 261
Tunnelend reservoir 103-5, 114

Underhill, William 299-300
Union Furnace (Mon.) 265
Uppermill aqueduct 98
Upton, — 29

Varley, John 21-3, 61, 107-9

Waite, John 169, 170, 175
  John jun. 170-1, 175
Wakefield (Yorks.) 81
Walker, Thomas 184, 194, 211-12
  see also Barber & Walker

Wall, John 281
Walters, Walter 258, 270, 271
Wandsworth (Surrey) 316-17
Wapping (Middlesex) 314
Warner, John 242-6
Warwick & Birmingham Canal 282
Waterside aqueduct 98
Watkins, William 252-3
Watt, James jun. 91, 187
  see also Boulton & Watt
Weldon, Robert 292-6
Wellow (Somerset) 295, 302
  railway 307
Werneth Canal (Lancs.) 120
Wessendon reservoirs 101
West Hallam (Derbys.) 52
Whaley Bridge (Derbys.) 8, 122, 126, 128
Wheatcroft, German 143, 145
White, William 63
Whitehead, David 109, 129, 142, 255
  Robert 115
  Thomas 115
Whitmore, William 298-300, 304
Whitworth, Robert
  and Ashby de la Zouch Canal 221-4
  and Leeds & Liverpool Canal 157, 159, 166-8
  mentioned 44, 87n., 102-3, 276-7, 285
Wilkes, Joseph 183, 203-4, 212, 219-20, 230, 233, 240-1, 278, 285-7, 289
Willersley Castle: see Cromford, Willersley Castle in
Willesley (Derbys.) 227-8, 238, 241, 248
Williams, — 107
Willoughby, Henry, 5th Baron Middleton 13, 46
Wiltshire & Berkshire Canal 291
Wingerworth (Derbys.), ironworks at 72-3, 179, 199
Wirksworth (Derbys.) 10, 16
Wollaton (Notts.) 46, 198
Woodhouse, Jonathan 110, 139
Wootton, George 61-3
Worcester 278-81
Worcester & Birmingham Canal, railways 204, 277-81, 304
Worksop (Notts.) 5, 327
Worsley (Lancs.) 130
Worthington, George 76, 81, 106, 119, 122
Wragg, Samuel 188
Wright, Francis 197
  John 180-2, 196, 197, 320, 322, 324-5, 327, 331
Wye, River (Gloucs.), navigation 312

York, assizes 169-71, 173